Semantic Web Technologies for Intelligent Engineering Applications

Stefan Biffl · Marta Sabou
Editors

Semantic Web Technologies for Intelligent Engineering Applications

Editors
Stefan Biffl
TU Wien
Vienna
Austria

Marta Sabou
TU Wien
Vienna
Austria

ISBN 978-3-319-82367-6   ISBN 978-3-319-41490-4 (eBook)
DOI 10.1007/978-3-319-41490-4

Softcover reprint of the hardcover 1st edition 2016

Printed on acid-free paper

This Springer imprint is published by Springer Nature
The registered company is Springer International Publishing AG Switzerland

Foreword I

In the 1970s and early 1980s, the Benetton Group experienced extraordinary growth, increasing the sales from 33 billion lire in 1970 to 880 billion lire in 1985 (the latter figure is roughly equivalent to 1.2 billion euro in today's value), an increase of over 2,500 %.[1] There were several reasons for this commercial success, but arguably, a key reason was the introduction of innovative manufacturing processes, which supported flexible, data-driven product customization. In practice, what Benetton pioneered (among other things) was a model, where clothes were produced undyed and were only finalized as late as possible, in response to data coming from retail sales. This approach was supported by a sophisticated (for the time) computing infrastructure for data acquisition and processing, which supported a quasi-real-time approach to manufacturing. It is interesting that in this historical example of industrial success, we have the three key elements, which are today a foundation of the new world of flexible, intelligent manufacturing: innovative manufacturing technologies, which are coupled with intelligent use of data, to enable just-in-time adaptation to market trends.

The term *Industrie 4.0* is increasingly used to refer to the emergence of a fourth industrial revolution, where intelligent, data-driven capabilities are integrated at all stages of a production process to support the key requirements of flexibility and self-awareness. Several technologies are relevant here, for instance the *Internet of Things* and the *Internet of Services*. However, if we abstract beyond the specific mechanisms for interoperability and data acquisition, the crucial enabling mechanism in this vision is the use of data to capture all aspects of a production process and to share them across the various relevant teams and with other systems.

Data sharing requires technologies, which can enable interoperable data modeling. For this reason, *Semantic Web technologies* will play a key role in this emerging new world of cyber-physical systems. Hence, this is a very timely book,

---

[1]Belussi F. (1989) "Benetton: a case study of corporate strategy for innovation in traditional sectors" in Dodgson M. (ed) Technology Strategies and the Firm: Management and Public Policy Longman, London.

which provides an excellent introduction to the field, focusing in particular on the role of Semantic Web technologies in intelligent engineering applications.

The book does a great job of covering all the essential aspects of the discussion. It analyzes the wider context, in which Semantic Web technologies play a role in intelligent engineering, but at the same time also covers the basics of Semantic Web technologies for those, who may be approaching these issues from an engineering background and wish to get up to speed quickly with these technologies. Crucially, the book also presents a number of case studies, which nicely illustrate how Semantic Web technologies can concretely be applied to real-world scenarios. I also liked very much that, just like an *Industrie 4.0* compliant production process, the book aims for self-awareness. In particular, the authors do an excellent job at avoiding the trap of trying to 'market' Semantic Web technologies and, on the contrary, there is a strong self-reflective element running throughout the book. In this respect, I especially appreciated the concluding chapter, which looks at the strengths and the weaknesses of Semantic Web technologies in the context of engineering applications and the overall level of technological readiness.

In sum, I have no hesitation in recommending this book to readers interested in engineering applications and in understanding the role that Semantic Web technologies can play to support the emergence of truly intelligent, data-driven engineering systems. Indeed, I would argue that this book should also be a mandatory read for the students of Semantic Web systems, given its excellent introduction to Semantic Web technologies and analysis of their strengths and weaknesses. It is not easy to cater for an interdisciplinary audience, but the authors do a great job here in tackling the obvious tension that exists between formal rigor and accessibility of the material.

I commend the authors for their excellent job.

April 2016

Prof. Enrico Motta
Knowledge Media Institute
The Open University
Milton Keynes, UK

Foreword II

The engineering and operation of cyber-physical production systems—used as a synonym for *Industrie 4.0* in Germany—need an adequate architectural reference model, secure communication within and in between different facilities, more intuitive and aggregated information interfaces to humans as well as intelligent products and production facilities. The architectural reference model in Germany is RAMI (ZVEI 2015) enlarged by, for example, agent-oriented adaptation concepts (Vogel-Heuser et al. 2014) as used in the *MyJoghurt* demonstrator (Plattform Industrie 4.0: Landkarte Industrie 4.0 – Agentenbasierte Vernetzung von Cyber-Physischen Produktionssystemen (CPPS) 2015). In the vision of *Industrie 4.0*, intelligent production units adapt to new unforeseen products automatically not only with changing sets of parameters but also by adapting their structure. Prerequisites are distinct descriptions of the product to be produced with its quality criteria including commercial information as well as a unique description of the required production process to produce the product, of the production facilities and their abilities (Vogel-Heuser et al. 2014), i.e., the production process it may perform (all possible options). Different production facilities described by attributes may offer their services to a market place. The best fit and most reliable production unit will be selected through matching the required attributes with the provided ones and subsequently adapts itself to the necessary process. There are certainly many challenges in this vision: a product description is required to describe especially customer-specific, more complex products adequately. Different formalized descriptions of production processes and resources are available, e.g., formalized process description (VDI/VDE 2015) or MES-ML (Witsch and Vogel-Heuser 2012), but structural adaptivity is still an issue.

Given that these attributes characterizing product, process and resource were available in a unique, interpretable, and exchangeable way, Semantic Web technologies could be used to realize this vision.

This coupling of proprietary engineering systems from different disciplines and different phases of the lifecycle is already well known since the Collaborative Research Centre SFB 476 IMPROVE running from year 1997 to year 2006 (Nagl and Marquardt 2008). CAEX has been developed in a transfer project of this collaborative research area at first only targeting at a port to port coupling of proprietary engineering tools during the engineering workflow of process plants. The idea is simple and still working: modeling the hierarchy of the resource (plant) in the different disciplinary views and mapping parts of the different discipline specific models to each other. Behavioral descriptions were added with PLCopen XML and geometric models with Collada, resulting in AutomationML, still under continuous and growing development. The future will show whether and how variability and version management—one of the key challenges in system and software evolution—may be integrated in or related to AutomationML. To specify a production facility is already a challenge, but describing its evolution over decades in comparison with similar production facilities and the library for new projects is even worse (Vogel-Heuser et al.; DFG Priority Programme 1593).

The more or less manual mapping from one *AutomationML* criterion in one discipline to another one in the other discipline should be replaced by coupling the discipline specific local vocabularies (ontologies) to a global (joint) vocabulary.

Ontologies have been in focus for more than one decade now, but are still being evaluated in engineering regarding real-time behavior in engineering frameworks on the one hand and regarding dependability and time behavior during runtime of machines and plants.

Semantic Web technologies can help to couple the models from the multitude of disciplines and persons involved in the engineering process and during operation of *automated production systems* (aPS). APS require the use of a variety of different modeling languages, formalisms, and levels of abstraction—and, hence, a number of disparate, but partially overlapping, models are created during engineering and run time. Therefore, there is a need for tool support, e.g., finding model elements within the models, and for keeping the engineering models consistent.

Different use cases for Semantic Web technologies in engineering and operation of automated production systems are discussed in this book, for example,

- To ensure compatibility between mechatronic modules after a change of modules by means of a Systems Modeling Language (SysML)-based notation together with the Web Ontology Language (OWL).
- To ensure consistency between models along the engineering life cycle of automated production systems: during requirements and test case design, e.g., by means of OWL and SPARQL, or regarding the consistency between models in engineering and evolution during operation (DFG Priority Programme 1593), making a flexible definition and execution of inconsistency rules necessary.
- To identify inconsistencies between interdisciplinary engineering models of automated production system and to support resolving such inconsistencies (Feldmann et al. 2015).
- To cope with different levels of abstraction is another challenge; therefore architectural models may be introduced and used to connect the appropriate levels with each other (Hehenberger et al. 2009).

Unfortunately, the key argument against an ontological approach based on Semantic Web technologies is the effort to develop the vocabularies and the mapping between discipline specific vocabularies as well as the rules to check inconsistencies between different attributes described with ontologies. Some researchers propose rule-based agents that map local ontologies to a global ontology (Rauscher 2015), but the domain-specific rules need to be formulated as a basis beforehand, which is a tremendous effort.

For example for more than 15 years, academia and industry are trying to develop a joint vocabulary for automated production systems being a prerequisite for self-aware service-oriented *Industrie 4.0* systems. This process is now part of the *Industrie 4.0* platform activities, but as often, setting up such vocabularies is, similar to standardization activities, difficult, takes time and—because of evolution in technology and methods—never ends. Often such ambitious and theoretically applicable approaches fail due to underestimated effort, shortage of money to cope with the effort and lack of acceptance, i.e., decreasing support from involved companies or companies needed for a successful solution refusing to participate. There will be long-term support needed and tremendous effort from both industry and academia necessary until Semantic Web technologies will gain their full potential.

To extract this knowledge from existing models and projects is certainly worth trying, but requires examples/models of engineering best practices without too many exceptions fulfilling single customer requirements, e.g., in special purpose machinery.

Regarding automation, the key challenges remains: how to agree on a local vocabulary and on domain-specific rules in close cooperation from academia and industry.

January 2016

Prof. Birgit Vogel-Heuser
Chair of Automation and Information Systems
TU München
Garching, Germany

References

DFG Priority Programme 1593—Design for Future—Managed Software Evolution. http://www.dfg-spp1593.de/. Accessed 7 Jan 2016

Feldmann, S., Herzig, S.J.I., Kernschmidt, K., Wolfenstetter, T., Kammerl, D., Qamar, A., Lindemann, U., Krcmar, H., Paredis, C.J.J., Vogel-Heuser, B.: Towards effective management of inconsistencies in model-based engineering of automated production systems. In: 15th IFAC Symposium on Information Control in Manufacturing, Ottawa, Canada (2015)

Hehenberger, P., Egyed, A., Zeman, K.: Hierarchische Designmodelle im Systementwurf mechatronischer Produkte. In: VDI Mechatronik, Komplexität beherrschen, Methoden und Lösungen aus der Praxis für die Praxis (2009)

Nagl, M., Marquardt, W. (eds.): Collaborative and Distributed Chemical Engineering. From Understanding to Substantial Design Process Support – Results of the IMPROVE Project. Springer Berlin (2008)

Plattform Industrie 4.0: Landkarte Industrie 4.0 – Agentenbasierte Vernetzung von Cyber-Physischen Produktionssystemen (CPPS). http://www.plattform-i40.de/I40/Redaktion/DE/Anwendungsbeispiele/265-agentenbasierte-vernetzung-von-cyber-physischen-produktionssystemen-tu-muenchen/agentenbasierte-vernetzung-von-cyber-physischen-produktionssystemen.html (2015). Accessed 7 Jan 2016

Rauscher, M.: Agentenbasierte Konsistenzprüfung heterogener Modelle in der Automatisierungstechnik. In: Göhner, P. (ed.) IAS-Forschungsberichte 2015, 2

VDI/VDE: Formalised Process Descriptions. VDI/VDE Standard 3682 (2015)

Vogel-Heuser, B., Legat, C., Folmer, J., Rösch, S.: Challenges of Parallel Evolution in Production Automation Focusing on Requirements Specification and Fault Handling. Automatisierungstechnik, **62**(11), 755–826

Vogel-Heuser, B., Diedrich, C., Pantförder, D., Göhner, P.: Coupling Heterogeneous Production Systems by a Multi-agent Based Cyber-physical Production System. In: 12th IEEE International Conference on Industrial Informatics, Porto Alegre, Brazil (2014)

Witsch, M., Vogel-Heuser, B.: Towards a Formal Specification Framework for Manufacturing Execution Systems. IEEE Trans. Ind. Inform. **8**(2) (2012)

ZVEI e.V.: The Reference Architectural Model RAMI 4.0 and the Industrie 4.0 Component. http://www.zvei.org/en/subjects/Industry-40/Pages/The-Reference-Architectural-Model-RAMI-40-and-the-Industrie-40-Component.aspx (2015). Accessed 7 Jan 2016

Preface

This book is the result of 6 years of work in the Christian Doppler Laboratory "Software Engineering Integration for Flexible Automation Systems" (CDL-Flex) at the Institute of Software Technology and Interactive Systems, Vienna University of Technology.

The overall goal of the CDL-Flex has been to investigate challenges from and solution approaches for semantic gaps in the multidisciplinary engineering of industrial production systems. In the CDL-Flex, researchers and software developers have been working with practitioners from industry to identify relevant problems and to evaluate solution prototypes.

A major outcome of the research was that the multidisciplinary engineering community can benefit from solution approaches developed in the Semantic Web community. However, we also found that there is only limited awareness of the problems and contributions between these communities. This lack of awareness also hinders cooperation across these communities.

Therefore, we planned this book to bridge the gap between the scientific communities of multidisciplinary engineering and the Semantic Web with examples that should be relevant and understandable for members from both communities. To our best knowledge, this is the first book to cover the topic of using Semantic Web technologies for creating intelligent engineering applications. This topic has gained importance, thanks to several initiatives for modernizing industrial production systems, including *Industrie 4.0*[2] in Germany, the *Industrial Internet Consortium* in the USA or the *Factory of the Future* initiative in France and the UK. These initiatives need stronger semantic integration of the methods and tools across several engineering disciplines to reach the goal of automating automation.

We want to thank the researchers, the developers, the industry partners, and the supporters, who contributed to the fruitful research in the CDL-Flex, as a foundation for providing this book.

[2]Because the term *Industrie 4.0* is the name of a strategic German initiative, the term will be used in its German form, without translation to English.

Researchers who applied basic research to use cases provided by industry partners: Luca Berardinelli, Fajar Juang Ekaputra, Christian Frühwirth, Olga Kovalenko, Emanuel Mätzler, Richard Mordinyi, Thomas Moser, Jürgen Musil, Petr Novák, Marta Sabou, Stefan Scheiber, Estefanía Serral, Radek Šindelář, Roland Willmann, Manuel Wimmer, and Dietmar Winkler.

Developers, who developed and evaluated scientific prototypes: Stephan Dösinger, Christoph Gritschenberger, Andreas Grünwald, Michael Handler, Christoph Hochreiner, Ayu Irsyam, Lukas Kavicky, Xiashuo Lin, Christian Macho, Kristof Meixner, Markus Mühlberger, Alexander Pacha, Michael Petritsch, Andreas Pieber, Michael Pircher, Thomas Rausch, Dominik Riedl, Felix Rinker, Barabara Schuhmacher, Matthias Seidemann, Lukas Stampf, Christopher Steele, Francois Thillen, Iren Tuna, Mathijs Verstratete, and Florian Waltersdorfer.

Industry and research partners, who provided support and data: Georg Besau, Florian Eder, Dieter Goltz, Werner Hörhann, Achim Koch, Peter Lieber, Arndt Lüder, Vladimir Marik, Alfred Metzul, Günther Raidl, Ronald Rosendahl, Stefan Scheffel, Anton Schindele, Nicole Schmidt, Mario Semo, Heinrich Steininger, and Wolfgang Zeller.

Administrative support: Natascha Zachs, Maria Schweikert.

Guidance and financial support from the Christian Doppler Society, the Federal Ministry of Economy, Family and Youth, and the National Foundation for Research, Technology and Development in Austria, in particular: Brigitte Müller, Eva Kühn, Gustav Pomberger, and A. Min Tjoa.

Vienna, Austria
April 2016

Stefan Biffl
Marta Sabou

Contents

Contributors

Luna Alani Giessen, Germany

Stefan Biffl Institute of Software Technology and Interactive Systems, CDL-Flex, Vienna University of Technology, Vienna, Austria

Fajar Juang Ekaputra Institute of Software Technology and Interactive Systems, CDL-Flex, Vienna University of Technology, Vienna, Austria

Jérôme Euzenat INRIA & Univ. Grenoble Alpes, Grenoble, France

Stefan Feldmann Institute of Automation and Information Systems, Technische Universität München, Garching near Munich, Germany

Wolfgang Kastner Technische Universität Wien, Vienna, Austria

Konstantin Kernschmidt Institute of Automation and Information Systems, Technische Universität München, Garching near Munich, Germany

Olga Kovalenko Institute of Software Technology and Interactive Systems, CDL-Flex, Vienna University of Technology, Vienna, Austria

Arndt Lüder Otto-von-Guericke University/IAF, Magdeburg, Germany

Richard Mordinyi Institute of Software Technology and Interactive Systems, CDL-Flex, Vienna University of Technology, Vienna, Austria

Thomas Moser St. Pölten University of Applied Sciences, St. Pölten, Austria

Petr Novák Institute of Software Technology and Interactive Systems, CDL-Flex, Vienna University of Technology, Vienna, Austria

Marta Sabou Institute of Software Technology and Interactive Systems, CDL-Flex, Vienna University of Technology, Vienna, Austria

Estefania Serral Leuven Institute for Research on Information Systems (LIRIS), Louvain, Belgium

Radek Šindelář CDL-Flex, Vienna University of Technology, Vienna, Austria

Simon Steyskal Siemens AG Austria, Vienna, Austria; Institute for Information Business, WU Vienna, Vienna, Austria

Tania Tudorache Stanford Center for Biomedical Informatics Research, Stanford, CA, USA

Birgit Vogel-Heuser Institute of Automation and Information Systems, Technische Universität München, Garching near Munich, Germany

Roland Willmann Institute of Software Technology and Interactive Systems, CDL-Flex, Vienna University of Technology, Vienna, Austria

Manuel Wimmer Institute of Software Technology and Interactive Systems, TU Vienna, Vienna, Austria

Dietmar Winkler Institute of Software Technology and Interactive Systems, CDL-Flex, Vienna University of Technology, Vienna, Austria; SBA-Research gGmbH, Vienna, Austria

Abbreviations

3D	3 Dimensional
AAA	Anyone is Allowed to Say Anything About Any Topic
AML	AutomationML, Automation Markup Language
API	Application Programming Interface
AS	Automation Systems
ASB	Automation Service Bus
ASE	Automation Systems Engineering
ATL	Atlas Transformation Language
AutomationML	Automation Markup Language
BFO	Basic Formal Ontology
BOM	Bill of Material
CA	Customer Attribute
CAD	Computer-Aided Design
CAEX	Computer-Aided Engineering Exchange
CC	Common Concepts
CE	Conformité Européene, European Conformity
CIM	Computer Integrated Manufacturing
CPK	Process Capability Index
CPPS	Cyber-Physical Production System
CSV	Comma Separated Values
CWA	Closed World Assumption
DB	DataBase
DE	Domain Expert
DIN	Deutsches Institut für Normung e.V.
DL	Description Logics
DP	Design Parameter
ECAD	Electrical CAD, Electrical Computer-Aided Design
EDB	Engineering Data Base
EDOAL	Expressive and Declarative Ontology Alignment Language
EKB	Engineering Knowledge Base

EMF	Eclipse Modeling Framework
EngSB	Engineering Service Bus
EO	Engineering Organization, Engineering Object
ERP	Enterprise Resource Planning
ETL	Extraction, Transformation and Load
FB	Feed Backward control
FF	Feed Forward control
F-Logic	Frame Logic
FLORA-2	F-Logic translator
FOAF	Friend-of-a-Friend
FPY	First Pass Yield
FR	Functional Requirement
HDM	Hyper-graph Data Model
HTML	Hypertext Markup Language
HTTP	Hypertext Transfer Protocol
I/O	Input/Output
I4.0	Industrie 4.0
IC	Integrated Circuit
ICT	Information and Communication Technologies
IDEF	Integrated Definition Methods
IEA	Intelligent Engineering Application
IRI	International Resource Identifier
JSON	JavaScript Object Notation
JSON-LD	JSON for Linking Data
KCMA	Knowledge Change Management and Analysis
KE	Knowledge Engineer
LD	Linked Data
LED	Linked Enterprise Data
LOD	Linked Open Data
MBE	Model-Based Engineering
MBSE	Model-Based Software Engineering
MCAD	Mechanical CAD, Mechanical Computer-Aided Design
MDE	Model-Driven Engineering
MDEng	Multidisciplinary Engineering
MDWE	Model-Driven Web Engineering
MES	Manufacturing Execution System
MM	MetaModel
MOF	Meta Object Facility
MOFM2T	MOF Model To Text Transformation Language
nUNA	Non-Unique Name Assumption
OBDA	Ontology-Based Data Access
OBII	Ontology-Based Information Integration
ODP	Ontology Design Pattern
OKBC	Open Knowledge Base Connectivity
OMG	Object Management Group

OPC UA	OPC Unified Architecture
OWA	Open World Assumption
OWL	Web Ontology Language
OWL DL	Web Ontology Language—Description Logic
OWL2	Web Ontology Language 2
OWL-S	OWL Services
P&ID	Piping and Instrumentation Diagram
PDF	Portable Document Format
PLC	Programmable Logic Controller
PM	Project Manager
ppm	parts per million
PPR	Product–Process–Resource
PPU	Pick and Place Unit
PROV-O	Provenance Ontology
PV	Process Variable
QFD	Quality Function Deployment
QVT	Query/View/Transformation
QVTo	QVT Operational
R2R	Run to Run control
RCS	Relational Constraint Solver
RDB	Relational Database
RDF	Resource Description Framework
RDF(S)	RDF Schema
RDFa	Resource Description Framework in attributes
RML	RDF Mapping Language
RM-ODP	Reference Model of Open Distributed Processing
ROI	Return on Investment
RQ	Research Question
RUP	Rational Unified Process
SCADA	Supervisory control and data acquisition
SDD	Specification-Driven Design
SE	Software Engineering
SEKT	Semantic Knowledge Technologies
SHACL	Shapes Constraint Language
SHOE	Simple HTML Ontology Extensions
SKOS	Simple Knowledge Organization System
SPARQL	SPARQL Protocol and RDF Query Language
SPC	Statistical process control
SPIN	SPARQL Inferencing Notation
SQL	Structured Query Language
SUMO	Suggested Upper Merged Ontology
SW	Semantic Web
SWRL	Semantic Web Rule Language
SWT	Semantic Web technologies
SysML	Systems Modeling Language

TCP	Transmission Control Protocol
TGG	Triple Graph Grammar
UC	Use Case
UML	Unified Modeling Language
UNA	Unique Name Assumption
URI	Uniform Resource Identifier
URL	Uniform Resource Locator
VCDM	Virtual Common Data Model
VDE	Verband der Elektrotechnik Elektronik Informationstechnik e.V. (Association for Electrical, Electronic & Information Technologies)
VDI	Verband Deutscher Ingenieure (Association of German Engineers)
W3C	World Wide Web Consortium
WG	Working Group
WWW	World Wide Web
XML	Extensible Markup Language
XSD	XML Schema Definition

Chapter 1
Introduction

Stefan Biffl and Marta Sabou

Abstract This chapter introduces the context and aims of this book. In addition, it provides a detailed description of industrial production systems including their life cycle, stakeholders, and data integration challenges. It also includes an analysis of the types of intelligent engineering applications that are needed to support flexible production in line with the views of current smart manufacturing initiatives, in particular *Industrie 4.0*.

Keywords Industrie 4.0 · Industrial production systems · Intelligent engineering applications · Semantic Web technologies

1.1 Context and Aims of This Book

Traditional industrial production typically provides a limited variety of products with high volume by making use of mostly fixed production processes and production systems. For example, a car manufacturer traditionally produced large batches of cars with the same configuration following the same process and using the same factory (i.e., production system). To satisfy increasingly diverse customer demands, there is a need to produce a wider variety of products, even with low volume, with sufficiently high quality and at low cost and risk. This is a major change of approach from traditional production because it requires *increased flexibility of the production systems and processes*.

S. Biffl (✉) · M. Sabou
Institute of Software Technology and Interactive Systems, CDL-Flex, Vienna University of Technology, Vienna, Austria
e-mail: Stefan.Biffl@tuwien.ac.at

M. Sabou
e-mail: Marta.Sabou@ifs.tuwien.ac.at

S. Biffl and M. Sabou (eds.), *Semantic Web Technologies for Intelligent Engineering Applications*, DOI 10.1007/978-3-319-41490-4_1

The move toward more flexible industrial production is present worldwide as reflected by relevant initiatives around the globe. Introduced in Germany, *Industrie 4.0*[1] is a vision for a more advanced production system control architecture and engineering methodology (Bauernhansl et al. 2014). Similar initiatives for modernizing industrial production systems have been set up in many industrial countries such as the *Industrial Internet Consortium* in the USA or the *Factory of the Future* initiative in France and the UK (Ridgway et al. 2013). A modern, flexible industrial production system is characterized by capabilities such as

1. *plug-and-participate of production resources* (i.e., machines, robots used in the production systems), such as a new machine to be easily used in the production process;
2. *self-* capabilities of production resources*, such as automated adaptation to react to the deterioration of the effectiveness of a tool or product; and
3. *late freeze of product-related production system behavior*, allowing to react flexibly to a changing set of products to be produced (Kagermann et al. 2013).

Achieving such flexible and adaptable production systems requires major changes to the entire life cycle of these systems, which, as described in Sect. 1.2, are part of a complex ecosystem combining diverse stakeholders and their tools. For example, the first step of the life cycle, the process of designing and engineering production systems needs to be faster and to lead to higher quality, more complex plants. To that end, there is a need to streamline the work of a large and diverse set of stakeholders which span diverse engineering disciplines (mechanical, electrical, software), make use of a diverse set of (engineering) tools, and employ terminologies with limited overlap (Schmidt et al. 2014). This requires dealing with heterogeneous and semantically overlapping engineering models (Feldmann et al. 2015). Therefore, a key challenge for realizing flexible production consists in intelligently solving data integration among the various stakeholders involved in the engineering and operation of production systems both across engineering domain boundaries and between different abstraction levels (business, engineering, operation) of the system.

Knowledge-based approaches are particularly suitable to deal with the data heterogeneity aspects of engineering production systems and to enable advanced capabilities of such systems (e.g., handling disturbances, adapting to new business requirements) (Legat et al. 2013). Knowledge-based systems support "(1) the explicit representation of knowledge in a domain of interest and (2) the exploitation of such knowledge through appropriate reasoning mechanisms in order to provide high-level problem solving performance" (Tasso and Arantes e Oliveira 1998). *Semantic Web technologies* (SWT) extend the principles of knowledge-based approaches to Web-scale settings which introduce novel challenges in terms of data size, heterogeneity, and level of distribution (Berners-Lee et al. 2001). In such

[1]Because the term *Industrie 4.0* is the name of a strategic German initiative, the term will be used in its German form, without translation to English.

setting, SWTs focus on large-scale (i.e., Web-scale) data integration and intelligent, reasoning-based methods to support advanced data analytics.

SWTs enable a wide range of advanced applications (Shadbolt et al. 2006) and they have been successfully employed in various areas, ranging from pharmacology (Gray et al. 2014) to cultural heritage (Hyvönen (2012) and e-business (Hepp 2008). A comparatively slower adoption of SWTs happened in industrial production settings. A potential explanation is that the complexity of the industrial production settings hampers a straightforward adoption of standard SWTs. However, with the advent of the *Industrie 4.0* movement, there is a renewed need and interest in realizing flexible and intelligent engineering solutions, which could be enabled with SWTs.

In this timely context, this book aims to provide answers to the following research question:

> How can SWTs be used to create *intelligent engineering applications (IEAs)* that support more flexible production processes as envisioned by *Industrie 4.0*?

More concretely the book aims to answer the following questions:

- Q1: What are semantic challenges and needs in *Industrie 4.0* settings?
- Q2: What are key SWT capabilities suitable for realizing engineering applications?
- Q3: What are typical Semantic Web solutions, methods, and tools available for realizing an IEA?
- Q4: What are example IEAs built using SWTs?
- Q5: What are the strengths, weaknesses, and compatibilities of SWTs with other technologies?

To answer these questions, this book draws on several years of experience in using SWTs for creating flexible automation systems with industry partners as part of the Christian Doppler Laboratory "Software Engineering Integration for Flexible Automation Systems": (CDL-Flex).[2] This experience provided the basis for identifying those aspects of *Industrie 4.0* that can be improved with SWTs and to show how these technologies need to be adapted to and applied in such *Industrie 4.0* specific settings. Technology-specific chapters reflect the state of the art of relevant SWTs and advise on how these can be applied in multidisciplinary engineering settings characteristics for engineering production systems. A selection of case studies from various engineering domains demonstrates how SWTs can enable the creation of IEAs enabling, for example, defect detection or constraint checking. These case studies represent work of the CDL-Flex Laboratory and other research groups.

We continue with a more detailed description of industrial production systems including their life cycle, stakeholders, and data integration challenges (Sect. 1.2). This is then followed by an analysis of what IEAs are needed to support flexible

[2]CDL-Flex: http://cdl.ifs.tuwien.ac.at/.

production in line with *Industrie 4.0* views (Sect. 1.3). We conclude with a readership recommendation and an overview on the content of this book in Sects. 1.4 and 1.5, respectively.

1.2 Industrial Production Systems

Industrial production systems produce specific kinds of *products*, such as automobile parts or bread, at high quality, low cost, and sufficiently fast (Kagermann et al. 2013). The design of the product to be produced in a production system (e.g., a factory, a manufacturing plant) defines the *production process*, i.e., the steps of production (e.g., gluing smaller parts together or drilling holes into a part), with their inputs and outputs (e.g., the raw input parts and the glued or drilled output part).

Figure 1.1 shows a small part of a production process for making bread. The process starts with a semifinished product, the *bread body*, which is input to the first

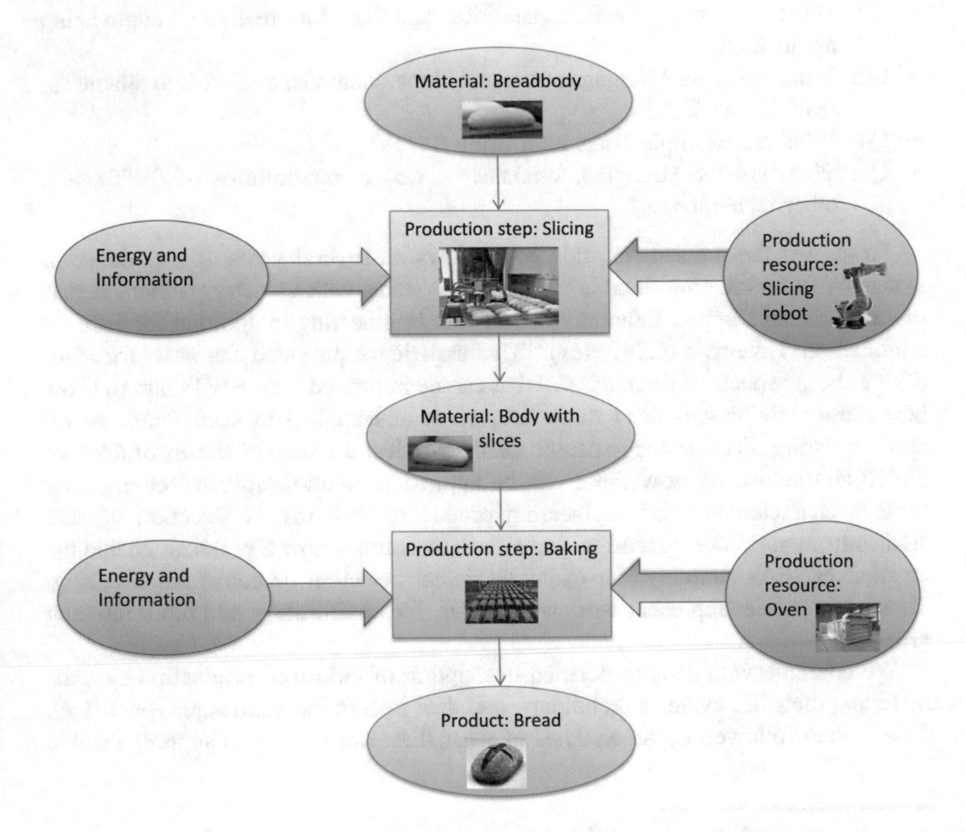

Fig. 1.1 Part of the production process for making bread

production step of slicing the top of the bread body. The output of this production step, *bread body with slices*, is the input to the next production step, baking the bread, which results in the final product, the *bread*, ready for packaging and delivery to customers. In an industrial production process context, each production step is supported with *production resources*, such as a robot with capabilities for slicing and an industrial oven for baking. The production process and resource need energy and they need to be controlled by programs based on information coming from sensors and human machine interfaces.

In general, the production process can be represented as a network consisting of several input parts and production steps that provide refined outputs and, in the end, the final product. The production steps require *production resources*, such as machines, that have the necessary *capabilities* to conduct the production activity, such as gluing or drilling, including support capabilities, e.g., handling the work piece during production (Tolio 2010).

Production resource capabilities can be provided by humans or machines. Figure 2.9 in Chap. 2 shows the example of a lab-size production system. Chapter 2 provides a more detailed view on industrial production systems and the engineering process of these production systems.

Figure 1.2 illustrates the engineering and operation of an industrial production system (Dorst 2015). There is an important distinction to be made between the two key phases in the life cycle of a production system. First, *the engineering phase* (left-hand side) concerns the planning and design of the production system. The engineering process starts on the top left-hand side with the business manager providing the business requirements to the engineers. During the engineering process representatives from several engineering disciplines, the customer, and project management need to design and evaluate a variety of engineering artifacts. Engineering artifacts include, but are not limited to: (1) the mechanical setup and function of the product and production system; (2) the electrical wiring of all

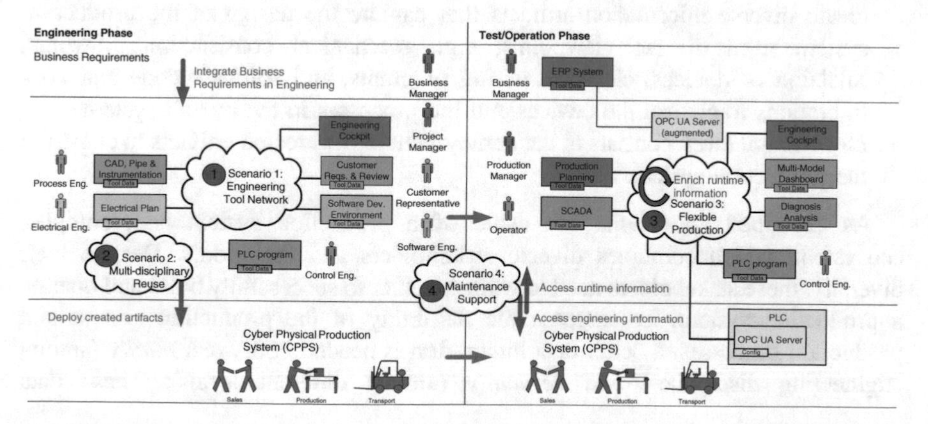

Fig. 1.2 Life cycle of industrial production systems: stakeholders, processes and Industrie 4.0-specific scenarios that enable increased production flexibility

devices used in the production system, such as sensors, motors, or actuators, and (3) the software to control the activities of all devices and to orchestrate the contributions of all devices into the overall desired production process. The safety of the production process is an important consideration during the design and evaluation of a production system. The production system design is the input to the construction and deployment of the system in the test and operation phase.

Second, *the test/operation phase* (right-hand side of Fig. 1.2) concerns the running production system, which can be tested, commissioned for production, and will eventually be monitored, maintained, and changed. A business manager uses an *enterprise resource planning* (ERP) system to schedule customer orders for production, based on a forecast of the available production capabilities in the system. On the production system level, the production manager and operator use manufacturing execution systems (MES) for production planning and control; and *supervisory control and data acquisition* (SCADA) systems to orchestrate the independent devices, which have to work together to conduct meaningful and safe production activities. Additionally to planning, other important functions in the test/operation phase are: diagnosis, maintenance, and reorganizing the production system. For example, OPC UA[3] servers provide data from the field level for integration with production planning to support the diagnosis of the current state of the production system.

Figure 1.2 also illustrates important *levels* over an industrial production system as well as the various stakeholders involved in these levels. These levels include (from top to bottom):

- *Business level*: the business manager determines the business requirements, e.g., which products shall be produced at what level of volume, which production process capabilities will be needed;
- *Engineering level*: the project manager, customer representative, and domain experts conduct the engineering process, in which experts from several domains work together to design the production system. During their work, engineers create diverse information artifacts that capture the design of the production system from diverse viewpoints, e.g., mechanical construction drawings, selection of devices, electrical wiring diagrams, and software code and configurations to control the devices and the processes in the overall system;
- *Deployment level*: consists of the deployment of the created artifacts to construct the production system.

As described above, the life cycle of a production system is a complex ecosystem, which combines diverse stakeholders and their tools. Despite their diversity, these stakeholders need to work together to successfully build and operate a production system. To increase the flexibility of the production system and production processes, a better data integration is needed both *horizontally* (among engineering disciplines) and *vertically* (among different levels). These data

[3]OPC UA: https://opcfoundation.org/about/opc-technologies/opc-ua/.

integration processes lay the foundation for IEAs both during the engineering and the test/operation of industrial production systems, as we describe next.

- *Horizontal data integration* includes the data exchange between different engineering disciplines, e.g., mechanical, electrical, and software engineering, which use different terminologies, methods, and tools. Such data exchange is challenging because typical engineering applications are software tools for a specific domain, which know only little about the production system engineering process as a whole or other engineering domains. There is a need for *IEAs* that can build on data integrated over several domains, e.g., to allow searching for similar objects in the plans of several engineering domains, even if terminologies differ.
- *Vertical data integration* also covers the data exchange between systems used to manage the different levels of a production system: business systems, engineering systems, and systems deployed in the field. Traditionally, the data formats on these levels differ significantly and make it hard to communicate changes between the business and field levels leading to applications that are limited to using the data on a specific level. There is a need for *IEAs* that can build on data integrated over several levels, e.g., for fast replanning of the production system operation in case of disturbances in the production system or changes in the set of product orders from customers.

These life cycle views provide the context to consider the contributions of engineering applications and how these can benefit from SWTs.

1.3 Intelligent Engineering Applications for Industrie 4.0

The *Industrie 4.0* vision addresses the question of how to provide sufficient flexibility at reasonable cost by representing the major process steps for the life cycle of a product and the life cycle of a production system, which allows producing the product, such as bread or automobiles, as input for the analysis of dependencies between product and production system. The upper half of Fig. 1.3 presents the relevant *life cycle phases of the product* while the lower half depicts the *life cycle phases for the production system* to be considered (VDI/VDE 2014a). The arrows crossing the line between the upper and lower halves provide a focus for the integrated consideration of product and production systems engineering (see also Fig. 2.1).

In Fig. 1.3, the *product life cycle* considers the engineering of products, such as a variety of bread types and automobile types, to be produced in a production system based on customer orders and the development and maintenance of the product lines containing the products. These product lines will impact the required capabilities of the production system. Based on possible products, marketing and sales will force product order acquisition.

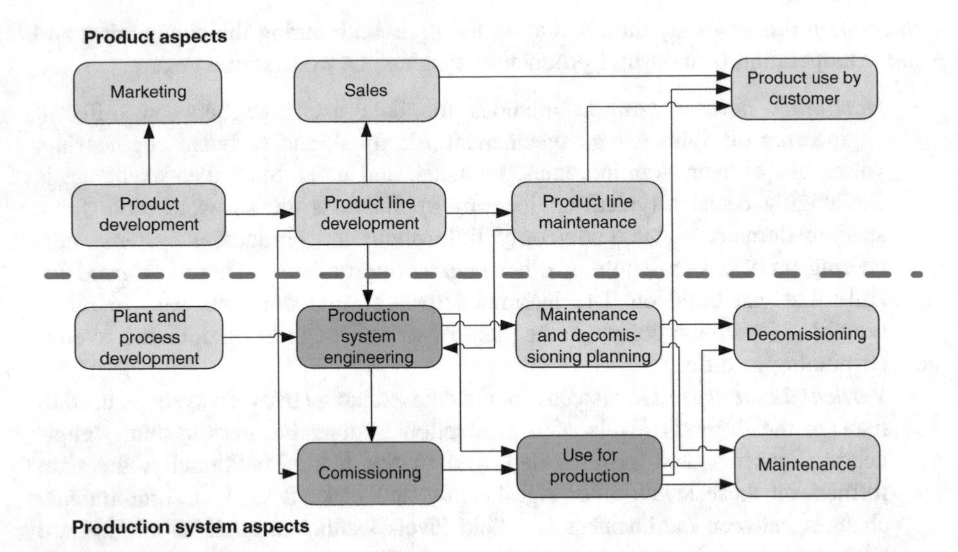

Fig. 1.3 Value-chain-oriented view on the product and production system life cycle based on (VDI/VDE 2014a)

The *life cycle of production systems* covers the main phases: *Plant and process development*, *Production system engineering*, *Commissioning*, *Use for production*, *Maintenance and decomposition planning*, *Maintenance*, and *Decommissioning*. In these phases, information related to products, production system components, and orders are required and processed leading to a network of information processing entities including humans using engineering tools and automated information processing within machines.

In summary, the current considerations in *Industrie 4.0* require that *information processing has to be enhanced toward a semantically integrated approach*, which allows data analysis on data coming both from product and production system lifecycle processes. In production system engineering, the current focus on data processing has to be moved on to information processing of semantically enriched data.

The vision of *Industrie 4.0* is much broader than creating flexible production systems, as described above. In fact, *Industrie 4.0* envisions the meaningful integration of life cycles relevant for production systems. These life cycles include the important step of engineering (i.e., designing and creating) industrial production systems. The main starting point of *Industrie 4.0* is the integrated consideration of production system life cycles (VDI/VDE 2014a), which include the engineering of industrial production systems.

In this context, *an engineering application is a software tool or a set of software tools for supporting engineering activities*, e.g., for product design and evaluation, e.g., of an automobile or production system part. An *intelligent engineering application* provides functionalities that seem intelligent, e.g., complex analytics for the optimization of product or production process properties, which are hard to automate. IEAs are a foundation to enable effectively and efficiently key engineering capabilities in industrial production systems, including plug-and-participate of production resources, such as a new machine to be used in the production process (Kagermann et al. 2013).

Figure 1.3 shows that IEAs can depend on information from a wide variety of sources in the engineering process, such as

- the bill of materials, e.g., for describing the materials needed for production,
- the production floor topology, e.g., the layout of production resources,
- the mechanical structure of a set of machines, e.g., robots in a manufacturing cell,
- the wiring plan, e.g., information cables between production resources and control computers, and
- the behavior plan, e.g., software controlling production process of a machine or the orchestration of a complex production process with many steps and sources of disturbances.

Unfortunately, there are many heterogeneous data models used in these information sources, for example, geometric and kinematic models, wiring plans, behavior specifications, and software programs in various representations. The *variety of data sources* is a major challenge that may prevent the sufficiently effective and efficient data exchange between engineering applications and their users.

To enable the engineering and production processes for flexible production systems, integrated information processing intends to ensure the lossless exchange and correct (meaningful) application of engineering and run-time information of a production system to gain additional value and/or to avoid current limitations of production system engineering and use.

In Fig. 1.2, the production system engineering process starts on the top left-hand side with providing the business requirements to the engineers. During the engineering process representatives and tools from several engineering disciplines, the customer, and project management need to design and evaluate a variety of engineering artifacts. These activities run in parallel and may include loops, which may lead to a complex flow of artifact versions in the network of tools used by the project participants. The semantics of engineering data have to be clarified in such a tool network to enable the systematic definition of processes that can be automated to support the domain experts in achieving their goals. SWTs have been shown to be a very good match for addressing the aspects of heterogeneity in data processing for a variety of fields due to their capability to integrate data intelligently and flexibly on a large scale (Shadbolt et al. 2006).

In Chap. 2, we discuss four scenarios (see the red numbered circles in Fig. 1.2) to illustrate the needs for Semantic Web capabilities in industrial production systems engineering and operation.

The first scenario, "*Discipline-crossing Engineering Tool Networks,*" explains in details the goals, challenges, and needs for Semantic Web capabilities in the context of the engineering phase of a single engineering project. This scenario considers the capability to interact appropriately within an engineering network covering different engineering disciplines, engineers, and engineering tools. The scenario further highlights the need for a common vocabulary over all engineering disciplines involved in an engineering organization creating a production system to enable fault free information propagation and use.

The second scenario, "*Use of existing Artifacts for Plant Engineering,*" has a focus on knowledge reuse (and protection) within engineering organizations. This scenario considers the problem of identification and preparation of reusable production system components within or at the end of an engineering project and the selection of such components within engineering activities. Here, the focus is on the required evaluation of component models to decide about the usability of the component within a production system. IEAs can help to analyze candidate components for reuse to support the engineer in evaluating reuse benefits and risks of a large number of candidate components.

The third scenario, "*Flexible Production System Organization,*" details the problem of run-time flexibility of production systems. Here, requirements following the intention of integration of advanced knowledge about the production system and the product within the production system control at production system runtime are sketched. Traditional production systems are fixed and hard to extend, e.g., for including new equipment for monitoring. For a flexible production system, an information system is needed to flexibly integrate production run-time data with engineering knowledge. This facilitates the automation of production planning on the business level, e.g., planning of feasible order volume in a given period, and production scheduling level, e.g., production resource availability and status of production jobs.

The fourth scenario, "*Maintenance and Replacement Engineering,*" describes situations where engineering and run-time information of a production system are combined toward improved maintenance capabilities of production system components. In traditional production systems engineering, the outcomes of the plant engineering process are printed documents on paper or as PDF files, not the engineering models created during the engineering phase. This practice may be insufficient for a flexible production system, if the stakeholders during operation need to reconfigure the production system, e.g., add components with new capabilities. A key question is how to provide engineering knowledge from the engineering phase on the left-hand side in Fig. 1.2 to the operation phase on the right-hand side in Fig. 1.2: what kind of engineering knowledge, made explicit in engineering models, will be needed, and what data exchange format is likely to be most useful?

From these scenarios, the authors of Chap. 2 derive four groups of needs for engineering data integration capabilities:

- engineering knowledge/data representation, integration, and analytics;
- efficient access to semi-structured data in the organization and on the Web;
- flexible and intelligent engineering applications and process knowledge support;
- provision of integrated engineering knowledge at production system run time.

The current approaches for modeling engineering knowledge have shortcomings that SWTs can help overcome. For example, major semantic challenges come from the need to provide tool support for processes that build on heterogeneous terms, concepts, and models used by the stakeholders in production system engineering and operation. Also, most of the knowledge is only implicitly given within the engineering and run-time artifacts of a production system, and has to be modeled and made explicit for further (re-)use. Improved support for the modeling of the semantics of engineering artifacts is required. Chapters 2 and 3 will introduce SWTs and their suitability to address important needs coming from engineering processes, which should be supported with advanced IEAs.

1.4 Who Should Read This Book and Why?

This book aims to bridge the communities of industrial production on one hand and Semantic Web on the other. Accordingly, stakeholders from both communities should find this book useful in their work.

Engineers and managers from engineering domains will be able to get a better understanding of the benefits and limitations of using SWTs. Moreover, thanks to the overviews of available technologies as well as the provided best practices for using these, engineers will be enabled to select and adopt appropriate SWTs in their own settings more effectively. Researchers and students interested in industrial production-related issues will get an insight into how and to what extent SWTs can address these issues.

Semantic Web researchers will gain a better understanding of the challenges and requirements of the industrial production domain especially in the light of the emerging *Industrie 4.0* requirements. This will support and guide Semantic Web researchers in developing new technologies and solutions for this important application area more effectively and efficiently.

1.5 Book Content and Structure

This book is structured in four parts, as follows.

Part I: Background and Requirements of *Industrie 4.0* for Semantic Web Solutions. Part I provides the necessary background information for understanding the rest of the book and covers questions Q1 and Q2 (see Sect. 1.1). Concretely, Chap. 2 describes the problem setting of engineering complex industrial production systems that match the *Industrie 4.0* vision. Chapter 3 introduces SWTs as a solution alternative for addressing the challenges raised by *Industrie 4.0* settings.

Part II: Semantic Web-Enabled Data Integration in Multidisciplinary Engineering. A main conclusion from Part I is that the engineering of complex industrial production systems that match the *Industrie 4.0* requirements happens in highly heterogeneous settings and that SWTs are, by their design, well suited for dealing with such heterogeneity through data integration approaches. Therefore, Part II focuses on how SWTs can be used for data integration in heterogeneous, multidisciplinary engineering settings typical in the creation of flexible production systems. Chapter 4 introduces the general data integration framework called the *Engineering Knowledge Base* (EKB), while the subsequent chapters focus on methods and tools for addressing the various aspects in this overall framework, namely: semantic modeling of engineering knowledge by using ontologies and the transformation of engineering knowledge elements into semantic data (Chap. 5); creating mappings between the semantic data derived from different engineering disciplines (Chap. 6); and managing changes in engineering data (Chap. 7). As such, this part covers question Q3.

Part III: Creating Intelligent Applications for Multidisciplinary Engineering. While Part II focuses on a set of methods necessary for data integration in multidisciplinary engineering settings, Part III demonstrates how the integrated engineering data can be used to support the creation of IEAs in line with question Q4. Chapter 8 describes the technical implementation of the data integration framework introduced in Chap. 4. The subsequent chapters focus on presenting IEAs that are enabled by and built on top of the integrated engineering data, namely: product ramp-up (Chap. 9) and industrial simulation (Chap. 10).

Part IV: Related and Emerging Trends in the use of Semantic Web in Engineering. Part II and Part III focus on a particular use of SWTs for creating IEAs as developed within the *CDL-Flex* research laboratory. Part IV complements these two previous parts with an outlook on the broader spectrum of approaches that make use of SWTs to support engineering settings. Chapter 11 provides an overview of the field and concludes with a synthesis of emerging trends in this area. As such this chapter places the work performed in *CDL-Flex* within the landscape of related research and motivates the rest of the chapters in part IV. These chapters contribute insights into how SWTs were used in the automotive industry (Chap. 12), for configuration management (Chap. 13), and in the domain of automated production systems (Chap. 14), thus providing further answers to question Q4.

Chapter 15 concludes the book with answers to question Q5 and an outlook on future opportunities in applying SWTs for creating IEAs in the setting of flexible industrial production systems.

Acknowledgments This work was supported by the Christian Doppler Forschungsgesellschaft, the Federal Ministry of Economy, Family and Youth, and the National Foundation for Research, Technology and Development in Austria.

References

Bauernhansl, T., ten Hompel, M., Vogel-Heuser, B. (eds.): Industrie 4.0 in Produktion, Automatisierung und Logistik. Springer (2014)

Berners-Lee, T., Hendler, J., Lassila, O.: The Semantic Web, pp. 29–37. Scientific American (2001)

Dorst, W. (ed.): Umsetzungsstrategie Industrie 4.0—Ergebnisbericht der Plattform Industrie 4.0. https://www.bitkom.org/Publikationen/2015/Leitfaden/Umsetzungsstrategie-Industrie-40/150410_Umsetzungsstrategie_0.pdf (2015). Accessed Nov 2015

Feldmann, S., Herzig, S.J.I., Kernschmidt, K., Wolfenstetter, T., Kammerl, D., Qamar, A., Lindemann, U., Krcmar, H., Paredis, C.J.J., Vogel-Heuser, B.: Towards effective management of inconsistencies in model-based engineering of automated production systems. In: Proceedings of IFAC Symposium on Information Control in Manufacturing (INCOM) (2015)

Hyvönen, E.: Publishing and Using Cultural Heritage Linked Data on the Semantic Web. Series: Synthesis Lectures on Semantic Web, Theory and Technology. Morgan and Claypool (2012)

Hepp, M.: GoodRelations: An ontology for describing products and services offers on the web. In: Proceedings of the 16th International Conference on Knowledge Engineering and Knowledge Management (EKAW), vol. 5268, pp. 332–347. Springer LNCS, Acitrezza, Italy (2008)

Gray, A.J.G., Groth, P., Loizou, A., Askjaer, S., Brenninkmeijer, C., Burger, K., Chichester, C., Evelo, C.T., Goble, C., Harland, L., Pettifer, S., Thompson, M., Waagmeester, A., Williams, A.J.: Applying linked data approaches to pharmacology: architectural decisions and implementation. Semant. Web J. **5**(2), 101–113 (2014)

Kagermann, H., Wahlster, W., Helbig, J. (eds.): Umsetzungsempfehlungen für das Zukunftsprojekt Industrie 4.0—Deutschlands Zukunft als Industriestandort sichern, Forschungsunion Wirtschaft und Wissenschaft, Arbeitskreis Industrie 4.0 (2013)

Legat, C., Lamparter, S., Vogel-Heuser, B.: Knowledge-based technologies for future factory engineering and control. In: Borangiu, T., Thomas, A., Trentesaux, D. (eds.) Service Orientation in Holonic and Multi Agent Manufacturing and Robotics, Studies in Computational Intelligence, vol. 472, pp. 355–374. Springer, Berlin (2013)

Ridgway, K., Clegg, C.W., Williams, D.J.: The Factory of the Future, Government Office for Science, Evidence Paper 29 (2013)

Shadbolt, N., Berners-Lee, T., Hall, W.: The semantic web revisited. IEEE Intell. Sys. **21**(3), 96–101(2006)

Schmidt, N., Lüder, A., Biffl, S., Steininger, H.: Analyzing requirements on software tools according to functional engineering phase in the technical systems engineering process. In: Proceedings of the 19th IEEE International Conference on Emerging Technologies and Factory Automation (ETFA). IEEE (2014)

Tasso, C., Arantes, E., Oliveira, E. (eds.): Development of Knowledge-Based Systems for Engineering. Springer, Wien (1998)

Tolio, T.: Design of Flexible Production Systems—Methodologies and Tools. Springer, Berlin (2010)

References

Part I
Background and Requirements of Industrie 4.0 for Semantic Web Solutions

Chapter 2
Multi-Disciplinary Engineering for Industrie 4.0: Semantic Challenges and Needs

Stefan Biffl, Arndt Lüder and Dietmar Winkler

Abstract This chapter introduces key concepts of the *Industrie 4.0* vision, focusing on variability issues in traditional and cyber-physical production systems (CPPS) and their engineering processes. Four usage scenarios illustrate key challenges of system engineers and managers in the transition from traditional to CPPS engineering environments. We derive needs for semantic support from the usage scenarios as a foundation for evaluating solution approaches and discuss Semantic Web capabilities to address the identified multidisciplinary engineering needs. We compare the strengths and limitations of Semantic Web capabilities to alternative solution approaches in practice. Semantic Web technologies seem to be a very good match for addressing the aspects of heterogeneity in engineering due to their capability to integrate data intelligently and flexibly on a large scale. Engineers and managers from engineering domains can use the scenarios to select and adopt appropriate Semantic Web solutions in their own settings.

Keywords Engineering process · Systems engineering · Cyber-Physical production systems · Semantic Web · Data integration

S. Biffl · D. Winkler (✉)
Institute of Software Technology and Interactive Systems,
CDL-Flex, Vienna University of Technology, Vienna, Austria
e-mail: dwinkler@sba-research.org; dietmar.winkler@tuwien.ac.at

S. Biffl
e-mail: stefan.biffl@tuwien.ac.at

A. Lüder
Otto-von-Guericke University/IAF, Magdeburg, Germany
e-mail: arndt.lueder@ovgu.de

D. Winkler
SBA-Research gGmbH, Vienna, Austria

S. Biffl and M. Sabou (eds.), *Semantic Web Technologies for Intelligent Engineering Applications*, DOI 10.1007/978-3-319-41490-4_2

2.1 Introduction

Production systems of any kind have to face two main drivers for evolution, (1) technical developments related to useable technologies and (2) customer requirement developments. The latter driver typically leads to product diversification. Related to technical development, especially development in the IT industry, increased communication capabilities have an important impact (Jacob 2012) and lead to the need of faster evolution of production systems (Kagermann et al. 2013).

Production systems are means for the creation of products. Following Grote and Feldhusen (Grote and Feldhusen 2014), the aim of a production system is the value-generation-based creation of goods, i.e., the creation of products. Therefore, the starting point of understanding of the engineering needs of production systems is the product. A *product* is defined as an item that ideally satisfies customer needs related to its application (Grote and Feldhusen 2014). These needs are considered within the product design as input providing main requirements to product design. In addition, product design has to consider technical capabilities of production systems as bordering conditions. Examples of products are items, which are of interest for an end user, such as cars, washing machines, mobile phones, clothes, or food, with product-related customer expectations, such as passenger safety, low energy and water consumption, internet access, style, or taste. In addition, electrical drives, cement, or oxygen need to be seen as products with more technical customer requirements, such as integrability into a production system, speed of setting, or usability in medical systems.

Within the *product design*, which has to be seen as a creative process of engineering the product: on the one hand the structure, visual nature, behavior, or functionality of the product are defined; on the other hand the way to create the product, i.e., the process, needs to be developed. This includes the definition of materials to be used and the definition of production steps to be executed. Materials cover all physical and nonphysical elements to be purchased for production, such as steel plates for cars or agriculture objects for food production. Production steps cover all required value-adding actions needed within the production process. Examples are welding of steel plates, mounting of components in car manufacturing, or cleaning and cooking in food processing.

Figure 2.1 models these dependencies between the product and the production system. Based on the defined way of product creation, the *production system* can be engineered. Therefore, under consideration of technological and economical possibilities, the set of required production steps is translated by a team of engineers coming from different engineering disciplines into a set of *production system resources* that are able to execute the production steps on the defined materials. Examples of such resources are welding robots required for steel plate welding, human workers required for the mounting of components in car manufacturing, a washing belt for agriculture product cleaning, and a steamer for cooking in food processing. These resources will be engineered in detail, implemented and finally used to create the products.

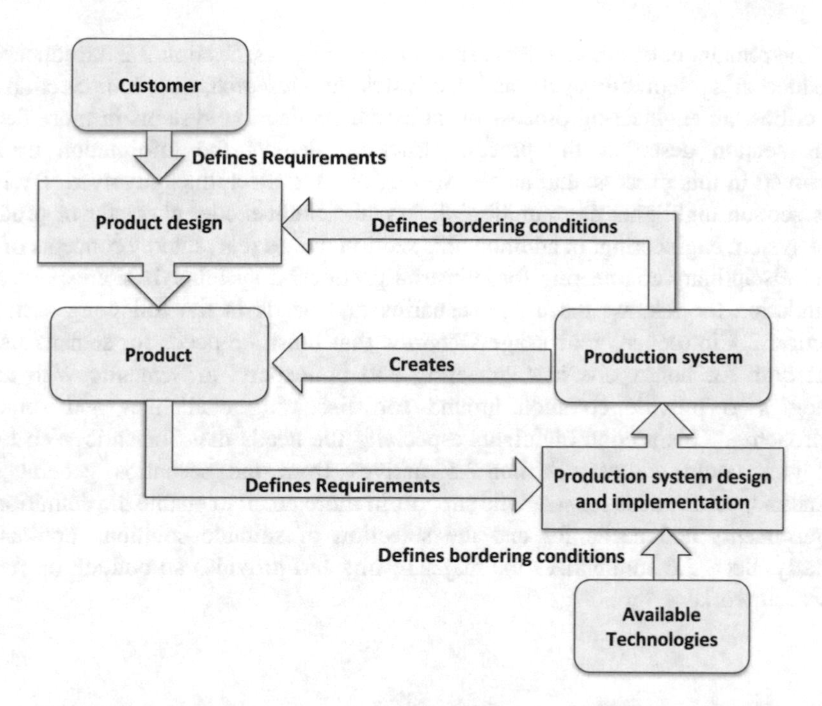

Fig. 2.1 Relations between product and the production system

The design and implementation of industrial production systems require the collaboration of several engineering disciplines. Most relevant among them are industrial plant planning, production process planning, and mechanical, electrical, and software engineering. Chapter 1 introduced a general notion of the basic concepts of *industrial production systems* and the need for *intelligent engineering applications* when moving from traditional to more flexible industrial production systems. This chapter builds on these general concepts to discuss semantic challenges and needs coming with the heterogeneous data models used in the multi-disciplinary engineering of industrial production systems.

The different engineering disciplines involved in the engineering of production systems apply engineering methodologies, tools, and data sets tailored to the needs of these engineering disciplines. These experts from different domains in production systems engineering are aware of the challenges coming from heterogeneous data models in the tool landscapes they use and want to better understand their options for a roadmap for technology adoption to effectively and efficiently move toward sufficient data integration. However, most production systems engineers are not experts in Semantic Web technologies and only few Semantic Web technology experts have a deep understanding of industrial production systems, their design, and evaluation. Therefore, this chapter intends to provide a foundation for the discussion between production systems engineering experts and Semantic Web technology experts on semantic challenges, needs, and options.

The remainder of this chapter is structured as follows. Section 2.2 introduces the production system life cycle and motivates the research question. Section 2.3 describes the engineering process of industrial production systems in more detail. This section describes the process structure, depicts the information usually involved in this process, and names the engineering disciplines involved. By that, this section highlights the multidisciplinary and multi-model character of production system engineering. In addition, this section names relevant key concepts of the multidisciplinary engineering for industrial production systems for nonexperts as a foundation for relating the usage scenarios and needs in the following sections. Section 2.4 introduces four usage scenarios that illustrate needs for semantic support both for nonexperts in engineering and nonexperts in Semantic Web technologies to provide common ground for discussing challenges and solution approaches. This section highlights especially the needs that Semantic Web technologies could address. Section 2.5 derives from the scenarios general and domain-specific needs for semantic support in more detail to enable the definition of requirements and needs for and the selection of suitable solution approaches. Finally, Sect. 2.6 summarizes the major results and provides an outlook on future research work.

2.2 Production Systems Life Cycle

Input to the production system engineering are the production steps and involved material required to produce the products intended and the technological and economical possibilities for production. Based on this input, in a first step the set of *production system resources* is selected. Here for example the type of welding robot required for the welding process of a special car is identified and named. To each production process step at least one (production system) resource is assigned. All selected resources will be put into a sequence to roughly define the production system. This assignment and sequencing is validated against economic conditions. If the resource assignment is successfully verified, each of the production resources is designed in detail in the next step of production system engineering. For the welding robot for example the welding gun is selected by the process engineer, and the mounting position is defined by the mechanical engineer, the energy supply is engineered by the electrical engineer, and the motion path is programmed by the robot programmer. The consistency of the overall engineering can be validated in the virtual commissioning using simulations. Once the detailed engineering is completed, the production system can be installed (i.e., set up physically) and commissioned (i.e., filled with control programs and started sequentially for the first time). If the commissioning was successful, the production system can be used for processing products. Over time, each running production system needs to be maintained to ensure a safe and long living operation. If the production system is not required anymore (for example the products cannot be sold anymore), it can be redesigned or decommissioned. Figure 2.2 shows this general process.

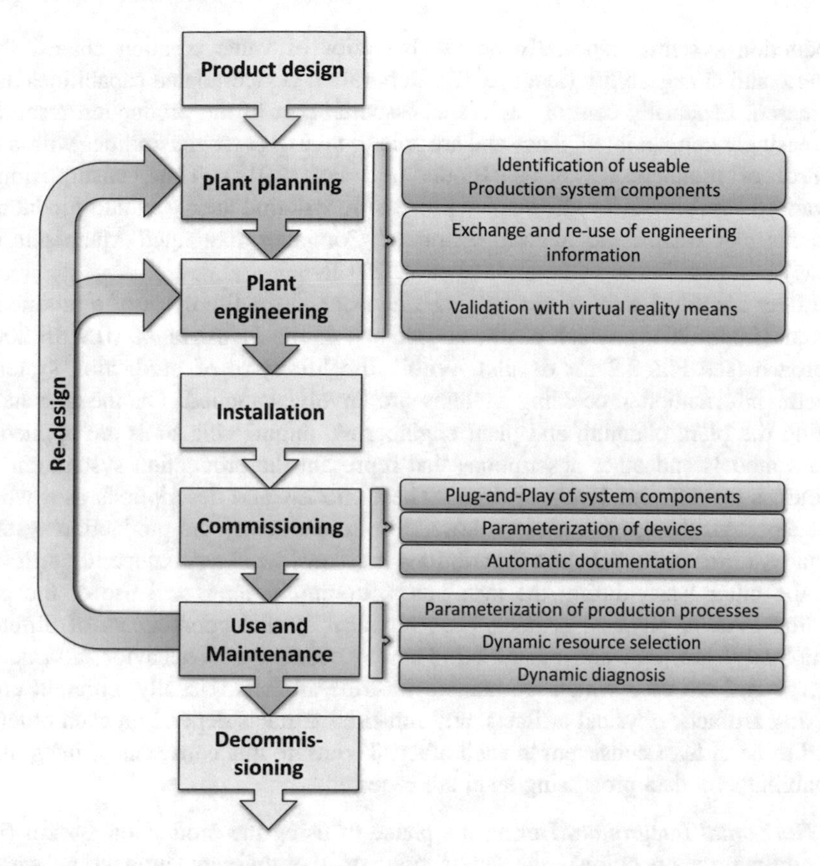

Fig. 2.2 Production system life cycle processes based on (Lüder 2014)

All technologies applicable in production systems can evolve. These developments enable new technical possibilities for the design and application of production systems. Envisioned possibilities shall include

- *plug-and-participate capabilities* of production resources (i.e., the integration and use of new or changed production resources during production system use without any changes within the rest of the production system),
- *self-* capabilities* of production resources (e.g., self-programming of production process control, self-maintenance in case of technical failures, or self-monitoring for quality monitoring), and
- *late freeze of product-related production system behavior* (i.e., fixing the characteristics of an ordered product at the latest possible point before production step execution, e.g., enabling to change the ordered color of a car until the start of painting) to name a few examples (Kagermann et al. 2013).

One step ahead of multidisciplinary engineering, information science and information technology have reached a point enabling a wide-ranging impact on production systems, especially on the behavior of value creation chains, their design, and changeability (Jacob 2012). In parallel, device-related capabilities have increased. Especially control devices as essential part of the production resources increasingly contain intelligence and are able to take over responsibilities within the control of production systems (Binder and Post 2012). Thus, unsurprisingly, advanced capabilities for information processing will find their way into production systems and realize the historic vision of Computer Integrated Manufacturing (CIM) in a new fashion (Kagermann et al. 2013).

Lüder identified a set of challenging situations in the life cycle of a production system (Lüder 2014), which can be tackled within the *Industrie 4.0* (BMBF 2007) approach (see Fig. 2.2 for details). Within the life cycle of production systems, several information processing systems are involved/applied. On the one hand, during the plant planning and plant engineering, engineering tools are applied to create models and other descriptions that represent the production system and its resources on different levels of detail. These models and descriptions as a whole will represent the specification on how to build and apply the production system. Usually, different models and descriptions are considered as engineering artifacts. On the other hand, during the installation, commissioning, and use of the production system, physical artifacts (i.e., physical system components of different complexity) are used and controlled. Thereby, control and behavior information emerge and are used which are named run-time artifacts. Usually, different engineering artifacts, physical artifacts, and run-time artifacts depend on each other or need to be at least consistent to each other. To ensure this consistency, integration capabilities on data processing level are required.

- *Horizontal Integration*. During the phase of using the production system (i.e., runtime or operation), the interaction of the different production system resources (possibly located at different geographical locations) and its control systems as well as its interaction with the production system environment (e.g., delivery of materials, energy consumption, waste disposal, or product delivery) need to be coordinated. This is considered as *horizontal integration* within a value chain network. Horizontal integration shall enable the automatic integration of new production resources within a production system, in the same way as today USB devices are integrated into a PC system by using plug-and-play mechanisms. It also shall enable the automation of routine tasks, such as process documentation or diagnosis of system components.
- *Vertical Integration*. During the development phase of the production system, starting with the plant planning until the use of the production system and its maintenance, it is of interest to coordinate the different steps of artifact creation and to ensure the availability of all necessary information/artifacts developed in previous engineering process steps. Therefore, an integration of engineering activities, engineering tools, and engineering disciplines is required, enabling the exchange and possibly reuse of developed information. This is named *vertical integration*. Vertical integration shall enable the automatic application

of the correct parameters out of the engineering information to configure a production resource or to dynamically select the right production resource for a special product order.

Figure 2.2 depicts an assignment of the challenges to life cycle phases. These challenging situations can only be tackled by enabling the adequate application of information and knowledge about the product, the product structure, and required production processes, the production system architecture and resources, their properties, and their possible production processes within the engineering and control of the production system. However, this adequate application is not possible yet. Most of the knowledge is only implicitly given within the engineering and run-time artifacts of a production system, and have to be modeled and made explicit for further use. Improved support is required for the modeling of the semantics of engineering artifacts as well as the subsequent usage of these semantics.

To deal with these challenges and to start a discussion between the two communities of production system engineering and information sciences related to these topics topic, this chapter intends to address research question (RQ) on the "**need for semantic support in multidisciplinary production system engineering**." Domain experts in production systems engineering typically are not also experts in Semantic Web technologies. However, these experts can provide usage scenarios in multidisciplinary engineering, which illustrate needs for semantic support, which pose challenges to the domain experts in their daily work. We aim at identifying recurring needs for semantic support in these scenarios to provide a foundation for communicating these challenges to Semantic Web technology experts, who typically are not also experts in multidisciplinary engineering.

To address this research question, we applied the following *methods for research*. We synthesized representative usage scenarios from literature study and own experience. Following the process of conceiving, instantiating, and changing a production system during engineering and operation phases, we identified recurring engineering activities that involve significant challenges regarding heterogeneous and/or implicit expert knowledge, which is necessary to conduct and possibly automate the engineering activity, as needed for CPPS engineering. We analyzed the needs for semantic support in these scenarios and identified common needs. The set of needs was validated and extended in discussions with domain experts, who adapted the generic scenarios to their own engineering contexts.

This chapter provides the following contributions for scientific communities and target audiences of the book. Engineers and managers from engineering domains will be able to get a better understanding of the benefits and limitations coming from using Semantic Web technologies. Engineers and managers from engineering domains can use the scenarios to select and adopt appropriate Semantic Web solutions in their own settings. Semantic Web researchers can use the scenarios and needs to get a better understanding of the challenges and requirements of the multidisciplinary engineering domain, which will support and guide the researchers in developing new technologies and solutions for this important application area more effectively and efficiently. Within this chapter, key challenges for semantic modeling exploited in production system engineering are identified. Therefore, selected important *Industrie 4.0* challenges are aggregated in four main scenarios of the application of ideas of *Industrie 4.0*: two scenarios related to engineering and two scenarios related to production system usage phases in the overall life cycle of a production system. These scenarios are recurring scenarios not only discussed within the collected requirements for *Industrie 4.0* (see Kagermann et al. 2013) but also addressed in several research activities.[1]

(1) The first scenario "*Discipline-crossing Engineering Tool Networks*" considers the capability to interact appropriately within an engineering network covering different engineering disciplines, engineers, and engineering tools. This scenario highlights the need for a common vocabulary and technical interfaces between engineering tools over all engineering disciplines involved in an engineering organization creating a production system to enable a fault-free information propagation and use.
(2) The second scenario "*Use of existing Artifacts for Plant Engineering*" has a focus on knowledge reuse and protection within engineering organizations. This scenario considers the problem of identification and preparation of reusable production system components within or at the end of an engineering project, and the selection of such components within engineering activities. Here, the focus is on the required evaluation of component models to decide about the potential usability of the component within a production system.
(3) The third scenario, "*Flexible Production System Organization*," discusses the problem of run-time flexibility of production systems. This scenario sketches requirements following the intention of integration of advanced knowledge about the production system and the product within the production system control at production system runtime.
(4) Finally, scenario four, "*Maintenance and Replacement Engineering*," combines engineering and run-time information of a production system toward improved maintenance capabilities of production system components.

These scenarios allow researchers and practitioners from the Semantic Web area to better understand the challenges of production system engineers and managers, to define goals and assess solution options using Semantic Web technologies. Major semantic challenges come from the need to provide tool support for processes that build on heterogeneous terms, concepts, and models used by the stakeholders in production system engineering and operation. From the challenges illustrated in the four scenarios, we derive needs for semantic support, including the support for multidisciplinary engineering process knowledge from the usage scenarios, as a foundation for evaluating solution approaches.

[1]AutomationML projects: https://www.automationml.org/o.red.c/projects.html.

2.3 Engineering of Industrial Production Systems

In Sect. 2.2, the life cycle of production systems and, especially, its engineering phase has been described on a high level. To enable the detailed evaluation of the multidisciplinary character of production system engineering, we will describe in this section the engineering process of production systems in more detail. Following the Engineers Council for Professional Development (Science 1941), an engineering process is defined as a sequence of activities that creatively apply scientific principles to design or develop structures, machines, apparats, or manufacturing processes; all with respect to reach an intended function for economic and safe operation. Thus, an engineering process is based on a sequence of design decisions to be taken. Engineering processes are executed by engineering organizations (EOs). Following the VDI regulations (VDI 2010), an EO can be an engineering company, an engineering subunit of a company, or a combination of different companies and/or subunits executing one or more engineering processes for a limited time and providing/containing the necessary resources for process execution. Thus, each EO will execute at least one engineering process described by a procedure model. The procedure model determines the sequence of engineering activities, covering design decisions with required input and output information, tools to be used, and responsible roles. The procedure model formalizes the engineering process and reflects the technical content of an engineering process in an organization.

Figure 2.3 shows the structure of an engineering activity, which can be also seen as a design decision or a set of design decisions. Within an engineering process for each engineering activity, the following requirements should be met: to enable the execution of the engineering activity, (1) a set of predecessor engineering activities has to be finished, (2) specific information has to be available, and (3) appropriate engineering tools should be usable. Humans with the required skills and knowledge should execute the engineering activities. The engineering activity will create information reflecting the taken design decisions and provide them for successive design decisions.

In addition, engineering activities establish a network of activities. As described by Lindemann (Lindemann 2007), the set of engineering activities containing the design decision execution has a hierarchical structure (see Fig. 2.4). On the highest level of the engineering process, design decisions can be considered as engineering process phases. These phases contain engineering activities, like control programming, which can be separated into sub-activities, like hardware configuration and code design. Again, these sub-activities can be divided into sub-sub-activities, like I/O configuration. Finally, the hierarchical decomposition of engineering activities ends with elementary actions, e.g., naming input and output variables of controls, defining the wiring path between a control and a sensor, or specifying the rotation speed of a drive for a special situation.

This hierarchical structure of engineering processes is reflected in nearly all engineering process models in literature (Lüder et al. 2011). Most engineering

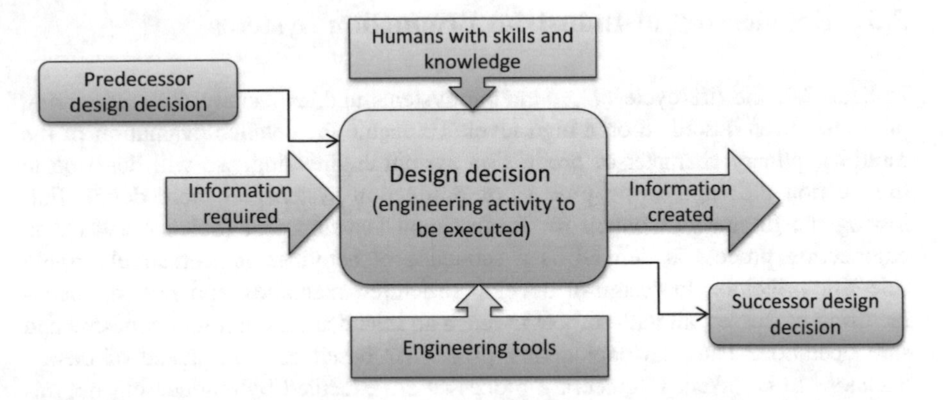

Fig. 2.3 Impact on/of design decisions within an engineering activity based on (Schäffler et al. 2013)

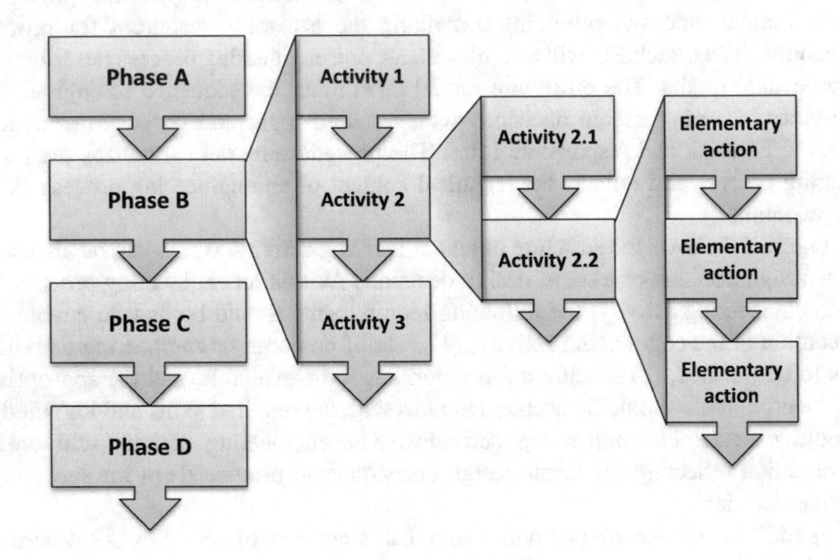

Fig. 2.4 Engineering activity hierarchy based on (Lindemann 2007)

processes have been developed for different purposes in the area of production systems engineering, resulting in considerable process variety. However, the described engineering processes can be distinguished with respect to the type of business model used, the industrial domain addressed, the completeness with respect to the life cycle of products and production systems, and the involved engineering disciplines and stakeholders. In the following, engineering processes will be distinguished with respect to the business model they follow. Of course, the set of processes considered in this chapter is not complete. From a business point of view, there are main business models as follows: *solution providers*, *component providers*, and *product developers*.

- The *solution provider business* addresses the engineering of a production system, e.g., a car manufacturing plant or a steel mill, as a unique system developed to produce specific products in a special way. Usually, solution providers cover plant planning, plant engineering, implementation, commissioning (i.e., acceptance test), and use of production systems. Thereby, these processes are project oriented and in some industrial domains—they are called "*project business*." Examples of such engineering processes are the engineering processes specified in the VDI 5200 Guideline "Factory planning" (VDI 2009a), the AutomationML reference engineering process (Drath et al. 2008), the engineering process proposed by Schnieder (1999), and the engineering process presented by Kiefer (2007).
- The *component provider business* addresses the development of reusable production system components, e.g., a welding robot or a type of steel press, which are typically intended for one industrial domain and a special set of production process steps they are applied within. These components are mainly developed based on best practices in solution business projects, market analysis, and technological possibilities. Therefore, the engineering processes typically contain the component planning, component engineering, implementation, validation, and the clear documentation of the developed engineering artifacts. Examples of such engineering processes are the AQUIMO engineering process (Aquimo 2010) and the Domain Engineering process (Maga et al. 2010).
- The *product development business* addresses the development of products usable within several industrial domains, e.g., a robot, an electrical drive or a sensor. Here it needs to be kept in mind that products in this business model are not consumer products but objects applicable within different technical settings by adapting them to the application case. The results of the engineering processes of this business type are intended to be multipurpose devices and construction elements. Facing the very different types of product-related engineering processes like (Ulrich and Eppinger 2007; Schäppi et al. 2005; VDI 2004), the product business mostly follows an engineering process covering the phases requirement analysis, product planning, product engineering, implementation, and validation.

There are some newer approaches considering different businesses in parallel or its combination. Here, two representative engineering processes are the engineering process of VDI Guideline 3695—"Plant engineering" (VDI 2010)—and the engineering process of the VDI Guideline 4499—"Digital factory" (VDI 2008, 2009b). The different engineering processes described above provide the possibility to be combined into one generalized engineering process. Figure 2.5 shows the generalized engineering process, which combines the product, component, and solution provider businesses using the knowledge about production system components, their functionalities, and use. The generalized engineering process consists of three subprocesses: (a) *product development business* for the design of products usable as mechatronic units within production systems and its components; (b) *component provider business* for the design of components usable as mechatronic units within

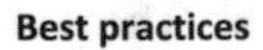

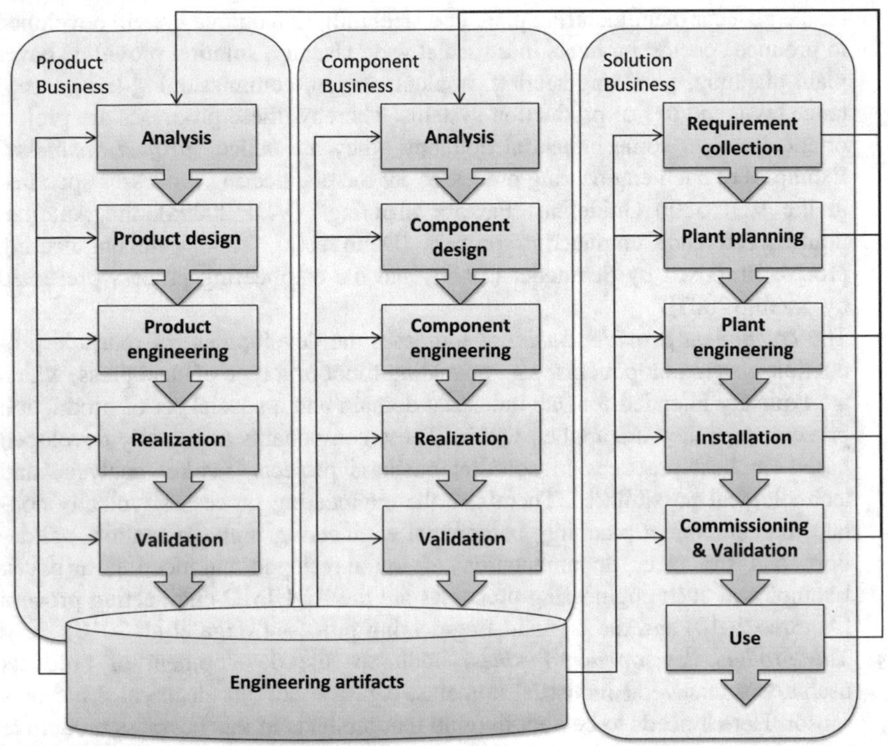

Fig. 2.5 Generalized engineering process based on (Lüder et al. 2011)

production systems; and (c) *solution provider business* for the production system design and implementation exploiting predefined mechatronic units. These businesses define subprocesses with their phases that cover all businesses needed to engineer a production system. Depending on the developed system, each process/phase can proceed with more or less effort.

Within the five phases of the subprocess for *product development business* the best practice of the production system design, implementation and use, the knowledge from components design, and the market and technology conditions are exploited as starting points and basic requirements. The results of the subprocess are mechatronic units and their information sets containing all engineering artifacts developed within the complete process and describing all relevant engineering knowledge (Lüder et al. 2010, Hundt et al. 2010) which can be used in various application cases not limited by industrial domains.

Similar to the subprocess for product development business, the subprocess for *component provider business* with its five phases starts also with an analysis of the best practice and requirements of the production system design, implementation, and use as well as the available mechatronic units, their provided functionalities,

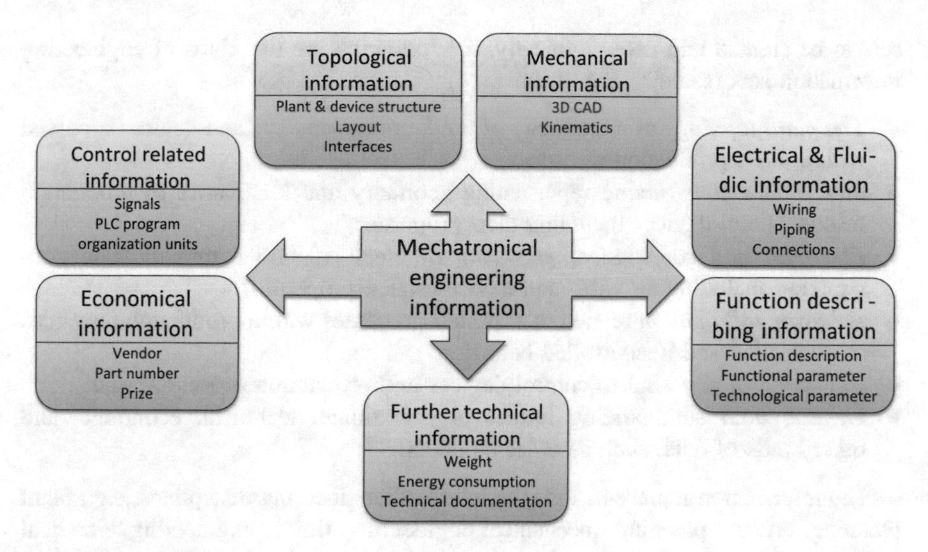

Fig. 2.6 Mechatronic engineering information sets based on (Lüder et al. 2010)

fulfilled requirements, and available engineering knowledge. This subprocess results in the provisioning of mechatronic units and their describing information sets with a higher complexity compared to the product development business. The mechatronic units are developed to be used in application cases limited to one industrial domain. Thus, the component- and product-related subprocesses will both feed a library of mechatronic units usable within components and production system-related subprocesses (see Fig. 2.6).

The subprocess for production system design and implementation (i.e., *solution provider business*) with its six phases exploits the provided set of mechatronic units as reusable components set to set up the production system resources. This subprocess starts with the requirement collection phase to collect all product-related and further requirements of the intended production system, e.g., for manufacturing a specific type of car. Based on these requirements, the production system is planned, engineered, implemented, commissioned, and used as described above. Beneath the running production system, the production system-related subprocess will result in best practices for the different industrial domains usable within the product and component design processes.

Note that the described process structure is idealized. Depending on the EO executing the engineering process and the industrial domain addressed, there is a wide variety of realizations and particularizations of this process. Typical production systems developed by the named processes are production systems like car body welding shops, combustion-turbine-based power plants or process plants for polyvinyl chloride generation. Within these production systems, components and devices such as welding robots, conveyor systems, gas turbines, pipes, vessels, PLCs, inductive sensors, and level indication switches, and drives are combined. Looking at these engineering processes, there are different engineering information sets to be created and used. Typically, the following are the kinds of engineering information sets (see also Fig. 2.6):

- The *plant topology* as a hierarchy of production resources and devices involved within these production resources;
- *Mechanical construction* representing geometry and kinematics of production resources and devices including their properties;
- *Electrical and communication wiring* of electrical and communication construction including all wire- and device-related properties;
- *Behavior information* related to expected processes within production resources as controlled and uncontrolled behavior;
- *PLC code* of any kind to control devices and production processes; and
- *Generic data* summarizing further organizational, technical, economic, and other kinds of data, such as order numbers.

The information represents also the involved engineering disciplines, e.g., plant planning, process planning, mechanical engineering, fluidic engineering, electrical engineering, communication system engineering, control engineering, simulation, and procurement. Depending on the application case, also other special engineering disciplines or natural sciences can be involved, such as chemistry in process plants or nuclear physics in nuclear power plants.

Figure 2.7 shows typical stakeholders involved within the engineering of a production system:

At first, there is the *plant owner* (1). He is responsible for the economic success of the production system and, therefore, defines the products to be produced, the capabilities of the production system to produce the products, and bordering conditions to the production system. Within this task, he exploits information gained by the product *customers* related to the intended product portfolio and certain

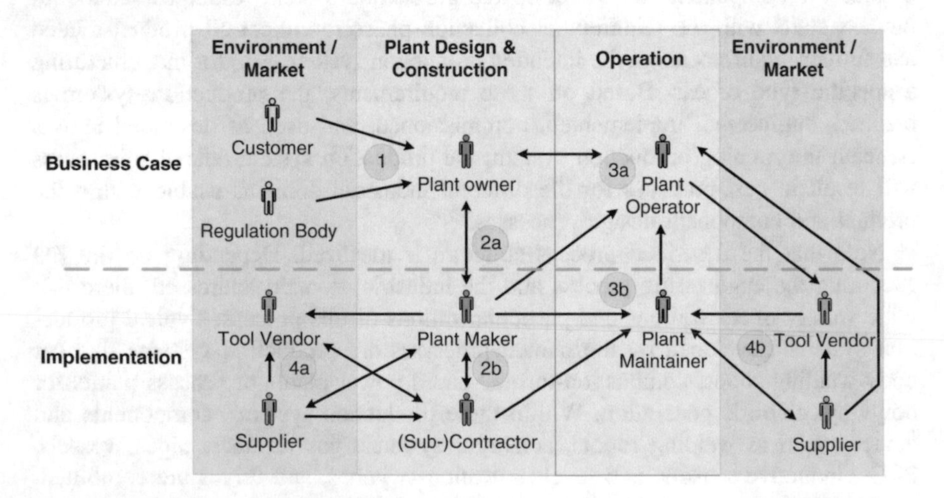

Fig. 2.7 Stakeholders in added-value chains related to industrial plant engineering

regulation bodies defining bordering conditions to the production system related to system safety or environmental protection.

The *plant owner* will contract a *plant maker* (or *system integrator*) to create the production system following the defined intentions and bordering conditions given in a technical and functional requirement specification (2a). The *plant maker* will take up these basic documents and will ensure the engineering, installation, and commissioning of the production system. Within the necessary activities, he is assisted by *subcontractors* (2b) taking over some of the engineering, installation, and commissioning activities and by *component and device suppliers* (following the device and component provider business) providing information for the engineering and physical objects for installation.

After the production system has been commissioned and handed over to the *plant owner*, the *plant owner* will authorize a *plant operator* to run the production system and to produce the intended products (3a). As production systems, like all technical systems, will degrade over use time, the *plant operator* gets technical support from *plant maintainers* (3b) who are responsible to ensure the availability of the production system resources and devices.

The *tool vendor* has essential impact on engineering (4a), use, and maintenance (3b) activities related to a production system. He provides engineers with software tools that are needed to create engineering artifacts, access plant components at commissioning and runtime, and enable device diagnosis. The capabilities and quality of the software tools have essential impact on the efficiency and quality of the activities they are applied within. Therefore, the software tool users (*plant maker*, *plant operator*, *plant maintainer*, *supplier*, and *subcontractor*) will provide the *tool vendor* with requirements toward tool capabilities resulting from their special application conditions. The described network of interactions between the different stakeholders is depicted in Fig. 2.7, which represents also the flow of information between the different stakeholders.

The discussion of the complete flow of information goes far beyond the scope of this chapter. Therefore, we will focus on discussing selected illustrative examples: *Customer* and *regulation bodies* will provide requirements for the product and the production system to the *plant owner*. The *customers* will define the product characteristics (e.g., product type and quality, amount) resulting in the definition of the production process to be executed on the production system. The regulation bodies will define safety-related product requirements (like CE conformity) and production system-related safety and workforce requirements (like protection guidelines from the employers' mutual insurance associations). All these kinds of information are related to function-related information or further technical information.

- *Plant owner* and *plant maker* will exchange both requirements to the production system and production system realization information. Usually, the *plant owner* will provide a requirements specification including functional and nonfunctional requirements within a tender document (German: *Lastenheft*) and the *plant maker* will reply with a technical specification (German: *Pflichtenheft*). Finally, the *plant maker* will provide a complete plant engineering documentation. Thus, the communication between *plant owner* and *plant maker* will cover all types of information shown in Fig. 2.7.

- *Plant maker* and *subcontractors* will exchange the same type of information like *plant owner* and *plant maker* but on a more detailed level covering only the parts of the technical system the *subcontractor* should contribute to.
- *Suppliers* will provide the *plant maker* and *(sub-)contractors* with information about their components or devices. Usually, *suppliers* will provide technical documentation covering user manuals, installation guidelines, and basic engineering information covering all information sets of Fig. 2.7.
- *Tool vendors* will get from all involved stakeholders (except for *customer* and *regulation bodies*), the information/requirements on engineering and data exchange capabilities of the tools to be used in engineering. Here, the most interesting information are the types of artifacts to be created in the engineering process (types of diagrams, such as P&ID or wiring diagrams, types of lists, such as device lists or terminal stripe assignments) and the capabilities for efficient engineering.
- *Plant operator* and *plant maintainer* will get the "plant as is" documentation from the *plant maker*. Also, here the exchanged information will cover all information sets of Fig. 2.7. This information is required to appropriately understand, run, and maintain the plant.

Note that the share of effort contributed by the different stakeholders may vary based on company size, industrial setting, and application case. For example, the engineering effort share of the *plant owner* and *plant maker* differs according to the company size. Larger *plant owners* within the automotive industry, e.g., Volkswagen or Daimler, will usually execute the basic engineering of the production system until the definition of the production resources to be applied and afterwards hand over to the *plant maker*. In contrast, small *plant owners* in consumer goods industries, such as a furniture manufacturer, only define the product and the *plant maker* will make the complete engineering. Similar sharing patterns can be found in maintenance activities, where, e.g., in the process industry special device *suppliers* and/or *subcontractors* will take over responsibilities for maintenance activities, while in automotive industry the *plant owner* will take the complete burden of maintenance.

2.4 Usage Scenarios that Illustrate Needs for Semantic Support

After having detailed the engineering process of production systems and illustrated its multidisciplinary and multi-model character, this section will introduce relevant scenarios of information application and information creation within production system engineering. Thereby, this section will provide a deeper view on needs for

Fig. 2.8 Lab size production system as running example

semantic integration during typical steps for the creation and application of data and information within the life cycle of production systems. These scenarios will cover different life cycle phases of the production system to represent the typical engineering knowledge and integration needs introduced in Sect. 2.2. For illustration, the scenarios will be discussed in the context of a lab-size production system that is hosted at Otto-v.-Guericke University Magdeburg.[2] The production system consists of a set of three multipurpose machines, eight turntables, and ten conveyors, and is wired with Field-IOs to Raspberry-Pi-based controllers as depicted in Fig. 2.8.

Each of the machines, turntables, and conveyors is equipped with the same types of sensors and actuators. There are inductive proximity switches to detect materials, position indicator switches for position sensing of the turntables and machine heads, and drives to move conveyor belts, rotate turntables, rotate tools, and move the machine heads. All control devices are connected to three modular Field-IOs by wire. Two Ethernet switches, a controller, and the Field-IOs establish a Modbus TCP network to enable access to the physical inputs and outputs. Finally, the Raspberry-Pi-based controller runs a PLC program that controls all devices.

[2]Institute of Ergonomics, Manufacturing Systems and Automation at Otto-v.-Guericke, University Magdeburg: Equipment Center for Distributed Systems, http://www.iaf-bg.ovgu.de/technische_ausstattung_cvs.html, last access March 2016.

2.4.1 Scenario 1—Discipline-Crossing Engineering Tool Networks

As mentioned in Sect. 2.2 the engineering of production systems is multidisciplinary engineering involving different engineering activities executed by engineers coming from different engineering disciplines with different engineering tools, where the engineering activities establish a network of dependencies. Thus, efficient engineering depends on efficient networking. For example, the *Industrie 4.0* value chain processes "Plant and process planning" and "Production system engineering" can be improved significantly with consistent engineering data exchange between the involved engineering activities (and therefore, the involved tools and engineers) This is also visualized in Fig. 2.5 as the exchange of engineering information through the mechatronic component library that contains engineering artifacts.

At the beginning of the life cycle of a production system, the plant owner will contract the plant maker to create a production system following a set of defined requirements. The plant maker will start an engineering process, as described in Sect. 2.2, involving subcontractors for special engineering purposes. The main aim of the plant maker (and also of the involved subcontractors) is the efficient creation of all required engineering artifacts (e.g., plans and programs) required to install and commission the production system with sufficient quality, following the aims of the plant owner and the requirements of subsequent phases of the production system life cycle.

As described in Sect. 2.2, the production system engineering process involves the execution of engineering activities in different disciplines, like mechanical, electrical, and control/software engineering. Each discipline has to specify discipline-specific parts of the production system. Looking at the example plant during the engineering, the different required plant parts have to be selected and instantiated resulting in a list of system components to be used. Within mechanical engineering, these system components are detailed, leading to the identification of devices (e.g., sensors, drives, and controllers) and other buyable parts and their location and application in the production system (and the production process). Within electrical engineering, the used devices are integrated into wiring networks. In control engineering, the software programs for the controllers are implemented, driving the devices to achieve the required system behavior. Therefore, each of the involved disciplines and engineers may deal with the same objects (e.g., plant sensors, drives, and controllers) from different viewpoints. To enable the mutual understanding of the different engineers, they require across all disciplines a common set of concepts (e.g., signals), which have a discipline-related incarnation (e.g., the position of a plug in mechanical engineering, a wiring point in electrical engineering, and a program variable in control engineering).

This set of *common concepts* is currently not explicitly represented in discipline-specific and isolated tools. The main reason for this problem is the existence and development of discipline-specific engineering tools by tool vendors. Usually, tool vendors focus their business on one engineering discipline or on related engineering disciplines. Thus, during the last decades engineering tools have been developed and highly optimized for the execution of discipline-specific engineering activities, like MCAD tools, ECAD tools, and control programming tools. These tools apply engineering-discipline-specific concepts and languages that evolved in parallel to the engineering tools following the understanding and acting in the engineering discipline. However, the focus on one engineering discipline led to neglecting the consideration of other engineering disciplines involved in the overall engineering process and of higher level participants, such as project managers or system integrators.

Within recent years, the need for better integration of engineering tools along the complete engineering tool chain (or tool network) has been identified and has received prominent attention as foundational requirement for the *Industrie 4.0* approach (Wegener 2015). Thus, there is a need for the identification of common concepts between and within the different engineering disciplines and to represent and integrate these common concepts to enable lossless data and information exchange along the engineering tool chains/networks. The identification of common concepts has to be supported by tool users (especially the plant maker) and tool vendors in cooperation.

Exploiting the developed common concepts within the network of engineering activities, two main benefits can be reached. On the one hand, the involved engineers can improve the quality of the developed system architecture by mutual discussions and understanding needs of the different involved engineering disciplines. Thereby, the overall quality of the developed production system can be improved. On the other hand, the information flow within the network can be improved. The common concepts will improve the capabilities of implementation and application of appropriate tool interfaces enabling lossless data exchange and a fault-free data application in the data-receiving tool.

Security is an issue in this scenario. The information exchange can be executed between different involved legal parties providing engineers to the engineering process. In Fig. 2.7, it is easy to see that on the one hand the involved stakeholders are interested in protecting the information exchange against access from outside of the engineering network. On the other hand, they might be interested in securing information against access of different partners within the engineering network to protect vital knowledge. For example, one subcontractor may require knowledge protection with respect to other subcontractors of a supplier to ensure that component-specific models can only be applied by special subcontractors (e.g., in case of virtual commissioning or simulation).

To ensure the necessary quality assurance and the high degree of parallel work in engineering, several needs for semantic capabilities arise. The knowledge on engineering artifacts needs to be explicitly represented (N1) to make them accessible and analyzable; the views of several engineering disciplines need to be integrated (N2); the quality assurance, e.g., consistency checks of engineering model views across disciplines, and automation of engineering, e.g., the generation of derived artifacts, require functions for engineering knowledge access and analytics (N3). In addition to structured data stored in databases or knowledge bases, the use of semi-structured data, e.g., technical fact sheets that include natural language text, or linked data, e.g., component information from a provider, in the organization, and on the Web can help improve the automated support for reuse processes (N4). These required capabilities provide the foundation to improve the support for multidisciplinary engineering process knowledge like supporting the exchange of advice to amendments within the different created artifacts based on identified dependencies (N6).

2.4.2 *Scenario 2—Use of Existing Artifacts for Plant Engineering*

As mentioned above, the frequency of activities in the life cycle of production systems is increased significantly (especially by reducing the time span of use phase) following the reduction of the life cycle of products. *Industrie 4.0* faces this trend by enforcing the integration of the different value chains of production systems, its components, and its devices (as discussed in Sect. 2.3). An example of this integration is given in the VDI/VDE guideline (VDI/VDE 2014). It considers the integration of manufacturers of PLCs, screws, metal sheets, and washing machines (see Fig. 2.9). Here, it gets visible that components used in technical systems

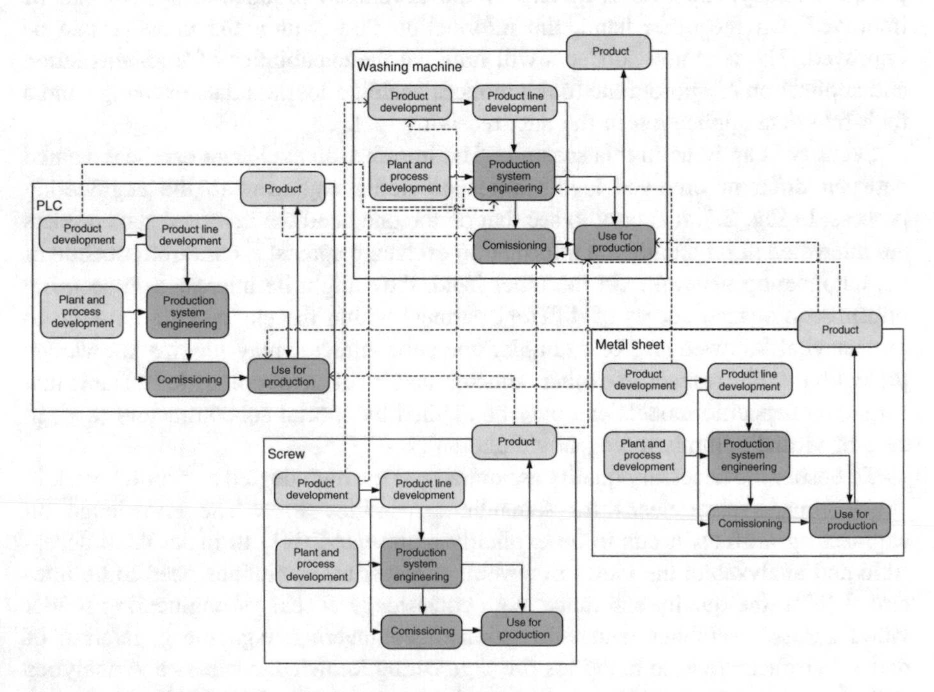

Fig. 2.9 Example for integration of value chains based on (VDI/VDE 2014)

(production system and products) can be reused. Thus, also their "digital shadows" developed/exploited in engineering, i.e., the data shown in Fig. 2.6, can be reused. As an example, the *Industrie 4.0* value chain processes "Plant and process planning" and "Production system engineering" can be improved significantly with better automated support for the reuse of existing artifacts during plant engineering. This automation of support for reuse is currently hampered by a weak integration of the diverse knowledge needed from several roles in the engineering process for matchmaking between: (1) reuse requirements for production systems and (2) the capabilities of reusable devices and components.

The plant maker is strongly interested in a most efficient engineering process as the engineering process efficiency has a major impact on the economic effect of the engineering project. Therefore, the plant maker intends to reuse existing engineering artifacts coming from various sources. These sources can be the plant maker (and his subcontractors) exploiting libraries of engineering artifacts developed in prior projects (reuse of engineering artifacts), or suppliers of devices or components providing libraries with detailed engineering information for these devices and components. To make this reuse possible, the engineering artifacts have to be developed and prepared for integration into a library, need to be identified as appropriate for an application case, and be adequately incorporated in the artifact set to be established by the plant maker. This reuse can be prepared for each discipline separately. However, reuse preparation that covers several disciplines based on a leading discipline (VDI 2010) is probably more efficient. An example for such a reusable artifact is a conveyor belt. It contains a set of automation devices like a drive chain and material sensors. For each conveyor belt, the mechanical engineering defines mechanical parts, the involved automation devices, and their assembly; the electrical engineering specifies the wiring of the devices including active and passive infrastructure; and the control engineering develops appropriate control functions for the belt motions. All these contributions can be developed once and reused in each project requiring a conveyor belt with these specified characteristics.

The preparation of artifacts for reuse in other engineering projects assumes the existence of an overall system architecture (within one discipline or across disciplines) with types of applicable system objects (the example in Fig. 2.9 includes machines, turntables, conveyors with drives and sensors, which are interconnected by wires), and system dependencies and interrelations (in the example, wiring relations, neighborhood relations, or control dependencies). The definition of system-object types requires well-defined semantics for production system components in the production system hierarchy, see, e.g., in Kiefer (2007; Hundt et al. 2010). The definition of system dependencies and interrelations requires a well-defined semantics for dependencies between components (within and across engineering disciplines), and a unique mapping of discipline-dependent artifacts to components. For example, the drive controller within the drive chain of the conveyor belt is considered in mechanical, electrical, and control engineering. Within electrical engineering, the set of control inputs and outputs is connected to wires enabling the information exchange between the drive controller and the overall controller for the conveyor belt. Within the control programming, there is a control variable assigned to each input and output of the drive controller. Consistency requires to ensure that for each connected input and output there is a variable with an appropriate type. The definition of these well-defined semantics is an activity of the plant maker in cooperation with his subcontractors and suppliers. Therein, the plant maker has to consider his specific production system architecture, while suppliers have to ensure the possibility to apply their components/devices within different plant architectures and engineering processes. For example, the plant maker will focus on the requirements of the specific product like a car body, while the provider of conveyors as components or drives as devices will focus on the applicability of these conveyors and drives within application cases going beyond welding shops including for example logistic systems for conveyors and elevators for drives.

If libraries of artifacts exist, the plant maker has to be able to identify the appropriate library entry (i.e., component or device) for his application case (i.e., the specific production system) within the library. This means, that the plant maker (especially the plant planner, the mechanical engineer, the electrical engineer, and the control engineer) will select appropriate production system components and devices (in the example, machines, turntables, and conveyors, drives, sensors, communication switches, and wire types) to execute the necessary production processes for the intended products. All engineering roles have to select the relevant components/devices based on requirements coming from the customer regarding the product, the production system owner regarding business performance capabilities, the engineers of the other engineering disciplines involved regarding production system interfaces. To enable the selection of appropriate components and devices, three elements are needed: suitable descriptions of the requirements to be fulfilled by the component, suitable descriptions of components and their capabilities, and a methodology to map the requirements to the reusable components. Usually, the requirements will address the required functionalities of the component/device (e.g., drilling function of a machine, material identification function of a sensor) in combination with the application conditions (e.g., drilling depth needs, material types that shall be identified, environmental conditions of sensor use).

Currently, descriptions of components and devices are very heterogeneous. For devices, first classification standards, like eCl@ss[3], exist to provide device type classifications following the functional requirements (like servo-stepper motor or a pointed profile mill) and device properties that enable the evaluation of the device application range (e.g., maximal torque of a motor or the maximum cutting depth and the cutting diameter of the profile mill). Even through the existence of first component models towards classification standards, there is no commonly accepted classification of components available. Examples of available models are eCl@ss or vendor-dependent classifications, such as the drive chain classification of Lenze

[3]eCl@ss: www.eclass.de.

(Götz 2013). Such a classification shall focus on the executed production process (or its parts) provided by the component following, for example, the DIN 8580 standard series. As the requirements to components and devices may be very different (coming from different sources and describing very different subjects), a discipline-crossing general representation of required and provided component capabilities, which can be compared, shall be developed based on a well-defined semantics of objects. In this task, the plant owner (knowing best the required production processes), the plant maker (who has to make the mapping), and suppliers (who know the component/device capabilities) have to be involved to bring in existing knowledge of their disciplines, possibly defined in standards for aspects of the application domain.

Exploiting the developed discipline-crossing general representation of required and provided component capabilities within the network of engineering activities again two main benefits can be reached. By exploiting completely engineered, tested, and validated library elements, the involved engineers can improve the quality of the developed system by preventing engineering errors and exploiting the integrated components/devices in the most appropriate way. In parallel, engineering efficiency can be increased by reusing developed structures, plans, and other engineering artifacts, so these artifacts do not need to be "redrawn" over and over for each new project.

Security issues in this scenario have a similar level compared to Scenario 1. The developed libraries contain sensitive knowledge of the device and component vendors. They contain device- and component-related information which shall only be accessible to a limited set of roles. Hence, access rights and encryption are part of the libraries and the component/device models. Beyond this knowledge-related issue, there is also a market-competition-related issue. Each vendor has a vital interest in making his products more attractive for users than the products of the competitors. Therefore, the vendor may make his product easier to find and use or make other vendors' components less likely to score well with requirements. These comparisons can be supported by a well-defined model semantics.

Against the background of the required quality of engineering artifacts and the assurance and high degree of parallel work in engineering, several needs for semantic capabilities can be identified. The knowledge on system requirements, system architecture, and reusable components/devices need to be explicitly represented (N1), the heterogeneous views of system procurers and several engineering disciplines need to be integrated, in particular, there has to be a common view on the system architecture, e.g., a best-practice architecture in an application area, such as for printing machines or welding cells (N2), the mapping of system requirements to component/device capabilities requires functions for engineering knowledge access and analytics, e.g., the capability to explore the system architecture model at design time (N3). In addition to structured data in databases or knowledge bases, the use of semi-structured data, e.g., technical fact sheets that include natural language text, or linked data, e.g., component information from a provider, in the organization and on the Web can help to improve the automated support for reuse processes (N4). Based on these required capabilities, software systems to support reuse in an engineering environment can be designed (N5) to improve the support for multidisciplinary engineering process knowledge (N6).

2.4.3 Scenario 3—Flexible Production System Organization

Within Scenario 2, the integration of value chains has been considered related to engineering activities. Looking at Fig. 2.9, integration can also be related to the *Industrie 4.0* value chain processes "Commissioning" and "Use for production." These processes require information from the engineering phases of the life cycle for the production system in scope, but also from the "Use for production" phase of the life cycles of relevant system components and devices used in the production system in scope. The value chain processes "Commissioning" and "Use for production" can be improved significantly with better automated support for flexible production system organization. This automation of support is currently hampered by a weak integration of the diverse knowledge coming from several roles in the engineering process with the flexible organization of a production system at runtime.

The plant owner and, forced by the plant owner, the plant operator are interested in a most flexible application of the production system. There are various types of relevant flexibility regarding produced products, production processes, and production resources to automate production processes (Terkaj et al. 2009). Most often, the plant owner wants more production-resource-related flexibility (i.e., the ability to integrate and change production system components, such as adding another multipurpose machine and a new cycle of turntables and conveyors in the example plant. These changes aim at increasing production resource availability. A plant owner may want to change a machine head to enable other milling or turning processes without negatively impacting the overall production system functionality). A plant owner may aim at increasing product-portfolio-related flexibility (i.e., the ability to change the product portfolio, like adding a new product type without negatively impacting the overall production system functionality).

To achieve sufficient resource-related flexibility, *Industrie 4.0* envisions the dynamic integration or change of production system components within the production system at runtime (Kagermann et al. 2013). Integration or change is envisioned for small components, such as position indicator switches or drives as given in the running example, for larger components, such as complete machines or turntables, and for parts of production systems. To make this vision possible, two information processing features are required. First, the newly integrated or changed production system component has to provide information about (a) its capabilities, (b) its access paths, and (c) its control-related features. In the context of the example production system, an example sensor component shall provide information about (a) its possible measurement capabilities, (b) the fieldbus protocol, addresses and services to access the sensor values, and (c) the control-application code fragments, e.g., function blocks, to be applied in the control system for using the sensor component. Second, the overall control system of the production system must be able to reason about the information provided by the component and to integrate the component at runtime within the processes (especially production processes) to be executed. Concretely, the component capabilities need to be compared with the required capabilities for the production processes (e.g., a product instance shall be identified with inductive sensing capabilities) and, if appropriate, shall be utilized in the overall control architecture for the production control based on the provided access paths and control code fragments.

To enable product-related flexibility, *Industrie 4.0* assumes that each product shall know in a certain way the required material and processes for its creation (Kagermann et al. 2013). In the context of the example production system, the product shall know what types of machining steps are required for its manufacturing and the required machining parameters (e.g., tool speed, chip size, cooling). This information has to be used on the one hand for the identification of the applicable manufacturing resources and on the other hand to parameterize the processes executed on the resources by applying the related values within the control system. To enable this flexible approach, production processes need to be modeled and automated reasoning about required and provided capabilities has to be supported. As production system resources shall be applicable within various industries (e.g., the example production system could be applied for metal-part processing or wooden-part processing), the process description and reasoning shall be independent of the application industry.

For both types of required flexibility, a common process description based on the product–process–resource concept (Pfrommer et al. 2013) is desirable. This description requires expressing in semantically well-defined and comparable ways, the concepts and relationships of (a) the needs for capabilities of production system components and devices; (b) the component/device use conditions (access path and control); and (c) product-related processing requirements.

Again, this is a cooperative task for the plant owner (knowing best the product and its required production processes), the plant operator (running the control system), the plant maker (knowing the overall system architecture including the control architecture), and suppliers (who know the component/device capabilities). Exploiting the common process description, the plant owner and the plant maker can reduce the effort for adapting the production system to the changing needs and can increase in parallel the reachable quality of the changes toward the changed requirements. As an example, a metal part of a car needs to be drilled in a certain way before welding. A model of this drilling process and its dependencies to the product can be exploited to identify automatically the right drilling machine for this step by comparing it with the capabilities of the set of production resources. This will lead to faster adaptations and a better ROI.

In this scenario, security plays a completely different role than in Scenarios 1 and 2. Here, the intention is to exploit the modeled information automatically within the production system. Thus, beyond the security issue of preventing unauthorized access to the information or parts of it, safety issues of the production system have to be reflected. The provided information shall not harm the production system, staff, products, or the environment by resulting in malfunctions. Thus, beyond the encryption discussion of Scenarios 1 and 2, here the common process description needs to be validated with respect to the resulting behavior of the created system.

To enable the necessary definition of system requirements for components/devices, the matching to component/device capabilities, and the proper use of components/devices at runtime require several needs for semantic capabilities. The knowledge on system requirements, system architecture, and component/device capabilities needs to be explicitly represented (N1), the heterogeneous views of system requirements and component/device capabilities need to be integrated (N2), the matching of system requirements to component/device capabilities requires functions for engineering knowledge access and analytics, e.g., the capability to explore a system architecture model at runtime (N3). These required capabilities provide the foundation to improve the support for multidisciplinary engineering process knowledge (N6). The integrated engineering knowledge needs to be provided at production system runtime (N7) to enable decision support during the production system operation phase.

2.4.4 Scenario 4—Maintenance and Replacement Engineering

Scenario 4 will extend the application of information from plant engineering, commissioning, and use phases of the system life cycle toward the "Maintenance and decomposition planning" and "Maintenance" phases of the *Industrie 4.0* value chains in the life cycle. These phases can be improved significantly with better automated support for the assessment of the impact of changes to selected plant components or devices. This automation of change impact support is currently hampered by a weak integration of the diverse kinds of knowledge coming from several roles in the engineering process with the maintenance knowledge during production system operation.

Physical production system components do not necessarily remain stable over the complete lifetime of a production system. These components are subject to certain types of wear-out processes and sporadic faults, which may result in a production system component and (very often) production system that do not work as required. In the context of the example system, the fault of a drive in a conveyor may make parts of the production system inaccessible and therefore (temporarily) lost to the production process.

In practice, the plant maintainer tries to mitigate the risk of a degraded production system with appropriate maintenance and repair strategies, including predictive maintenance and component diagnosis. In case of predictive maintenance, the engineering knowledge about typical component wear-out processes is applied to identify the probability of a fault in the near future, to define component change or repair cycles based on the fault risk, and to ensure proper replacement engineering. In case of diagnosis, run-time measurements are applied to identify dangerous states of components. In the case of the example production system, the knowledge of the usual drive lifetime can be used to change the drive before it is likely to fail. For diagnosis, sound sensors can be installed to measure vibrations in the system and to identify dangerous vibrations announcing the breaking of drive chain parts.

New maintenance capabilities can be reached, if the run-time information can be combined with engineering knowledge. If the supplier of a drive and the plant operator can aggregate their knowledge on drive chain wear-out processes in correlation with drive chain use in the production system and the sound propagation of drive chains, they may enable the development of more sophisticated drive-chain diagnosis methodologies making predictive maintenance actions dependent on sensor measurements and, thereby, automatic control, to name an example.

In addition, existing engineering knowledge has to be applied to ensure the correct integration of new devices within replacement strategies. Here, the involved engineer has to ensure that the new system components fulfill the needs of the application, even if the new component is not an exact copy of the component being replaced.

Such scenarios require a common semantics of engineering and run-time information related to system components and devices. General behavior models of components are required, which exploit engineering information and specific system knowledge, and can be combined with run-time information coming from the production system. The creation of such a common component-behavior semantics is a cooperative task of all roles involved in the design and use of production systems (except for the customer). The plant owner, plant operator, and plant maintainer have to provide knowledge about the production system's run-time behavior. Plant maker, subcontractor, and supplier have to provide knowledge about the production system components and devices and the systems engineering. With the common component-behavior semantics, the plant maintainer can improve his diagnosis and planning skills resulting in an increase of the production system availability.

Similar to Scenario 3, security issues are related to knowledge protection and production system safety leading to the same type of encryption and model validation requirements.

To enable the necessary definition of component-fault-risk models, production system-risk models, and change impact analysis for production system components, several needs for semantic capabilities arise. The knowledge on system requirements, fault risks, and change impact possibilities need to be explicitly represented (N1), the heterogeneous views of system engineers, operators, and maintainers need to be integrated (N2), the modeling of risks and change impact analysis require functions for engineering knowledge access and analytics (N3). In addition to structured data in databases or knowledge bases, the use of semi-structured data, e.g., technical fact sheets that include natural language text, or linked data, e.g., component information from a provider, in the organization and on the Web can help improve the automated support for risk and change impact analysis (N4). Based on these required capabilities, software systems to support risk and change impact analysis in an engineering and operation environment can be designed (N5) to improve the support for multidisciplinary engineering process knowledge (N6). The integrated engineering knowledge needs to be provided at runtime (N7) to enable decision support during diagnosis and replacement in the maintenance phase.

2.5 Needs for Semantic Support Derived from the Scenarios

Major semantic challenges come from the need to provide tool support for processes that build on the heterogeneous terms, concepts, and models used by the stakeholders in production system engineering and operation. Based on the selected scenarios, this section will derive a set of needs that illustrate key capabilities that semantic solutions for multidisciplinary engineering should address. Usage scenarios can be addressed with intelligent engineering applications supported with Semantic Web approaches, e.g., for engineering data integration and storage, for consistency checking, or for the organization and recommendation of engineering components for reuse in a specific target project. Table 2.1 lists the production system engineering usage scenarios (UCx) and derived needs for engineering knowledge modeling and integration capabilities (Nx), similar to the discussion on enterprise service ecosystems in (Oberle 2014).

N1—Explicit engineering knowledge representation. Domain experts in production systems engineering routinely use a wide variety of engineering models, which represent certain aspects of engineering knowledge explicitly (Newen et al. 2011). However, in many cases, the formality/expressiveness of the modeling approaches used does not support the sufficiently complete expression of

Table 2.1 Production system engineering usage scenarios (UCx) and needs for engineering knowledge modeling and integration capabilities (Nx)

Production System Engineering Needs & Use Cases		UC1	UC2	UC3	UC4
N1	Explicit engineering knowledge representation	✓	✓	✓	✓
N2	Engineering data integration	✓	✓	✓	✓
N3	Engineering knowledge access and analytics	✓	✓	✓	✓
N4	Efficient access to semi-structured data in the organization and on the Web	✓	✓		✓
N5	Flexible and intelligent engineering applications		✓		✓
N6	Support for multidisciplinary engineering process knowledge	✓	✓	✓	✓
N7	Provisioning of integrated engineering knowledge at production system runtime			✓	✓

knowledge needed to automate engineering processes for production system engineering. Therefore, there is a need for knowledge representation support, which allows analyzing the requirements for the knowledge to represent and providing the domain experts with appropriate tools for knowledge representation and design.

In all use cases (UC1 to UC4), example knowledge representations are engineering models; also the representation of semi-structured data, such as external documentation like technical fact sheets or component provider data on the Web; access to domain expertise, like expert networks, and collective intelligence systems, like recommender systems for reusable components. For multidisciplinary engineering, common concepts of several stakeholders across disciplines need to be represented to allow sharing engineering artifact views between engineering disciplines. In UC1, engineering process steps require the exchange of engineering tool results, responsibilities of project participants, and progress states of engineering objects. If domain experts use a variety of tools that have only limited knowledge of the engineering project and process, an integrated plant model is very useful to provide a complete view on the project progress. The plant model is based on the capability to store versions of engineering models and tool data to support parallel engineering processes. In UC2, production system engineering project and reusable engineering artifact characteristics support the selection and adaptation of candidate artifacts for reuse. In UC3 and UC4, access to run-time data in the semantic context of engineering models enables the automated interpretation of run-time data elements, even if the production system structure or processes are adapted.

N2—Engineering data integration (common concepts model). The engineering tool network in a typical production system engineering environment contains a collection of tools with heterogeneous data models, which use different terms and data formats for similar concepts (Biffl et al. 2012). Due to this semantic heterogeneity it is difficult, costly, and error prone to provide a consistent production system plant model for parallel engineering. In particular, in multidisciplinary engineering there is the need for an engineering data integration approach, which provides an integrated data model of the common concepts of stakeholders across disciplines to enable the linking of engineering knowledge across disciplines.

Example engineering data integration requirements are in all use cases (UC1 to UC4) including a process for identifying common concepts and for transforming data for the versioned exchange of engineering results in a tool network, e.g., identifying which parts of an electrical engineering plan are semantically linked to part of a mechanical engineering plan. Use cases UC1 and UC2 require data integration between business and engineering/automation levels, e.g., linking an engineering model version to a requirement in the engineering project plan as foundation for checking the consistency of these planning views. In UC2, an integrated plant view is useful to assess the impact of reusing a specific engineering artifact, e.g., understanding which components in different disciplines need to be considered, if a part of a previous system should be extracted for reuse. In UC3 and UC4, engineering plan knowledge has to be linked to run-time data access configurations, e.g., an automated process step to an OPC UA variable, to enable the effective and consistent adaptation of a production system and its associated engineering models at runtime.

N3—Engineering knowledge access and analytics. Knowledge access and analytics in production system engineering builds on the availability of formally represented (N1) and integrated (N2) engineering data in a multidisciplinary engineering environment. Domain experts and managers need basic functions operating on common data model, e.g., reports and analyses based on data coming from an integrated plant model to check the project progress and the quality of the results from parallel engineering. Therefore, effective and efficient mechanisms are needed (a) for querying of engineering models, including versions and changes; and (b) for defining and evaluating engineering model constraints and rules across several views. Example engineering knowledge management requirements in all use cases (UC1 to UC4) include analysis functions for an integrated plant model. In UC1, domain experts can collaborate more efficiently when supported by checks that reveal missing or inconsistent engineering results between two or more disciplines, e.g., a new device that has not yet been addressed in a partner discipline. In UC2, the recommendation of components for reuse needs the capability to analyze both target engineering plans and reuse candidate engineering artifacts from a set of reusable projects and to perform a matchmaking operation between needs and offers. In UC3 and UC4, business managers can plan production services more effectively, if they can quickly integrate engineering knowledge on the current state of the production system with external web resources, e.g., the production capability of the plant with changing customer orders and updated input from component providers and backup producers that all come in via web services or from web pages. The needs N1 to N3 are basic needs of production system engineering derived from the use cases in Sect. 2.4. Addressing the need for explicit knowledge representation (N1) is the foundation for addressing the need for engineering data integration (N2), which in turn is a foundation for addressing the need for engineering knowledge access and analytics (N3). Addressing the basic needs N1 to N3 is the foundation for addressing the advanced needs N4 to N7.

N4—Efficient access to semi-structured data in the organization and on the Web. Engineering process automation today is mostly based on structured data, e.g., in databases or documents that follow a metamodel. In addition to structured data in databases or knowledge bases, the use of semi-structured data, e.g., technical fact sheets that include natural language text, or linked data, e.g., component information from a provider, in the organization and on the Web can help improve the automated support for reuse processes. Therefore, there is a need for more efficient access to semi-structured data in the organization and on the Web. Example requirements for efficient access to semi-structured or Web data are found in UC1, UC 2, and UC4. In UC2, the tool support for reuse depends in many cases on information that comes from outside the organization, e.g., vendor information on components or issues and recommendations posted in software engineering discussion forums. In a similar way during maintenance in UC4, when changing a component for a different component with similar capabilities, domain experts can use the information on the experience of others to assess the impact of a change on the overall system.

N5—Flexible and intelligent engineering applications (automation of engineering). Assuming the capability of knowledge access and analytics (N3) on an integrated production system plant model, intelligent engineering applications, such as defect detection and constraint checking can be designed. In a production system context, these engineering solutions need to be flexible to adapt to the changes in the production system both at design time and at runtime. An intelligent engineering application goes beyond hard-coded programs and is driven by the description of the production system plant. Production systems, in general, are mission- and safety-critical systems, which require a very high level of correctness and completeness of results from the engineering applications. Therefore, the benefits of domain experts depend on the extent to which they can rely on the results coming from the engineering application. Important nonfunctional requirements are highly intuitive user interfaces and scalability. User interfaces have to be easy to understand and use for practitioners, e.g., many engineers want to continue using their well-known best-practice tools, which should be augmented with the knowledge they need in a project. Scalability becomes a major requirement for typical large engineering projects if several projects in a company need to be analyzed together.

In UC2, a reuse system in a company is an example for an engineering application that has to be easy to extend as new or changed types of production systems, components, and devices have to be considered. In UC4, requirements for an engineering application for defect detection and constraint checking are to enable the simple addition of another engineering discipline to extend the range of constraint checking, e.g., new kinds of constraints coming from a specific kind of simulation model. At runtime, flexibility can mean the change of the production system or the addition of new disciplines, such as plant maintenance.

N6—Support for multidisciplinary engineering process knowledge. A major goal of the stakeholders in production system engineering is improving the productivity of the engineering project in a repeatable way, e.g., by avoiding unnecessary work. Therefore, there is a need to support increasing the quality and efficiency of the multidisciplinary engineering process by representing engineering process responsibilities and process states linked to the production system plant model. This need extends N3, which focuses on knowledge regarding engineering artifacts, with respect to knowledge on engineering processes.

Example engineering process requirements can be found in all use cases (UC1 to UC4). In UC1, the maturity state of engineering results and the responsibilities of domain experts and organizational units enable more efficient planning and monitoring of the engineering process. In UC2, both projects and project-independent reuse processes need to be defined and need access to engineering knowledge for effective reuse recommendations. In UC3 and UC4, the inclusion of all relevant stakeholders from production system planning, engineering, operation, and maintenance, enables more effective system engineering and overall production system process management.

N7—Provisioning of integrated engineering knowledge at system runtime. In a flexible production system context, domain experts need engineering knowledge at runtime to assess in a situation, which needs changing the system, the set of options for a successful change. In addition, changes have to be documented in a way that supports future change analysis. Therefore, there is a need for providing integrated engineering knowledge at system runtime beyond simple pdf printouts of engineering plans. The knowledge has to be available in a sufficiently timely manner to support applications that depend on reacting in time to real-time processes.

Example requirements for engineering knowledge at system runtime can be found in UC3 and UC4. In UC3, if a flexible production system changes, the engineering knowledge on the system structure, available components, and wiring to signals has to be updated in a knowledge base to enable the correct linking of data from sensors to engineering objects. In UC4, a current engineering knowledge on the system is necessary to correctly assess the impact of a changed component on the overall system behavior.

The collection of needs is the foundation to investigate how well Semantic Web capabilities can address these needs for semantic support.

2.6 Summary and Outlook

This chapter introduced key elements of the life cycle processes of engineering production systems toward *Industrie 4.0*, focusing on heterogeneity in the engineering of production systems. We introduced four usage scenarios to illustrate key challenges of system engineers and managers in the transition from traditional to cyber-physical production system engineering environments. These scenarios allow semantic researchers and practitioners to better understand the challenges of production system engineers and managers, to define goals and assess solution options using Semantic Web technologies. Major semantic challenges come from the need to provide tool support for processes that build on the heterogeneous terms, concepts, and models used by the stakeholders in production system engineering and operation. From the challenges illustrated in the four scenarios, we derived needs for semantic support from the usage scenarios as a foundation for evaluating solution approaches.

Outlook. Chapter 3 will provide a basic assessment on how well Semantic Web approaches seem suitable to address these needs compared to alternative approaches. Table 2.2 summarizes the assessment discussed in Chap. 3 to inform potential users on the relevant Semantic Web capabilities and their match to the needs identified in Sect. 2.5. Table 2.2 considers how important each capability (Cx) is to address the set of needs to enable selecting a suitable set of Semantic Web technologies as starting point for planning a solution strategy.

The qualitative evaluation of Semantic Web capabilities shows good coverage of the previously identified production systems engineering needs. The conclusion

Table 2.2 Semantic Web technology capabilities (Cx) and needs for engineering knowledge modeling and integration capabilities (Nx); strong (++) and moderate (+) support of a need by a capability

Semantic Web Capabilities & Needs		N1	N2	N3	N4	N5	N6	N7
C1	Formal and flexible semantic modeling	++	+	++	+	+	+	+
C2	Intelligent, web-scale knowledge integration	+	++	++	++	++	++	
C3	Browsing and exploration of distributed data set			+	++	+	+	+
C4	Quality assurance of knowledge with reasoning					++	++	++
C5	Knowledge reuse	+	+	++			++	+

Chap. 15 will compare the strengths and limitations of Semantic Web approaches with alternative approaches that are also available to production systems engineers. Engineers and managers from engineering domains can use the scenarios introduced in this chapter to get a better understanding of the benefits and limitations coming from using Semantic Web technologies in comparison to alternative approaches as foundation to select and adopt appropriate Semantic Web solutions in their own settings.

Semantic Web researchers and practitioners, who need to define goals for intelligent engineering applications and to assess solution options using Semantic Web technologies, can compare their application scenarios to the usage scenarios introduced in Sect. 2.4 to derive needs that can be addressed well by the capabilities of Semantic Web technologies, see Chaps. 3 and 15.

Acknowledgments Parts of this work were supported by the Christian Doppler Forschungsgesellschaft, the Federal Ministry of Economy, Family and Youth, and the National Foundation for Research, Technology and Development in Austria.

References

AQUIMO. aquimo—Ein Leitfaden für Maschinen- und Anlagenbauer. VDMA (2010)

Biffl, S., Mordinyi, R., Moser, T.: Anforderungsanalyse für das integrierte Engineering—Mechanismen und Bedarfe aus der Praxis. ATP Edition **54**(5), 28–35 (2012)

Binder. J., Post, P.: Innovation durch Interdisziplinarität: Beispiele aus der Industriellen Automatisierung. In: Anderl, M., Eigner, U., Sendler, R., Stark, R. (eds.) Smart Engineering, acatech DISKUSSION, pp. 31–43 (2012)

BMBF. Produktionsforschung—57 erfolgreiche Projekte für Menschen und Märkte. Bundesministerium für Bildung und Forschung (2007)

Drath, R., Lüder, A., Peschke, J., Hundt, L.: An evolutionary approach for the industrial introduction of virtual commissioning. In: Proceedings of the 13th IEEE International Conference on Emerging Technologies and Factory Automation (ETFA). IEEE (2008)

Götz, O.: AutomationML—Effizienz für den Engineeringprozess. In: SPS-Magazin, 5/2013, pp. 43–45 (2013)

Grote, K.-H., Feldhusen, J. (eds.) Dubbel—Taschenbuch für den Maschinenbau. 24. aktualisierte Auflage. Springer Vieweg (2014)

Hundt, L., Lüder, A., Estévez, E.: Engineering of manufacturing systems within engineering networks. In: Proceedings of the 15th IEEE International Conference on Emerging Technologies and Factory Automation (ETFA). IEEE (2010)

Jacob, M.: Management und Informationstechnik. Eine Kompakte Darstellung. Springer (2012)

Kagermann, H, Wahlster, W, Helbig, J. (eds.): Umsetzungsempfehlungen für das Zukunftsprojekt Industrie 4.0—Deutschlands Zukunft als Industriestandort sichern. Forschungsunion Wirtschaft und Wissenschaft. Arbeitskreis Industrie 4.0 (2013)

Kiefer, J.: Mechatronikorientierte Planung automatisierter Fertigungszellen im Bereich Karosseriebau. Dissertation. Universität des Saarlandes. Schriftenreihe Produktionstechnik Band 43 (2007)

Lindemann, U.: Methodische Entwicklung technischer Produkte: Methoden flexibel und situationsgerecht anwenden. 3rd edn. Springer (2007)

Lüder, A., Hundt, L., Foehr, M., Wagner, T., Zaddach, J.J.: Manufacturing System Engineering with Mechatronical Units. In: Proceedings of the 15th IEEE International Conference on Emerging Technologies and Factory Automation (ETFA). IEEE (2010)

Lüder, A., Foehr, M., Hundt, L., Hoffmann, M., Langer, Y., Frank, S.: Aggregation of engineering processes regarding the mechatronic approach. In: Proceedings of the 16th IEEE International Conference on Emerging Technologies and Factory Automation (ETFA). IEEE (2011)

Lüder, A.: Integration des Menschen in Szenarien der Industrie 4.0. In: Bauernhansl, T., Vogel-Heuser, B., ten Hommel, M. (eds.) Industrie 4.0 in Produktion, Automatisierung und Logistik. Springer Fachmedien, pp. 493–507 (2014)

Maga, C., Jazdi, N., Göhner, P., Ehben, T., Tetzner, T., Löwen, U.: Mehr Systematik für den Anlagenbau und das industrielle Lösungsgeschäft—Gesteigerte Effizienz durch Domain Engineering. In: at—Automatisierungstechnik 9, pp. 524–532

Newen, A., Bartels, A., Jung, E.-M.: Knowledge and Representation. Centre for the Study of Language & Information (2011)

Oberle, D.: Ontologies and reasoning in enterprise service ecosystems. Inform. Spektrum **37**(4) (2014)

Pfrommer, J., Schleipen, M., Beyerer, J.: PPRS: production skills and their relation to product, process, and resource. In: Proceedings of the 18th IEEE Conference on Emerging Technologies and Factory Automation (ETFA). IEEE (2013)

Science. Engineers Council for Professional Development. Science **94**(2446), 456 (1941)

Schäffler, T., Foehr, M., Lüder, A., Supke, K.: Engineering process evaluation—evaluation of the impact of internationalization decisions on the efficiency and quality of engineering processes. In: Proceedings of the 22nd IEEE International Symposium on Industrial Electronics (ISIE). IEEE (2013)

Schäppi, B., Andreasen, M., Kirchgeorg, M., Rademacher, F.-J.: Handbuch Produktentwicklung. Hanser Fachbuchverlag (2005)

Schnieder, E.: Methoden der Automatisierung. Vieweg, 380 pp. (1999)

Terkaj, W., Tolio, T., Valente, A.: Focused flexibility in production systems. In: Hoda, A., ElMaraghy (eds.) Changeable and Reconfigurable Manufacturing Systems. Springer Series in Advanced Manufacturing, Springer, pp. 47–66 (2009)

Ulrich, K., Eppinger, S.: Product Design and Development. McGraw-Hill Higher Education (2007)

VDI. Entwicklungsmethodik für mechatronische Systeme. VDI-Richtlinie 2206. Verein Deutscher Ingenieure e.V (2004)

VDI. Digitale Fabrik—Grundlagen. VDI-Richtlinie 4499–1. Verein Deutscher Ingenieure e.V (2008)

VDI. Fabrikplanung. VDI-Richtlinie 5200. Verein Deutscher Ingenieure e.V (2009)

VDI. Digitale Fabrik—Blatt 2. VDI-Richtlinie 4499-1. Verein Deutscher Ingenieure e.V (2009)

VDI. Engineering of industrial plants. Evolution and Optimizations. Fundamentals and Procedures. International Standard. ICS 35.240.50 Part 1–Part 4. Verein Deutscher Ingenieure e.V.. VDI/VDE 3695, November 2010 (2010)

VDI/VDE. Industrie 4.0—Wertschöpfungsketten. VDI/VDE Gesellschaft Mess- und Automatisierungstechnik. Statusreport, April 2014 (2014)

Wegener, D.: Industrie 4.0—Chancen und Herausforderungen für einen Global Player. In: Bauernhansl, T., ten Hompel, M., Vogel-Heuser, B. (eds.) Industrie 4.0 in Produktion, Automatisierung und Logistik. Springer (2015)

Chapter 3
An Introduction to Semantic Web Technologies

Marta Sabou

Abstract The process of engineering cyber-physical production systems (CPPS) relies on the collaborative work of multiple and diverse teams of engineers who need to exchange and synchronize data created by their domain-specific tools. Building applications that support the CPPS engineering process requires technologies that enable integrating and making sense of heterogeneous datasets produced by various engineering disciplines. Are Semantic Web technologies suitable to fulfill this task? This chapter aims to answer this question by introducing the reader to key Semantic Web concepts in general and core Semantic Web technologies (SWT) in particular, including ontologies, Semantic Web knowledge representation languages, and Linked Data. The chapter concludes this technology overview with a specific focus on core SWT capabilities that qualify them for intelligent engineering applications. These capabilities include (i) formal and flexible semantic modeling, (ii) intelligent, web-scale knowledge integration, (iii) browsing and exploration of distributed data sets, (iv) knowledge quality assurance and (v) knowledge reuse.

Keywords Semantic Web technologies · Ontologies · Semantic Web languages · Reasoning · Linked Data · Semantic Web technology capabilities

3.1 Introduction

The analysis of typical CPPS engineering scenarios in Chap. 2 led to conclude that these are knowledge- and data-intensive settings. They are characterized by the need to integrate and make sense of large-scale, evolving, and heterogeneous engineering data. In this chapter, we build on the findings of Chap. 2 and investigate how Semantic Web technologies (SWTs) can facilitate innovative CPPS engineering applications.

M. Sabou (✉)
Institute of Software Technology and Interactive Systems, CDL-Flex,
Vienna University of Technology, Vienna, Austria
e-mail: Marta.Sabou@ifs.tuwien.ac.at

S. Biffl and M. Sabou (eds.), *Semantic Web Technologies for Intelligent Engineering Applications*, DOI 10.1007/978-3-319-41490-4_3

To make the technical discussion in this chapter accessible to audiences that are not familiar with the Semantic Web, we adopt a gradual approach starting with a high level, nontechnical introduction of this technology followed by a more detailed presentation of core SWTs. This chapter lays the groundwork for a comprehensive understanding of the materials covered in the remainder of this book.

The three main goals of this chapter are as follows:

1. *Providing a high-level introduction to the Semantic Web*. The present chapter starts with providing an overall introduction to the Semantic Web, including the motivation behind it, the typical SWTs developed and their uptake in enterprises (Sect. 3.2). This introduction provides a high-level understanding of SWTs to motivate their relevance for solving the needs of engineering applications as briefly discussed in Sect. 3.2.4.
2. *Providing details of core SWTs*. The subsequent sections (Sects. 3.3–3.6) detail the various SWT elements that were determined as relevant for addressing the needs of engineering applications, including ontologies, Semantic Web knowledge representation languages, reasoning, and Linked Data.
3. *Identifying SWT capabilities particularly useful in engineering applications.* Section 3.7 revisits the question of how SWTs can be applied in the engineering domain by identifying core SWT capabilities. It also discusses how these capabilities can in principle support the various needs of engineering applications which were introduced in Chap. 2.

Section 3.8 concludes the chapter with a summary.

3.2 The Semantic Web: Motivation, History, and Relevance for Engineering

In this section, we provide a brief introduction to the Semantic Web. We start with a motivating scenario for the emergence of the Semantic Web research area (Sect. 3.2.1). We then define the Semantic Web and describe typical elements of a Semantic Web-based approach (Sect. 3.2.2). We demonstrate how the use of SWTs diffused from the original web-based setting into enterprises (Sect. 3.2.3) and conclude by motivating the relevance of concrete SWT elements to address the needs of engineering applications (Sect. 3.2.4).

3.2.1 Why Was the Semantic Web Needed?

Research on SWTs was motivated by the explosive growth of the *World Wide Web* (Berners-Lee 1999) to levels where finding information became a nontrivial task. Document retrieval on the Web (that is, Web search) was particularly affected as

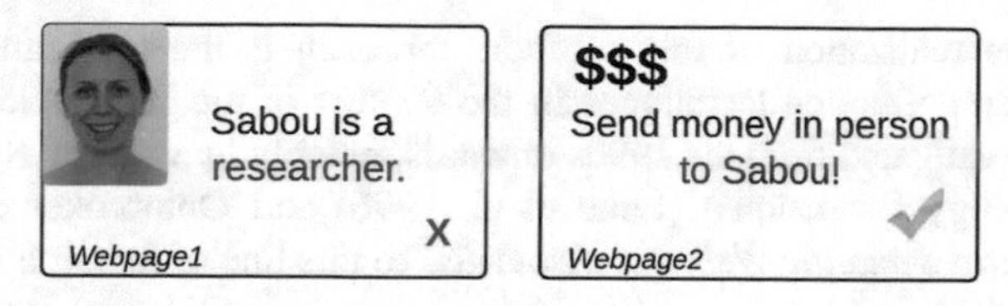

Fig. 3.1 Example false negative (Webpage1) and false positive (Webpage2) results for the "Sabou person" query

keyword-based web search algorithms could reliably identify all web pages that contain a given set of keywords (i.e., high recall) but also returned many irrelevant results (i.e., low precision).

Assume, for example, that you have briefly met one of the editors of this book at an event and that you later on try to find information about her. Suppose you remember only her family name, *Sabou*. Ideally, Web search can point you to her homepage (*Webpage 1* in Fig. 3.1) within the first page of the search result list.

A search for *Sabou,* however, will return over 300,000 pages including webpages about persons (mostly soccer-players), geographic locations (a town in Burkina Faso), or music albums. Narrowing the search by specifically querying persons with "*Sabou person*" will still yield a substantial number of search results (25,000). The websites returned are those that contain these two keywords explicitly. This includes a website of a "*Send money in Person*" service, in particular its webpage referring to the African town of Sabou (represented as *Webpage 2* in Fig. 3.1) although, obviously this webpage does not refer to a person. Therefore, in information retrieval terminology, it is a false positive (Baeza-Yates and Ribeiro-Neto 1999). *Webpage 1* is not retrieved because it does not contain the term "person" it is a false negative (Baeza-Yates and Ribeiro-Neto 1999).

How can search engines be made more effective? How can they sieve out wrong results from a result list? The Semantic Web aimed to solve such issues, as we detail in Sect. 3.2.2.

3.2.2 *The Semantic Web in a Nutshell*

To tackle the limitations of search and other applications that aimed to process the vast, textual Web, the idea emerged to augment Web information with a formal (i.e., machine-processable) representation of its meaning. In other words, the idea was to add additional information to webpages that would more clearly describe the meaning of their content (e.g., that they are about a person or a town) and therefore avoid typical mistakes made by computer programs as exemplified above. A direct benefit of this machine-processable meaning is the enhancement and potential automation of several information management tasks, such as search or data integration. For example, search engines would deliver more appropriate results if they could easily tell whether a webpage is about a town or a person.

A concrete realization of this solution approach is the application of formal knowledge representation techniques in the context of the Web. Such approaches have been investigated from the 1990s onwards, notably by work on SHOE (Simple HTML Ontology Extensions) (Luke et al. 1996) and Ontobroker (Fensel et al. 1998). The term *Semantic Web* was associated to this line of research in 2001 when it was defined as:

> The Semantic Web is an extension of the current Web, in which information is given well-defined meaning, better enabling computers and people to work in cooperation. (Berners-Lee et al. 2001)

Well-defined meaning is provided by semantic descriptions, often referred to as *metadata* (i.e., data about data). In Fig. 3.2 we revisit the motivating scenario and add the elements of SWTs to it. The bottom layer of the figure consists of our example webpages, which constitute the *Data* layer. A *Metadata* layer is then added which clarifies that *Webpage1* is about an entity *Sabou* who has a *Job*, and declares that *Researcher* is a kind of *Job*. The metadata of *Webpage2* states that the webpage refers to an entity *Sabou* which is a *Town*.

By itself, solely adding descriptions of (i.e., *metadata* about) webpages as text will not solve the aforementioned issues as search engines will make the same interpretation mistakes that they make on current textual webpages. Instead, appropriate technologies must be used to make these metadata descriptions less ambiguous and hence interpretable by computer programs.

To achieve this, a number of principles must be followed. First, metadata should describe information with terms that have clear meaning to machines and also reflect the agreement of a wide community. For our example, terms such as "person," "town," or "job" are important. It is also important to convey how these terms can be related to state, for example, that persons can have jobs, that towns are

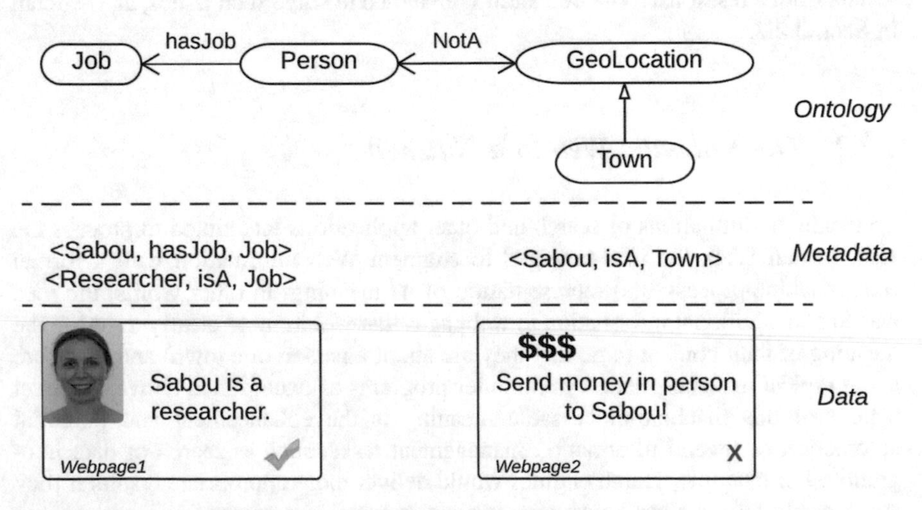

Fig. 3.2 The Semantic Web idea in a nutshell

geographic locations, and that the set of persons and geographic locations is disjoint (i.e., there is no entity which is both a person and a geographic location). A collection of terms and their relations forms a domain vocabulary. SWTs allow specifying these domain vocabularies through *formal, shared domain models* i.e., *ontologies* (Gruber 1993). For example, the top part of Fig. 3.2 shows a basic ontology that captures the information we described above. The *Metadata* layer uses ontology terms to describe the content of the webpages in the *Data* layer.

Second, metadata should be expressed in a *representation language* that can be parsed and interpreted by computer programs. For example, HTML (Hypertext Markup Language), is a simple representation language that instructs a browser computer program *how* to display information on a webpage, e.g., any text included between the tags <b> and </b> will be shown in bold-face. To realize the vision of the Semantic Web, representation languages that describe *what* certain information means are needed. These languages can be used to specify which words refer to concepts and which to relations between concepts.

The meaning of semantic representation languages is grounded in logics, which entails two important advantages. First, it allows to unambiguously state complex facts such as: only persons can have jobs (i.e., anything that has a job is a person); nothing can be both a person and a geographic location (i.e., the set of persons and geographic locations are disjoint); a town is a kind of geographic location. Second, the logics-based semantics (meaning) can be leveraged to enable computer programs to derive new information, a process referred *to as inference or reasoning*.

Continuing our example, *Webpage1* refers to an individual *Sabou* who has a Job as a researcher. Because persons can have jobs according to the example ontology, it follows that *Sabou* in this web page refers to a *Person*. A search algorithm that can parse metadata and reason with knowledge described in ontologies can therefore deduce that this webpage should be returned as a result to the query for "*Sabou person*". Similarly, since *Webpage2* refers to *Sabou* that is a *Town*, a search algorithm can infer that this mention of *Sabou* also refers to a *GeoLocation* and therefore it cannot be about a *Person* as the example ontology explicitly states that these two concepts are disjoint. Therefore, *Webpage2* should not be returned as a result to the "*Sabou person*" query.

3.2.3 The Use of Semantic Web Technologies in Enterprises

Although initially developed to improve access to Web data, SWTs have proved to be beneficial in *enterprise settings* as well, especially in data-intensive domains where they facilitate integration of heterogeneous data sets (Shadbolt et al. 2006). A prime example is e-Science where ontologies can facilitate data interoperability between scientific communities in separate subfields around the world and allow them to share and communicate with each other, which may in turn lead to new scientific discoveries. In this scenario, ontologies are used as a "means of communicating and resolving semantic and organizational differences between biological databases" (Schuurman and Leszczynski 2008). Semantic integration of datasets is achieved using ontologies as mediators. Furthermore, based on the formal nature of ontology languages, automated reasoning can be used to derive new knowledge as well as to detect potential errors and inconsistencies in ontologies (Köhler et al. 2011).

Responding to these data integration needs, a new set of *Linked Data* technologies has evolved focusing on methods for creating links across semantic datasets to aid their meaningful integration (Heath and Bizer 2011). Wood (2010) discusses the adoption of these techniques in enterprises as Linked Enterprise Data. In the media domain, for instance, one of the earliest adopters of SWT is the BBC,[1] which uses Linked Data for integrating its diverse and disconnected internal datasets as well as to link to external semantic datasets (Kobilarov et al. 2009). One such external semantic dataset is DBpedia,[2] a Semantic Web representation of Wikipedia data (Lehmann et al. 2015).

3.2.4 How Are SWTs Relevant for Engineering Applications?

Similarly to the previously discussed e-Science and media industry settings, typical use cases of engineering of complex production systems (including CPPS) as discussed in Chap. 2 involve integrating and making sense of heterogeneous datasets produced by different engineering disciplines. To realize these use cases, the following needs must be fulfilled: abilities to explicitly represent engineering knowledge (need N1), to integrate engineering knowledge (N2), to provide access to and analytics on (the integrated) engineering knowledge (N3) as well as to provide efficient access to semi-structured data in the organization and on the Web (N4, see Table 2.1 in Chap. 2).

Since SWTs were deemed to be suitable for addressing such needs in other domains, this chapter will explore SWTs required for addressing typical needs when creating intelligent engineering applications (IEA). The explicit representation of engineering knowledge (N1) is achieved with ontologies (Sect. 3.3) and formal knowledge representation languages (Sect. 3.4). Knowledge access and analytics (N3) rely on reasoning functionalities made possible by the formal nature of ontologies (Sect. 3.5). Data integration (N2) internally to the enterprise as well as with external data is well supported by the Linked Data technologies discussed in Sect. 3.6. Section 3.7 revisits in more detail this initial analysis of how SWT capabilities can address the needs of IEAs.

---

[1]BBC: http://www.bbc.com/.

[2]DBpedia: http://wiki.dbpedia.org/.

3.3 Ontologies

As discussed in Sect. 3.2.2, an *ontology* is a technical artifact that acts as a centerpiece of any Semantic Web-based solution and allows the explicit and formal representation of knowledge relevant for the application at hand. Adopters of Semantic Web solutions therefore need to acquire an ontology either by creating it themselves or by reusing one from similar applications. This section defines ontologies, explains the main elements of an ontology and describes a set of characteristics to be considered when reusing ontologies. Chapter 5 expands on the topic of semantic modeling (i.e., ontology creation) in the context of the engineering domain.

Studer et al. (1998) define an ontology as "*a formal, explicit specification of a shared conceptualization*". In other words, an ontology is a domain model (*conceptualization*) which is *explicitly* described (*specified*). An ontology should express a *shared* view between several parties, a consensus rather than an individual view. Also, this conceptualization should be expressed in a machine-readable format (*formal*). As consensual domain models, the primary role of ontologies is to enhance communication between humans (e.g., establishing a shared vocabulary, explaining the meaning of the shared terms to reach consensus). As formal models, ontologies represent knowledge in a computer-processable format thus enhancing communication between humans and computer programs or two computer programs.

For example, an ontology in the mechanical engineering domain, such as the ontology snippet depicted in Fig. 3.3, could be used to explicitly record mechanical engineering knowledge necessary for semantically describing relevant information in an engineering project. This ontology could include concepts such as `Conveyer` or `Engine`. A *concept* represents a set or class of entities with shared

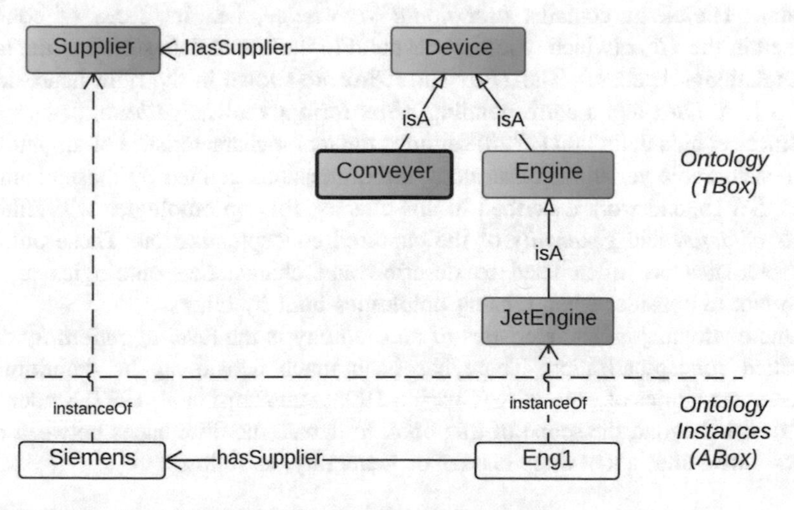

Fig. 3.3 Snippet from a Mechanical Engineering domain ontology

characteristics within the domain of the ontology. Alternative terms for referring to ontology concepts are *Classes*, *Types*, or *Universals*.

Entities (individuals, instances) represented by a concept are the things the ontology describes or potentially could describe. For example, entities can represent concrete objects such as a specific engine (`Eng1`) or a company (`Siemens`). Ontology instances are related with the *instanceOf* relation to the ontology concepts denoting their types.

A set of *relations* can be established between ontology concepts. An important relation is the *isA* relation, which indicates subsumption between two concepts. In other words, it connects more generic (sometimes also referred to as *parent*) concepts to more specific ones (or *child* concepts) thus providing means for organizing concepts into *taxonomies* (i.e., *inheritance hierarchies)*. For example, a `JetEngine` is more specific than an `Engine`, so a subsumption relation can be declared between `JetEngine` and `Engine`. Similarly, `Conveyer` and `Engine` are more specific than `Device`.

Ontologies can also define domain-specific relations. For example, `hasSupplier` relates a `Device` to its `Supplier`. Similarly, characteristics of instances denoted by a concept can be described with relations such as `hasWeightInKg` or `hasMaxWorkingHours` (these are not depicted in Fig. 3.3). These relations connect an instance of a concept with a data value (e.g., an integer or a string). Ontologies can also contain more complex information such as various constraints depending on the modeling capabilities offered by knowledge representation languages, which will be discussed in Sect. 3.4 in more detail.

Since current ontology representation languages are based on Description Logics (DL) (Baader et al. 2003), the DL terms of *ABox* and *TBox* can be used to characterize different types of knowledge encoded by an ontology. The *TBox* covers *terminological knowledge*, which includes ontology concepts and relations. It describes the structure of the data and therefore it plays a similar role as a database schema. The *ABox* contains *assertional knowledge*, i.e., instances of concepts defined in the *TBox* (which would correspond to the actual data stored in the tables of a relational database). The *TBox* and *ABox* are shown in the right hand side of Fig. 3.1. A *TBox* and a corresponding *ABox* form a *knowledge base*.

Studer et al.'s definition (1998) captures the major characteristics of an ontology, but considerable variations exist along the dimensions defined by these characteristics. SWTs, and work described in this chapter, rely on ontologies with different *levels of detail* and *generality* of the captured conceptualization. These ontology characteristics are often used to describe and characterize ontologies and are important to consider when reusing ontologies built by others.

One of the major characteristics of an ontology is the *level of generality* of the specified conceptualization. There has been much debate on the definitions of different categories of generality (Guarino 1998; van Heijst et al. 1997; Studer et al. 1998). It is beyond the scope of this book to debate the differences between these views—we rather adopt three classes of generality, as follows:

- **Foundational (or top-level) ontologies** are conceptualizations that contain specifications of domain and problem independent concepts and relations (such as space, time, matter) based on formal principles derived from linguistics, philosophy, and mathematics. The role of these ontologies is to serve as a starting point for building new ontologies, to provide a reference point for easy, and rigorous comparisons among different ontological approaches, and to create a foundational framework for analyzing, harmonizing, and integrating existing ontologies and metadata standards, in any domain, engineering included. Examples of such ontologies are DOLCE (Masolo et al. 2003), the *Suggested Upper Merged Ontology* (SUMO) (Pease et al. 2002), OpenCyc[3] and the *Basic Formal Ontology* (BFO) (Smith 2003). A comparison of these top-level ontologies is provided by Borgo et al. (2002).
- **Generic ontologies** contain generic knowledge about a certain domain such as medicine, biology, mathematics, or engineering. Domain-specific concepts of generic ontologies are often specified in terms of top-level concepts defined in foundational ontologies thus inheriting the general theories behind these top-level concepts. Examples of such ontologies are the OWL-S ontology (Martin et al. 2007), a generic vocabulary for describing web services in any domain; the *Good Relations* ontology (Hepp 2008), which provides a vocabulary for describing product offering data on webpages and could be used to describe e-commerce offerings of engineering specific products; or the Onto-CAPE ontology in the domain of the computed added process engineering (Marquardt et al. 2010).
- **Domain ontologies** are specific to a particular domain. For example, the *Friend-of-a-Friend* (FOAF) ontology[4] provides a vocabulary for describing personal information.

A second ontology classification criterion is the *level of detail* of the specification. The ontology community distinguishes between *lightweight* and *heavyweight* ontologies (Corcho et al. 2003). *Lightweight ontologies* are domain models that include a taxonomic hierarchy as well as properties between concepts. For example, the FOAF vocabulary would qualify as a lightweight ontology as it contains the definition of only a handful of concepts (`Person`, `Agent`, `Organization`, `Project`) and their relevant properties (e.g., `firstName`, `logo`, `knows`). *Heavyweight ontologies* contain axioms and constraints as well. For example, *OpenCyc* and DOLCE are heavyweight ontologies. Note that the distinction between lightweight and heavyweight ontologies is blurred as these are intuitive rather than fixed measures. While heavyweight ontologies are more difficult to build, they enable a larger range of reasoning tasks that can lead to more diverse functionalities of engineering applications using them.

---

[3]OpenCyc: http://www.opencyc.org.

[4]FOAF: http://www.foaf-project.org/.

Table 3.1 Web- and Semantic Web-specific standard namespaces

Namespace	Global URI for namespace
rdf	http://www.w3.org/1999/02/22-rdf-syntax-ns#
rdfs	http://www.w3.org/2000/01/rdf-schema#
owl	http://www.w3.org/2002/07/owl#
xsd	http://www.w3.org/2001/XMLSchema#

3.4 Semantic Web Languages

To represent Semantic Web-specific data, a set of Web-based knowledge representation languages has been developed. In this section, we provide an introduction to these languages including: RDF (*Resource Description Format*) in Sect. 3.4.1, RDF(S) (RDF Schema) in Sect. 3.4.2 and OWL (*Web Ontology Language*) in Sect. 3.4.3 (Table 3.1). Table 3.2 provides an overview of the key modeling constructs of these languages defined in this section. We conclude the section with an introduction to the SPARQL query language in Sect. 3.4.4, which allows querying semantic data represented in RDF and therefore plays a key role in many IEAs built using SWTs.

3.4.1 Resource Description Framework (RDF)

The *Resource Description Framework*[5] (RDF) is a language for describing *resources* on the Web and was adopted as the data interchange model for all Semantic Web languages. Resources can refer to anything including "physical things, documents, abstract concepts, numbers, and strings" (Cyganiak et al. 2014). RDF allows expressing relationships between two resources through *RDF statements*. RDF statements consist of three elements: a *subject*, a *predicate*, and an *object*, which are collectively referred to as *triples*. Objects in an RDF statement can also be represented by *literals*, which are used for representing values such as strings, numbers, and dates.

The subject and the object of an RDF statement denote the resources that are related, while the predicate is a resource itself denoting the relation that exists between the subject and the object. For example, the following triples declare that `Eng1` is an `Engine`, specify its weight and its maximum working hours, and relate it to a supplier:

[5]RDF: https://www.w3.org/RDF/.

Table 3.2 Overview of the key RDF/RDF(S)/OWL modeling constructs described in this chapter and their definitions adapted from the corresponding language reference documentation

Modeling construct	Definition according to language specification
`rdf:type`	States that a resource is an instance of a class
`rdf:Property`	Used to define an RDF property
`rdfs:Class`	Declares a resource as a class for other resources
`rdfs:subClassOf`	States that all the instances of one class are instances of another
`rdfs:domain`	Declares the class or datatype of the subject in triples whose second component is a certain predicate
`rdfs:range`	Declares the class or datatype of the object in triples whose second component is a certain predicate
`rdfs:subPropertyOf`	States that all resources related by one property are also related by another
`owl:equivalentClass`	Relates two classes whose class extensions contain exactly the same set of individuals
`owl:disjointWith`	Asserts that the class extensions of the two class descriptions involved have no individuals in common
`owl:intersectionOf`	Defines a class that contains the same instances as the intersection of a specified list of classes
`owl:unionOf`	Defines a class that contains the same instances as the union of a specified list of classes
`owl:complementOf`	Defines a class as a class of all individuals that do not belong to a certain specified class
`owl:ObjectProperty`	Defines a property that captures a relation between instances of two classes
`owl:DatatypeProperty`	Defines a property that captures a relation between instances of classes and RDF literals/XML Schema datatypes
`owl:inverseOf`	If a property, P1, is owl:inverseOf P2, then for all x and y: P1(x, y) iff P2(y, x)
`owl:TransitiveProperty`	If a property, P, is transitive then for any x, y, and z: P(x, y) and P(y, z) implies P(x, z)
`owl:ReflexiveProperty`	A reflexive property relates everything to itself
`owl:SymmetricProperty`	If a property P is symmetric then if the pair (x, y) is an instance of P, then the pair (y, x) is also an instance of P

Eng1 isA Engine.
Eng1 hasSupplier Siemens.
Eng1 hasWeightInKg 2.
Eng1 hasMaxWorkingHours 14000.

These triples are represented graphically in Fig. 3.4, with resources depicted by an ellipse and literals by a rectangle.

A set of RDF triples constitutes an *RDF graph*. An important principle of RDF is that individual triples can be merged whenever one of their resources is the same. The example triples in Fig. 3.2a all have resource `Eng1` as their subject. As a result,

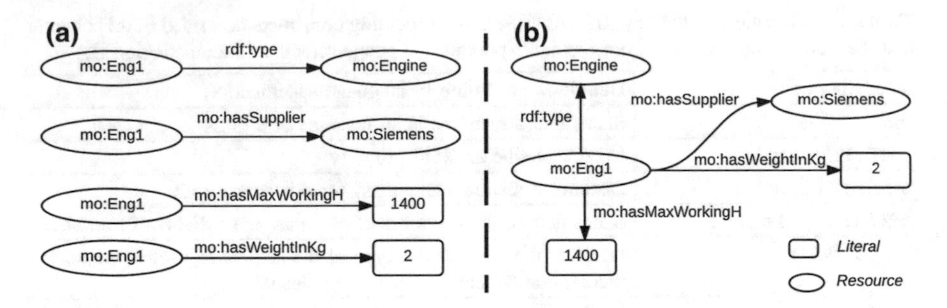

Fig. 3.4 RDF Triples and the corresponding RDF Graph. **a** RDF triples **b** RDF graph

they can be merged into a graph structure as shown in Fig. 3.4b. This characteristic of RDF facilitates tasks that require integrating data from various sources, for example, from different webpages that provide (potentially) different information about the same entity (for example, the same person or company). This characteristic differentiates the graph-based RDF data model from more traditional, relational data models.

Two RDF resources are considered same if they have the same unique identifier (or name). In RDF, each resource is identified using URLs (*Uniform Resource Locators*), which are well-established Web technologies for identifying webpages. For example, instance `Eng1` can be identified with the URL http://www.tuwien.ac.at/mechanics/Eng1. URLs, used as Web addresses, point to the exact location for accessing a resource on the Web (e.g., most commonly a webpage). In RDF, resource names do not necessarily locate a resource, it is sufficient that they universally identify that resource. Therefore, RDF relies on URIs (*Uniform Resource Identifiers*) to identify resources. Effectively, URLs are more specific URIs because they do not only identify a resource, but also provide information on how to access it. In practical terms, although URLs and URIs have the same format, URLs will always point to a Web-resource such as a webpage (i.e., they will locate this) whereas URIs do not necessarily point to a concrete resource. To enable the use of non-ASCII characters, IRIs (*International Resource Identifiers*) can be used. To conclude with a more precise definition of resource equality: two RDF resources are considered the same if they are identified by exactly the same URI.

While lengthy IRI strings can be handled well by computers, they are cumbersome to handle in print. As a solution, RDF relies on the mechanism of *qualified names* (or *qnames*) used in XML (*Extensible Markup Language*) namespaces. A *qname* consists of a *namespace* and an *identifier* within that namespace, separated by a colon. For example, consider `mo` as a namespace representing http://www.tuwien.ac.at/mechanics/. Then the respective *qname* for http://www.tuwien.ac.at/mechanics/Eng1 would be `mo:Eng1`.

The W3C (*World Wide Web Consortium*)[6] has defined namespaces for some of the major Web- and Semantic Web-specific standards, as shown in Table 3.1. These namespaces will be used throughout the chapters of this book.

The success of the Web and its exponential growth has been facilitated by its open nature. On the Web, "**A**nyone is allowed to say **A**nything about **A**ny topic," a feature that Allemang and Hendler (2011) refer to as the *AAA slogan*. This principle heavily influenced several decisions in the design of SWTs. In particular, the Semantic Web relies on the *Nonunique Naming Assumption*, which means that two syntactically different IRIs might refer to the same real-world entity. In other words, just because two IRIs differ, it does not necessarily follow that they refer to two different entities. This assumption is indispensable in a Web setting, where diverse content creators can use syntactically varying IRIs to refer to the same real-world entity.

Serializations. RDF graphs can be written down (i.e., serialized) using a variety of formats. These include the following:

- the Turtle family of RDF languages (N-Triples, *Turtle*);
- JSON-LD, which is based on JSON syntax;
- RDFa[7] (*Resource Description Framework* in attributes), which is suitable for embedding RDF content into HTML and XML; and
- RDF/XML, which is an XML Syntax for RDF.

Examples of these diverse serializations are provided in the RDF 1.1 Primer.[8] Throughout the present chapter we use *Turtle* for our examples. Listing 3.1 shows the previously discussed triples (see Fig. 3.4) in a *Turtle* serialization. Lines 1–2 contain the namespace declarations, while lines 3–5 contain the actual triples stating that `Eng1` is an `Engine`, and it has certain values of weight and maximum working hours.

Listing 3.1:

```
1. @prefix mo: <http://www.tuwien.ac.at/mechanics/>.
2. @prefix rdf: <http://www.w3.org/1999/02/22-rdf-syntax-ns#>.
3. mo:Eng1 rdf:type mo:Engine.
4. mo:Eng1 mo:hasWeightInKg "2".
5. mo:Eng1 mo:hasMaxWorkingHours "14000".
```

[6]W3C: https://www.w3.org/.

[7]RDFa: https://rdfa.info/.

[8]RDF 1.1 Primer: http://www.w3.org/TR/2014/NOTE-rdf11-primer-20140225/.

3.4.2 RDF Schema—RDF(S)

RDF provides basic means for data modeling for the Semantic Web, but it is limited in expressing what its resources are and how they relate. *RDF Schema*[9] (RDF(S)) is based on RDF and allows the definition of basic ontology elements (McBride 2004), such as those discussed in Sect. 3.3. RDF(S) provides a vocabulary to declare classes (sets) of objects in the world (classes correspond to ontology concepts) as well as class hierarchies. Classes are declared with `rdfs:Class`. One can declare that a class A is more specific than a class B using `rdfs:subClassOf`. This means that all resources (instances) belonging to class A also belong to class B (A being more specific than B). For example, `JetEngine` can be declared as a subclass of `Engine` because each jet engine instance can also be considered an instance of type engine.

Relations between ontology classes can be declared in RDF(S) using `rdf:Property`. Furthermore, it is possible to specify the domain and range of each property. The domain of a property (specified with `rdfs:domain`) specifies the type of any resource that is used as a subject for that property. Such information allows reasoners to deduce (i.e., infer) that any instance that acts as the subject of a property has the type of the domain declaration (even is such type information was not explicitly asserted for that instance). Similarly, the range of a property (specified with `rdfs:range`) defines the type of resources that can appear as objects of a property. Consider, for example, the property `mo:hasSupplier` for which `mo:Engine` can be declared as domain and `mo:Supplier` as a range. Asserting a triple of the form:

`mo:Eng1 mo:hasSupplier mo:Siemens.`

will lead an inference mechanism to deduce, based on the domain and range declarations of the `mo:hasSupplier` property, that `mo:Eng1` is of type `mo:Engine`, while the instance `mo:Siemens` is of type `mo:Supplier`. It is evident, that property domain and range declarations are a useful mechanism for deducing additional information about instances. Note that domain and range declarations cannot be used for constraint checking as typical for database systems. More information on constraint checking on the Semantic Web is available in Chap. 13.

As with classes, relations (or properties) can also be arranged into hierarchies using `rdfs:subPropertyOf`. If a property P is a subproperty of P', it means that resource pairs, which can be related with P, are also related with P'. Continuing the example above, some suppliers might only supply a company with services relevant for engines, for example, repair services. To capture this, a more specific property can be declared, namely `mo:hasServiceSupplier`, which is a subproperty of `mo:hasSupplier`.

`mo:hasServiceSupplier rdfs:subPropertyOf mo:hasSupplier.`

[9] RDF(S): https://www.w3.org/TR/rdf-schema/.

We conclude that, despite the limited number of modeling primitives, RDF(S) is well suited for expressing lightweight ontologies, consisting of concepts, relations, and their hierarchies.

3.4.3 *The Web Ontology Language (OWL)*

The *Web Ontology Language* (OWL)[10] was introduced as a more expressive ontology language than RDF(S). Since 2012, W3C recommends the use of OWL 2, an extended and improved version of OWL 1. OWL 2 enhances the expressivity of RDF(S) in several ways. We provide a few examples of OWL 2 constructs that are used in the chapters of the book, but refrain from a complete introduction to this language.

For classes, OWL 2 provides constructs beyond simple subsumption hierarchies declared with `rdfs:subClassOf`. One can specify, e.g., classes that have exactly the same set of instances with `owl:equivalentClass` or specify classes that share no instances at all with `owl:disjointWith`. Further, complex classes can be declared as intersections of (`owl:intersectionOf`), unions of (`owl:unionOf`), or complements of other (sets of) classes (`owl:complementOf`). The meaning of these modeling constructs is equivalent to that of the corresponding set theory operators.

Unlike RDF(S), OWL 2 distinguishes object properties (i.e., properties where the range is another instance identified by an IRI, declared with `owl:ObjectProperty`) and datatype properties (i.e., properties where the range is a literal—e.g., string, number, date—declared with `owl:DatatypeProperty`). For example, `mo:hasSupplier` is an object property, while the other two properties `mo:hasWeightInKg` and `mo:hasMaxWorkingH`, are datatype properties. For both types of properties, OWL 2 provides constructs to specify cardinality (e.g., min, max, exact) and value restrictions (e.g., universal and existential quantifiers). Cardinality constraints on a property P specify the minimum, maximum, or exact number of values that an instance will be connected to with property P.

Similarly to classes, a property can be equivalent to or disjoint with other properties. Object properties can have a set of property characteristics that add additional meaning to the ontological model. Some of the property characteristics used in this book are as follows:

- *Inverse* (`owl:inverseOf`)*:* a property P is an inverse of property P', if for any pair (x, y) connected by P', it is true that P connects (y, x). For example, for the property `mo:hasSupplier` we can declare an inverse property `mo:isSupplierFor`. In this case, for any triple such as <`mo:Eng1, mo:`

[10] OWL: https://www.w3.org/TR/owl2-overview/.

hasSupplier, mo:Siemens>, the following triple will be inferred: <mo:Siemens, mo:isSupplierFor, mo:Eng1>.

- *Transitive* (owl:TransitiveProperty*)*: a property P is transitive if from P (a,b) and P(b,c) we can deduce that P(a,c). Ancestry relations are classical examples of transitive relations.
- *Reflexive* (owl:ReflexiveProperty*)*: a reflexive property relates everything to itself. In mathematics, equality is a reflexive property as any number is equal to itself. In ontologies, a relation such as hasRelative is an example of a reflexive property since it can be said that everybody is a relative to him/herself. The inverse of owl:ReflexiveProperty is owl:IrreflexiveProperty and represents any property that cannot relate an individual to itself. For example, hasParent is irreflexive because nobody can be a parent to him/herself.
- *Symmetric* (owl:SymmetricProperty*)*: a property P is symmetric if for any P(x, y) P(y, x) also holds. In other words, these are properties where the direction of the property does not matter, or the property and its inverse are identical. An example in the mechanical engineering domain would be the isConnectedTo relation between two devices. Another intuitive example is the hasSpouse relation, where asserting that hasSpouse(x,y) has the same meaning as asserting that hasSpouse(y,x). Asymmetric properties (owl:AsymmetricProperty), on the other hand, are those properties P where asserting P(x, y) means that P(y, x) can never be asserted. A property hasParent relating a child to his parent, is a classic asymmetric property example. An example for asymmetric property from the mechanical domain is contains (one device contains another).

3.4.4 SPARQL (SPARQL Protocol and RDF Query Language)

As discussed in Sects. 3.4.2 and 3.4.3, RDF(S) and OWL are languages that allow describing semantic models, both in terms of ontologies and ontology instances. RDF(S) and OWL models follow RDF's graph structure. SPARQL (Harris and Seaborne 2013) is a W3C-endorsed query language for querying such graph-based data structures. Semantic Web datasets are can be made accessible through a *SPARQL Endpoint*, which enables querying that dataset via HTTP (Hypertext Transfer Protocol). For example, *DBpedia* offers a SPARQL endpoint at http://dbpedia.org/sparql.

SPARQL works on the principle of matching query patterns over an RDF data graph. Listing 3.2 shows a basic query that returns all the suppliers of mo:Eng1.

Listing 3.2:

```
1. PREFIX mo: <http://www.tuwien.ac.at/mechanics/>
2. SELECT ?supplier
3. WHERE {
4. mo:Eng1 mo:hasSupplier ?supplier
5. }
```

SPARQL queries start with namespace declarations in the PREFIX clause. The SELECT clause identifies the information that should be returned, in this case we specify a variable named `?supplier`. Variables in SPARQL always start with a "?" and can match any node (resource or literal) in an RDF data graph. The WHERE clause specifies *triple patterns* that should be matched. A *triple pattern* is similar to a triple, but any part of it can be replaced by a variable. In the example query (see Listing 3.2), we are interested in all triples of the form:

`mo:Eng1 mo:hasSupplier ?supplier`

In the case of the example dataset there is only one triple that matches this pattern, namely the triple:

`mo:Eng1 mo:hasSupplier mo:Siemens`

where the resource `mo:Siemens` will be bound to variable `?supplier`. Hence, the result set consists only of a single value, `mo:Siemens`.

A more complex example query is shown in Listing 3.3. This query returns not only the supplier but also the maximum working hours of `mo:Eng1`. Here, SELECT does not specify concrete variables that should be extracted, but rather states that the bindings for both specified variables should be returned in the result set (by "*"). Additionally, the WHERE clause now contains a graph pattern created through the conjunction of two triple patterns (through ".").

Listing 3.3:

```
1. PREFIX mo: <http://www.tuwien.ac.at/mechanics/>
2. SELECT *
3. WHERE {
4. mo:Eng1 mo:hasSupplier?supplier.
5. mo:Eng1 mo:hasMaxWorkingHours?maxH
6. }
```

Suppose that our example ontology also contains the datatype property `mo:hasActualWorkingHours` that records the amount of hours for which an engine was operational. Furthermore, an engine that has been used for more hours than specified by its maximum working hours property is considered a mo:WornOutEngine. With SPARQL it is possible to identify those engines that have been used for more hours than allowed. To this end, in Listing 3.4, the graph pattern that identifies the actual (?actH) and maximum (?maxH) working hours is extended with a filtering operator (FILTER) that selects only those ?eng entity resources for which the condition ?actH> ?maxH holds. SPARQL CONSTRUCT queries constitute a powerful mechanism for implementing rules and for deriving new knowledge from available information. CONSTRUCT clauses allow creating data graphs based on retrieved information: e.g., for every identified worn out engine resource, a corresponding triple is created stating that that engine resource is of type mo:WornOutEngine (a is a frequently used SPARQL shortcut for rdf:type).

Listing 3.4:

```
1. PREFIX mo: http://www.tuwien.ac.at/mechanics/
2. PREFIX rdfs: http://www.w3.org/2000/01/rdf-schema#
3. CONSTRUCT {
4. ?eng a mo:WornOutEngine
5. }
6. WHERE {
7. ?eng mo:hasActualWorkingHours ?actH.
8. ?eng mo:hasMaxWorkingHours ?maxH.
9. FILTER (?actH > ?maxH)
10. }
```

Additional explanations and examples of SPARQL queries can be found in Chap. 14.

3.5 Formality and Reasoning

Reasoners are inference engines that deduce (i.e., infer) new information from a given ontology model. They should be actively used while building an ontology for the early detection of undesired logical consequences that can be inferred from the ontology. Although several reasoners are currently available,[11] *Pellet* (Sirin et al. 2007), *HermiT* (Glimm et al. 2014), and *Fact ++* (Tsarkov and Horrocks 2006) are the most widely used reasoners thanks to their support of OWL 2 reasoning as well as integration with the widely popular Protégé ontology editing environment

[11]List of reasoners: http://owl.cs.manchester.ac.uk/tools/list-of-reasoners/.

(Musen et al. 2013).[12] A detailed comparison of the strengths and weaknesses of these (and other) reasoners when used for reasoning on large ontologies is available in (Dentler et al. 2011).

All Semantic Web languages are based on logics and possess an associated semantics. For example, OWL is based on *Description Logics* (Baader et al. 2003) and has model-theoretic semantics. These semantics enable certain inferences to be made about ontologies and their individuals. Inference (i.e., reasoning) techniques in the Semantic Web inherit from precursory technologies, e.g., logics, theorem provers, deductive databases, or semantic nets (Oberle 2014). Some typical types of reasoning tasks are subsumption checking, consistency checking and instance classification. We will now briefly discuss and exemplify some of the most common reasoning tasks in more detail.

Subsumption checking deduces whether subsumption relations can be established between ontology classes based on their definitions.

Consistency checking detects any logical contradictions within an ontological model. For example, if two classes A and B are declared disjoint, and then an instance is added that is of both types A and B, this leads to an inconsistent ontology. This could be the case for classes `mo:Pump` and `mo:Engine` in our example domain, which can be declared disjoint. If an instance is asserted as an instance of both these classes, then the ontology model becomes inconsistent.

Instance classification deduces a type (class) for an instance. In the ontology above, we declared the domain `mo:Engine` and the range `mo:Supplier` for the `mo:hasSupplier` property. If the `mo:hasSupplier` relation holds between instances `mo:E1a` and `mo:S1a`, then an inference engine will automatically deduce that the type of `mo:E1a` is `mo:Engine` (since the `mo:hasSupplier` property has its domain restricted to instances from this class) while `mo:S1a` is of type `mo:Supplier` (since the `mo:hasSupplier` property has its range restricted to instances from this class). Therefore, an instance classification was performed.

Open versus Closed world assumption. The openness of the Web, described by the AAA Slogan, has additional implications on Semantic Web applications beyond the *Nonunique Naming Assumption* discussed in Sect. 3.4.1. With respect to reasoning, Semantic Web systems are built on the assumption that the system has incomplete information and therefore adopt an *Open World Assumption* (OWA). Under the OWA, information that is not explicitly stated is considered unknown rather than false. For example, asking whether `mo:Eng1` has "Ford" as a supplier, will not return "No," but an empty value set. By contrast, database systems operate under a *Closed World Assumption* (CWA) and therefore assume complete information. In other words, information that is not known is considered false. The same query as above will return "No" as a result under CWA.

---

[12]Protégé ontology editor: http://protege.stanford.edu/.

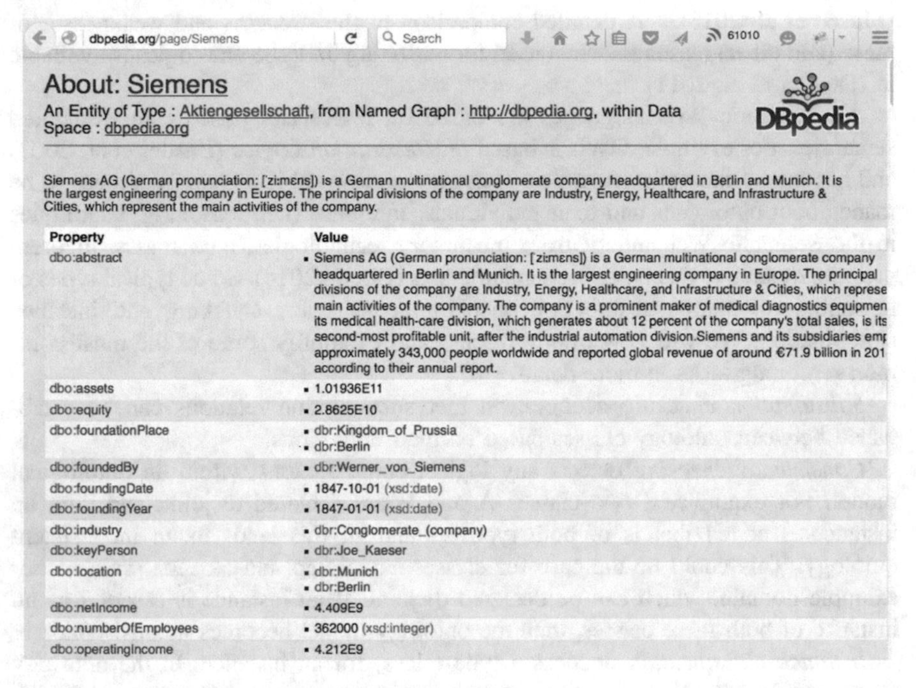

dbpedia.org/page/Siemens

About: Siemens

An Entity of Type : Aktiengesellschaft, from Named Graph : http://dbpedia.org, within Data Space : dbpedia.org

DBpedia

Siemens AG (German pronunciation: [ˈziːmɛns]) is a German multinational conglomerate company headquartered in Berlin and Munich. It is the largest engineering company in Europe. The principal divisions of the company are Industry, Energy, Healthcare, and Infrastructure & Cities, which represent the main activities of the company.

Property	Value
dbo:abstract	▪ Siemens AG (German pronunciation: [ˈziːmɛns]) is a German multinational conglomerate company headquartered in Berlin and Munich. It is the largest engineering company in Europe. The principal divisions of the company are Industry, Energy, Healthcare, and Infrastructure & Cities, which represe main activities of the company. The company is a prominent maker of medical diagnostics equipmen its medical health-care division, which generates about 12 percent of the company's total sales, is its second-most profitable unit, after the industrial automation division.Siemens and its subsidiaries emp approximately 343,000 people worldwide and reported global revenue of around €71.9 billion in 201 according to their annual report.
dbo:assets	▪ 1.01936E11
dbo:equity	▪ 2.8625E10
dbo:foundationPlace	▪ dbr:Kingdom_of_Prussia ▪ dbr:Berlin
dbo:foundedBy	▪ dbr:Werner_von_Siemens
dbo:foundingDate	▪ 1847-10-01 (xsd:date)
dbo:foundingYear	▪ 1847-01-01 (xsd:date)
dbo:industry	▪ dbr:Conglomerate_(company)
dbo:keyPerson	▪ dbr:Joe_Kaeser
dbo:location	▪ dbr:Munich ▪ dbr:Berlin
dbo:netIncome	▪ 4.409E9
dbo:numberOfEmployees	▪ 362000 (xsd:integer)
dbo:operatingIncome	▪ 4.212E9

Fig. 3.5 DBpedia page for Siemens (screenshot taken on 30.12.2015)

3.6 Linked Data

One of the original motivations for the Semantic Web was to enable Web-scale data integration. As detailed in Sect. 3.4.1, RDF's triple merging mechanism allows for easy integration of data graphs based on shared resource IRI's. In an open Web setting characterized by the AAA Slogan it is difficult to ensure that the same IRI is used across data providers to refer to the same entity. For example, the entity *Siemens* is assigned different IRIs in different datasets. In the New York Times dataset, *Siemens* has the IRI http://data.nytimes.com/N82918236209763785922. This dataset provides information about the number of Siemens-related articles as well as the dates when the company was mentioned first and last by the newspaper. The OpenEI (Open Energy Information) website[13] contains energy-related data as *Linked Open Data* (LOD). The OpenEI dataset identifies *Siemens* with the http://en.openei.org/lod/-page/wiki/Siemens IRI and contains information such as the power plants where *Siemens* provided a wind turbine. From this example, it is clear that data from the two datasets cannot be merged using the RDF graph merging capabilities because *Siemens* is identified differently in the two datasets (i.e., with two syntactically different IRIs).

[13]OpenEI website: http://en.openei.org/wiki/Main_Page.

Linked Data aims to overcome exactly this shortcoming of Web-scale data integration (Heath and Bizer 2011). The core idea of Linked Data is to create explicit links between datasets (similarly to hyperlinks in HTML pages). To this end, typically, an `owl:sameAs` relation is declared between resources from different datasets that represent the same real-life entity. A relation declared between entities from two datasets is also referred to as a "link." In other words, when an `owl:sameAs` link is specified between two resources, the triples containing these can be merged as if they were represented by two syntactically equal IRIs. For example, a single triple, as shown in Listing 3.5, is sufficient to declare that two IRIs in the *New York Times* and *OpenEI* datasets, respectively, refer to the same entity, i.e., Siemens:

Listing 3.5:

```
@prefix nyt: <http://data.nytimes.com/> .
@prefix ope: <http://en.openei.org/lod/page/wiki/Siemens> .
@prefix owl: <http://www.w3.org/2002/07/owl# > .

nyt:N82918236209763785922 owl:sameAs ope:Siemens.
```

The link between the two datasets established with the triple in Listing 3.5 enables computer programs to understand that the two IRIs represent the same resource and that data about those resources in the two datasets can be combined.

An alternative strategy to explicit link specification between two datasets is to integrate datasets by linking to a third dataset that acts as a "semantic bridge" between the two datasets (i.e., a common base of reference). A good example of such a dataset is *DBpedia* (Lehmann et al. 2015), which exposes the content of *Wikipedia* as structured data on the Web. For example, the *Wikipedia* page about *Siemens*[14] has a corresponding entity in *DBpedia* with the IRI http://dbpedia.org/resource/Siemens. By dereferencing this IRI, one can inspect all semantic metadata that is available for *Siemens* in *DBpedia* (see Fig. 3.5 for a partial screenshot of DBpedia's Siemens page). In addition to properties that describe the *Siemens* entity, DBpedia also provides links to other datasets that have their own data about Siemens, such as the *New York Times* and the *OpenEi* datasets. Given the breadth of Wikipedia, DBpedia contains the description of a broad range of entities. Therefore, it is often used as a "bridge" for integrating various datasets: two datasets do not necessarily need to create links between their entities directly, but can relate to DBpedia instead by linking their entities to the corresponding entity IRI in DBpedia.

The three datasets mentioned above (DBpedia, New York Times dataset, and OpenEI dataset) are examples of Linked *Open* Data because their data is

[14]Siemens Wikipedia page: https://en.wikipedia.org/wiki/Siemens.

represented in RDF, it is openly available on the Web, and they specify links to other datasets. In certain settings, however, the open publication of data might not be desirable. This does not hinder the use of Linked Data *within* the organizational network. RDF datasets that are linked to each other within an organization are referred to as Linked Data (LD) or *Linked Enterprise Data* (LED) (Wood 2010). Examples of *Linked Enterprise Data* adopters are the BBC (Kobilarov et al. 2009) and *NXP Semiconductors* (Walker 2015). These companies apply SWTs as a solution for enterprise-wide data integration problems.

3.7 Semantic Web Capabilities Relevant for Engineering Needs

As discussed in Chap. 2, the transition from traditional to cyber-physical production system engineering environments involves major challenges when building IEAs. These applications aim to support processes that build on the heterogeneous terms, concepts, and models used by the stakeholders in production system engineering and operation. Taking a step back from the technology details described in Sects. 3.3–3.6, we synthesize a set of SWT capabilities that are important for addressing the aspects of heterogeneity in production systems' engineering discussed in Chap. 2. We took the set of ontology-specific technology features identified by Oberle (2014) as a starting point for defining these capabilities and revised those from the perspective of the engineers' needs. Therefore, they are not always purely technical capabilities but rather useful functionalities that can support the various needs of engineering scenarios introduced in Sect. 2.4. The following Semantic Web capabilities were identified:

C1—Formal and flexible semantic modeling. Semantic (or conceptual) modelling is the capability to capture a universe of discourse. Similarly to other conceptual modeling approaches (e.g., UML—*Unified Modeling Language*), ontologies allow capturing a universe of discourse with their modeling primitives (classes, properties, and instances). In addition to what other semantic modeling approaches offer (e.g., UML), ontologies allow the explicit representation of knowledge as conceptual models in a formal and flexible way. The conceptual model described by ontologies is formulated by means of knowledge representation languages that are based on logics (see Sect. 3.4). The formality of these languages helps to assign unambiguous meaning to their modeling constructs. For example, the `rdfs:subClassOf` modeling element is mapped to the set theoretical subset operator by virtue of the model theory underlying OWL (Oberle 2014). Therefore, ontologies allow to formally describe the semantics of the vocabulary used in the dataset. As such they differ from databases where schemas merely describe how data is stored. By formally explaining the meaning of data, ontologies make data easier to understand to a wider range of users, both humans and machines.

The flexibility of semantic modeling with ontologies covers two aspects (Oberle 2014). First, the ability to evolve an ontology on the schema and instance levels at runtime, a characteristic called "*agile schema development,*" provides a high degree of flexibility. Second, *direct interaction* allows creating, changing, and populating semantic models by means of generic tools.

Formal and flexible semantic modeling with ontologies and Semantic Web languages is a strong foundation for addressing the need of production system engineering for explicit knowledge representation (N1[15]) and therefore it is marked with "++" in Table 2.2, Chap. 2. Reasoning and querying functionalities enabled by semantic models provide access to and analytics on engineering knowledge (N3). Furthermore, engineering knowledge that is captured in ontologies and formally represented facilitates data integration (N2) because meaningful relations between diverse datasets can be clearly specified. Such data integration can also be achieved with semi-structured data (e.g., technical fact-sheets including natural language) both within the organization and on the Web (N4) by specifying links to this data. Semantic models support the creation of IEAs (N5). Semantic modeling technique can help elicit and explicitly represent engineering process knowledge, such as engineering process responsibilities and process states (N6). Semantic representations of engineering models are a prerequisite for addressing the need of providing integrated engineering knowledge at system run time (N7).

C2—Intelligent, web-scale knowledge integration. The heterogeneity of engineering knowledge is a major challenge in typical production systems engineering scenarios. SWTs can tackle this heterogeneity by automatically and flexibly solving complex data integration problems on a large scale. This has already been successfully demonstrated in advanced applications that integrate many millions of data elements in domains as varied as pharmacology (Groth et al. 2014) or media publishing (Kobilarov et al. 2009).

The ability to define links and transformations between ontologies is a key enabler for the knowledge integration capability. *Ontology matching* techniques (Euzenat and Shvaiko 2013) are examples of such mechanisms for defining correspondences and links between ontologies. Chapter 6 of this book discusses in more detail how these techniques can be used in engineering settings. In addition to ontology matching and, as discussed in Sect. 3.5, *Linked Data* technologies aim to support data integration at Web-scale. To that end, they rely on the foundation of a stack of technical standards for publishing, querying, and annotating ontological information on the Web e.g., RDF(S), OWL, SPARQL. A key benefit for enterprises is the possibility to use *Linked Data* to seamlessly link internal and external data sets, i.e., data coming from the organization with external data that is available via the Web (Kobilarov et al. 2009). Additionally, organizations can provide third-party developers with an API to their data via a web site (i.e., a SPARQL endpoint), an approach that has been shown to be highly maintainable and flexible, e.g., in large-scale applications for publishing (Kobilarov et al. 2009).

[15]N" denotes the needs identified in Chap. 2.

Ontology matching and *Linked Data* mechanisms establish links between semantic models and data by making use of formal constructs, such as, for example `owl:sameAs`. Therefore, on one hand, they further extend capability C1 for semantic modeling by declaring knowledge *between* semantic models. On the other hand, due to their formal semantics, reasoning mechanisms can exploit these cross-model links to discover new knowledge that is only implicitly represented. For example, if the `mo:Device` concept of the example ontology is declared equivalent to the `Component` concept of another ontology, by virtue of reasoning, it can be deduced that any subconcept of `mo:Device` (e.g., `mo:Engine, mo:Conveyer`) is also a subconcept of `Component`. This ability to reason about cross-dataset links enables data integration applications to act intelligently.

In terms of addressing production system engineering needs, C2 provides strong support for engineering data integration (N2). Integrated data also enables knowledge access and analytics needs especially in settings where access to project-wide, integrated data is required (N3). Linked data plays a key role in facilitating efficient access to semi-structured data in the organization and on the Web by providing mechanisms to interlink these resources (N4). Flexible and intelligent engineering applications (N5) benefit from access to integrated knowledge. Finally, C2 provides support for multidisciplinary engineering process knowledge which is inherently distributed and must be integrated for a meaningful insight onto project-wide processes (N6). Capability C2 provides knowledge representational power to SWTs as it allows specifying links between models and therefore it clearly supports explicit engineering knowledge representation needs (N1).

C3—Browsing and exploration of distributed data sets. In the context of the Semantic Web, and in particular Linked Data technologies, user-friendly browsing and exploration of distributed data sets is a key enabler for the wide acceptance with nonexperts in software and data engineering (Hausenblas et al. 2014). For example, the *Linked Data* interface of *DBpedia* (see an example screenshot in Fig. 3.3) does not only allow easy browsing of the sematic data, but also allows users to navigate across datasets to which links are provided (for example, the OpenEI or NYT LD interfaces). Note that Linked Data interfaces are automatically created on top of Semantic Web data. In the context of engineering applications, this capability can be used to browse and explore both engineering models internal to an organization and external data sources, such as Web resources coming from third-party providers, in similar ways to (Gabrilovich and Markovitch 2007).

Browsing and exploration of distributed data sets is a core requirement for efficient access to semi-structured data in the organization and on the Web (N4). It also offers relevant support for addressing production system engineering need N3 for engineering knowledge access through browsing, exploration, and navigation. Browsing capabilities are important features of flexible and intelligent engineering applications (N5). By supporting sense-making and following links across engineering disciplines, capability C3 supports the increased productivity of multidisciplinary engineering processes (N6). Finally, navigational data access interfaces can provide access to integrated engineering knowledge during the production-system run time (N7).

C4—Quality assurance of knowledge with reasoning. In systems engineering, the quality assurance of engineering knowledge with advanced checks is highly relevant. Given the mission-critical character of engineering projects, inconsistencies, defects and faults among diverse engineering models should be discovered as early as possible. However, state of the art engineering models (such as those expressed in SysML) tend to capture only limited aspects of cross-domain interdependencies with links between models that are understandable by humans but not by machines (Kernschmidt and Vogel-Heuser 2013). Semantic Web mechanisms can improve the state of the art because they enable formally representing (C1) and interlinking (C2) engineering knowledge. The formal nature of the engineering models and their links allows their interpretation through reasoning mechanisms (detailed in Sect. 3.5), thus enabling the automation of a variety of quality assurance tasks. Semantic Web reasoning facilities can be used for supporting a wide range of tasks, but the ability to ensure quality (e.g., through consistency checks) is particularly important for the engineering domain.

Quality assurance of knowledge with reasoning is a key capability for creating flexible and intelligent engineering applications that both act intelligently (i.e., deduce new information from existing information by virtue of reasoning mechanisms) and reliably with respect to quality assurance (N5). Capability C4 allows detecting defects, faults, and inconsistencies and as such it increases the productivity of multidisciplinary engineering process (N6) while ensuring that high-quality engineering knowledge is provided at production-system run time (N7).

C5—Knowledge reuse. A basic motivation for the development of SWTs has been the need to describe key terms and relationships of domains of discourse in reference ontologies (Arp et al. 2015). These semantic reference models are the basis for sharing and reusing knowledge between domains and for building applications on top of these reference models (Arp et al. 2015). One implication of the declarative nature of ontologies is that knowledge represented in ontologies can be easily reused among different application use cases. This reusability of knowledge is one of the fundamental concepts underlying SWTs (Simperl 2009).

Therefore, knowledge reuse will strongly support the need for engineering knowledge access and analytics (N3). Reusing already specified knowledge (and therefore avoiding specifying this knowledge again) increases the productivity of multidisciplinary engineering processes (N6). Capability C5 also supports explicit engineering knowledge representation as existing knowledge sources might be reused to create the requested semantic models (N1). Data integration needs (N2) could also be supported by reusing already specified alignments between semantic models. Finally, the provision of integrated engineering knowledge at production-system run time can leverage on knowledge reuse (N7).

A qualitative evaluation of how well Semantic Web capabilities address the previously identified production systems engineering needs is presented in Table 2.2 in Chap. 2 of this book. Overall, it shows a good coverage of these needs. Chap. 15 presents a more detailed evaluation of how SWT capabilities address production systems engineering needs, based on the content of this book's chapters.

3.8 Summary

The present chapter provides an introduction to basic Semantic Web-related technology concepts that are necessary to understand the remainder of this book. We focused on explaining key elements of Semantic Web-based solutions and exploring the key SWT capabilities of particular relevance for intelligent engineering applications. The chapter introduced and discussed the following key topics:

- The *Semantic Web* (Sect. 3.2) is a vision for the extension of the HTML-based Web with formally represented knowledge (in the form of *semantic metadata*) that can be better understood by computer programs and therefore facilitates more effective solutions for data-intensive tasks (such as search).
- *Semantic Web technologies* have been designed to comply with the open nature of the Web captured by the "**A**nyone is allowed to say **A**nything about **A**ny topic" slogan (Allemang and Hendler 2011). Some implications of this compliance are the *Nonunique Naming Assumption* (Sect. 3.4.1) and the *Open World Assumption* (Sect. 3.5).
- *Ontologies* are conceptual models representing consensual domain knowledge (Sect. 3.3). They provide the terms (vocabulary) to create semantic metadata. Ontologies "explain" what domain terms mean and how they relate to each other in ways that are understandable to humans and machines alike.
- *Semantic Web Languages* (Sect. 3.4) are semantics-based knowledge representation languages used to represent ontologies in a formal way (i.e., in a way that can be parsed and interpreted by computer programs). RDF defines the basic data model of the Semantic Web based on concepts such as resources, triples, and graphs. RDF(S) allows defining lightweight ontologies. OWL is suitable for defining complex models. SPARQL allows querying ontologies and semantic metadata encoded in RDF, RDF(S) or OWL.
- *Reasoning* is the mechanism for deducing new knowledge from explicitly declared knowledge in ontologies and links between ontologies (Sect. 3.5). Reasoning enables the intelligent behavior of applications based on SWTs.
- *Linked Data technologies* (Sect. 3.6) are a type of SWTs that focus on creating links between semantic metadata and exposing it in browsable interfaces (and HTTP accessible query access points), both openly on the Web (Linked *Open* Data) and within enterprises (Linked *Enterprise* Data).
- *Semantic Web capabilities* of particular relevance for engineering applications (Sect. 3.7) are: (C1) formal and flexible semantic modeling; (C2) intelligent, web-scale knowledge integration; (C3) browsing and exploration of distributed data sets; (C4) quality assurance of knowledge with reasoning and (C5) knowledge reuse. A qualitative evaluation of how Semantic Web capabilities address production systems engineering needs is presented in Table 2.2 in Chap. 2, while the relation between needs and capabilities as present in the chapters of this book is discussed in Chap. 15.

Further SWT specific information is available in several chapters, as follows. Chapter 5 details how to create ontologies in general and specifically for engineering applications. Chapter 6 discusses approaches for creating mappings between ontologies and datasets that go beyond equivalence (as expressed with `owl:sameAs`). Chap. 7 focuses on methods for managing changes in Semantic Web data. Chap. 11 exemplifies the adoption of Semantic Web technologies to support engineering scenarios and Chaps. 9, 10, 12, 13, and 14 showcase the usage of SWTs to create a variety of IEAs for solving diverse engineering tasks.

Acknowledgments This work was supported by the Christian Doppler Forschungsgesellschaft, the Federal Ministry of Economy, Family and Youth, and the National Foundation for Research, Technology and Development in Austria.

References

Allemang, D., Hendler, J.: Semantic Web for the Working Ontologist, 2nd edn. Morgan Kaufmann (2011)

Arp, R., Smith, B., Spear, AD.: Building Ontologies with Basic Formal Ontology, p. 248. MIT Press (2015). ISBN 978-0262527811

Baader, F., Calvanese, D., McGuinness, D.L., Nardi, D., Patel-Schneider, P.F. (eds.).: The Description Logic Handbook: Theory, Implementation, and Applications. Cambridge University Press, New York (2003)

Baeza-Yates, R., Ribeiro-Neto, B.: Modern Information Retrieval. ACM Press, Addison-Wesley, New York (1999)

Berners-Lee, T.: Weaving the Web. Harpur, San Francisco (1999)

Berners-Lee, T., Hendler, J., Lassila, O.: The Semantic Web. Sci. Am. 29–37 (2001)

Borgo, S., Gangemi, A., Guarino, N., Masolo, C., Oltramari, A.: Ontology RoadMap. WonderWeb Deliverable D15 (2002)

Corcho, O., Fernandez-Lopez, M., Gomez-Perez, A.: Methodologies, tools and languages for building ontologies. Where is the meeting point? Data Knowl. Eng. **46**(1), 41–46 (2003)

Cyganiak, R., Wood, D., Lanthaler, M.: RDF 1.1 Concepts and Abstract Syntax. W3C Recommendation. http://www.w3.org/TR/rdf11-concepts/ (2014). Accessed 25 Feb 2014

Dentler, K., Cornet, R., ten Teije, A., de Keizer, N.: Comparison of reasoners for large ontologies in the OWL 2 EL profile. Semant. Web J. **2**(2), 71–87 (2011)

Euzenat, J., Shvaiko, P.: Ontology Matching. Springer (2013)

Fensel, D., Decker, S., Erdmann, M., Studer, R.: Ontobroker: or how to enable intelligent access to the WWW. In: Proceedings of the 11th Knowledge Acquisition for Knowledge-Based System Workshop (KAW98). Banff, Kanada (1998)

Gabrilovich, E., Markovitch, S.: Computing semantic relatedness using Wikipedia-based explicit semantic analysis. In: Proceedings of the 20th International Joint Conference on Artificial Intelligence (IJCAI'07). Morgan Kaufmann Publishers Inc., pp. 1606–1611 (2007)

Glimm, B., Horrocks, I., Motik, B., Stoilos, G., Wang, Z.: HermiT: An OWL 2 Reasoner. J. Autom. Reason. **53**(3), 245–269 (2014)

Groth, P., Loizou, A., Gray, AJG., Goble, C., Harland, L., Pettifer, S.: API-centric linked data integration: the open phacts discovery platform case study. Web Semantics, Science, Services and Agents on the World Wide Web, vol. 29, pp. 12–18, Dec 2014

Gruber, T.R.: A translation approach to portable ontology specifications. Knowl. Acquis. **5**(2), 199–220 (1993)

Guarino, N.: Formal ontology and information systems. In Guarino, N. (ed.) Formal Ontology in Information Systems. Proceedings of FOIS'98, pp. 3–15. IOS Press (1998)

Harris, S., Seaborne, A.: SPARQL 1.1 query language. W3C Recommendation 21 Mar 2013. http://www.w3.org/TR/2013/REC-sparql11-query-20130321/

Hausenblas, M., Ruth, L., Wood, D., Zaidman, M.: Linked Data, Manning, p. 336 (2014). ISBN 978-1617290398

Heath, T., Bizer, C.: Linked Data: Evolving the Web into a Global Data Space, 1st edn. Synthesis Lectures on the Semantic Web: Theory and Technology. Morgan & Claypool (2011)

Hepp, M.: GoodRelations: an ontology for describing products and services offers on the web. In: Proceedings of the 16th International Conference on Knowledge Engineering and Knowledge Management (EKAW). Springer LNCS, vol. 5268, pp. 332–347 (2008)

Kernschmidt, K., Vogel-Heuser, B.: An interdisciplinary SysML based modelling approach for analyzing change influences in production plants to support the engineering. In: Proceedings of the IEEE International Conference on Automation Science and Engineering (CASE), pp. 1113–1118 (2013)

Kobilarov, G., Scott, T., Raimond, Y., Oliver, S., Sizemore, C., Smethurst, M., Bizer, C., Lee, R.: Media meets Semantic Web—How the BBC uses DBpedia and linked data to make connections. In: Proceedings of the 6th European Semantic Web Conference on The Semantic Web: Research and Applications (ESWC). Springer, pp. 723–737 (2009)

Köhler, S., Bauer, S., Mungall, C.J., Carletti, G., Smith, C.L., Schofield, P., Gkoutos, G.V., Robinson, P.N.: Improving ontologies by automatic reasoning and evaluation of logical definitions. BMC Bioinformatics **12**, 418 (2011)

Lehmann, J., Isele, R., Jakob, M., Jentzsch, A., Kontokostas, D., Mendes, P.N., Hellmann, S., Morsey, M., van Kleef, P., Auer, S., Bizer, C.: DBpedia—a large-scale, multilingual knowledge base extracted from Wikipedia. Semant. Web J. **6**(2), 67–195 (2015)

Luke, S., Spector, L., Rager, D.: Ontology-based knowledge discovery on the World-Wide Web. In: AAAI Workshop on Internet-Based Information Systems (1996)

Martin, D., Burstein, M., Mcdermott, D., McIlraith, S., Paolucci, M., Sycara, K., McGuinness, D.L., Sirin, E., Srinivasan, N.: Bringing semantics to web services with OWL-S. World Wide Web **10**(3), 243–277 (2007)

Marquardt, W., Morbach, J., Wiesner, A., Yang, A.: OntoCAPE: A Re-usable Ontology for Chemical Process Engineering. Springer, Heidelberg (2010)

Masolo, C., Borgo, S., Gangemi, A., Guarino, N., Oltramari, A.: WonderWeb Deliverable D18—Ontology Library (Final Version). Working paper (2003)

McBride, B.: The Resource Description Framework (RDF) and its vocabulary description language RDFS. In: Staab, S., Studer, R. (eds.) Handbook on Ontologies, International Handbooks on Information Systems, pp. 51–66. Springer (2004)

Musen, M.: Protégé Team, Protégé Ontology Editor. In: Encyclopedia of Systems Biology, pp 1763–1765. Springer, New York (2013)

Oberle, D.: Ontologies and reasoning in enterprise service ecosystems. In: Informatik Spektrum, vol. 37, no. 4 (2014)

Pease, A., Niles, I., Li, J.: The suggested upper merged ontology: a large ontology for the Semantic Web and its applications. In: Working Notes of the AAAI Workshop on Ontologies and the Semantic Web. Edmonton, Canada (2002)

Schuurman, N., Leszczynski, A.: Ontologies for bioinformatics. Bioinform. Biol. Insights **2**, 187–200 (2008)

Shadbolt, N., Berners-Lee, T., Hall, W.: The Semantic Web revisited. IEEE Intell. Syst. **21**(3), 96–101 (2006)

Simperl, E.: Reusing ontologies on the Semantic Web: a feasibility study. Data Knowl. Eng. **68** (10), 905–925 (2009)

Sirin, E., Parsia, B., Cuenca Grau, B., Kalyanpur, A., Katz, Y.: Pellet: a practical OWL-DL reasoner. J. Web Semant. **5**(2), 51–53 (2007)

Smith, B.: Basic Formal Ontology. Technical report, Institute for Formal Ontology and Medical Information Science, University of Leipzig (2003)

Studer, R., Benjamins, V., Fensel, D.: Knowledge engineering: principles and methods. Data Knowl. Eng. **25**(1–2), 161–197 (1998)

Tsarkov, D., Horrocks, I.: FaCT++ description logic reasoner: system description. In: Furbach, U., Shankar, N. (eds.) Proceedings of the Third International Joint Conference on Automated Reasoning (IJCAR), pp. 292–297. Springer, Berlin (2006)

van Heijst, G., Schreiber, A.T., Wielinga, B.J.: Using explicit ontologies in KBS development. Int. J. Hum. Comput. Stud. **46**(2/3), 183–292 (1997)

Walker, J.: Is linked data the future of data integration in the enterprise? http://blog.nxp.com/is-linked-data-the-future-of-data-integration-in-the-enterprise/ (2015). Accessed 22 Oct 2015

Wood, D.: Linking Enterprise Data. Springer (2010)

Part II
Semantic Web Enabled Data Integration in Multi-disciplinary Engineering

Chapter 4
The Engineering Knowledge Base Approach

Thomas Moser

Abstract Systems and software engineering projects depend on the cooperation of experts from heterogeneous engineering domains using tools that were not designed to cooperate seamlessly. Current semantic engineering tool and data integration is often ad hoc and fragile, thereby making the evolution of tools and the reuse of integration solutions across projects unnecessarily inefficient and risky. This chapter describes the engineering knowledge base (EKB) framework for engineering environment integration in multidisciplinary engineering projects. The EKB stores explicit engineering knowledge to support access to and management of engineering models across tools and disciplines. The following Chaps. 5–7 discuss individual aspects of the EKB framework, which provides (1) data integration based on mappings between local and domain-level engineering concepts; (2) transformations between local engineering concepts; and (3) advanced applications built on these foundations, e.g., end-to-end analyses. As a result, experts from different organizations may use their well-known tools and data models and can access data from other tools in their syntax. Typical applications enabled by implementations of this framework are discussed in Chaps. 9 and 10.

Keywords Engineering knowledge base · Explicit knowledge · Data integration · Transformation · End-to-End analysis

4.1 Introduction

Software-intensive systems in industrial automation become increasingly complex due to the need for flexibility of business and engineering processes (Schäfer and Wehrheim 2007). Such systems and software engineering projects bring together experts from several engineering domains and organizations, who work in a heterogeneous engineering environment with a wide range of models, processes, and tools

T. Moser (✉)
St. Pölten University of Applied Sciences, Matthias Corvinus-Straße 15,
3100 St. Pölten, Austria
e-mail: thomas.moser@fhstp.ac.at

S. Biffl and M. Sabou (eds.), *Semantic Web Technologies for Intelligent Engineering Applications*, DOI 10.1007/978-3-319-41490-4_4

that were originally not designed to cooperate seamlessly, but specifically designed for a task or a single engineering discipline. A core question is how to integrate data across engineering tools and domain boundaries. Current semantic engineering environment integration is often ad hoc and fragile, making the evolution of tools and reuse of integration solutions across projects risky (Halevy 2005; Noy et al. 2005).

In order to reach the common goal of developing automation and control software in the engineering team, it is important to share the necessary knowledge between engineering domain experts (Schäfer and Wehrheim 2007). However, this knowledge is often only implicitly available and therefore inefficient to share, resulting in time-consuming repetitive tasks; often it is hard or even impossible to create and maintain common shared knowledge repositories. A method and platform for making expert knowledge explicit and efficiently shareable is needed in order to support quality and project managers in their data analyses based on engineering knowledge and concrete data in the engineering tool models, which is currently achieved using inefficient or fragile approaches.

This chapter proposes the engineering knowledge base (EKB) framework for supporting engineering environment integration in multidisciplinary engineering teams and projects. Since standards are hard to apply in projects where experts from different organizations participate, these experts may use their well-known local tools and data model, and additionally can access data from other tools in their local syntax. The EKB is located on top of a common data repository and stores explicit engineering knowledge to support access and management of engineering models across tools and disciplines by providing (a) data integration by exploiting mappings between local and common engineering concepts; (b) transformations between local engineering concepts following these mappings; and (c) advanced applications using these foundations, e.g., end-to-end analyses. As a result, experts from different organizations may use their well-known local tools and data model, and additionally can access data from other tools.

The research results have been evaluated in the industrial application domain of production automation systems, regarding effort, feasibility, performance, scalability, robustness, and usability. The evaluation is based on prototypes for a set of specific use cases of the industrial application domain, as well as on empirical studies. Major results of this work are the feasibility of the EKB framework, i.e., the process, method, and tool support is usable and useful across engineering domains, as well as better accuracy, effectiveness, and efficiency. In addition, defects are found earlier in the engineering process, resulting in risks like errors or inconsistent entries in data models being mitigated earlier and more efficiently. Initial evaluation results indicate an effort reduction for reuse in new engineering projects and finding defects earlier in the engineering process.

The remainder of this chapter is structured as follows: Sect. 4.2 summarizes background work on automation systems engineering and on semantic integration of tool data models and additionally introduces the research challenges for data and tool integration in multidisciplinary engineering environments. Section 4.3 explains differentiations to other technologies and approaches for engineering tool and process data integration, and describes related approaches. Section 4.4 introduces the engineering knowledge base (EKB) framework and describes the generic framework architecture. Section 4.5 summarizes the results of applying the EKB framework to the industrial application scenarios and presents the results of the EKB framework evaluation. Finally, Sect. 4.6 concludes and summarizes.

4.2 Background and Research Challenges

This section summarizes background work on automation systems engineering and on semantic integration of tool data models and introduces the research challenges for data and tool integration in multidisciplinary engineering environments.

4.2.1 Automation Systems Engineering

Automation systems (AS) depend on distributed software to control the system behavior. The behavior of AS must be testable and predictable to meet safety and quality standards. In automation systems engineering (ASE), software engineering depends on specification data and plans from a wide range of other engineering aspects in the overall engineering process, e.g., physical plant design, mechanical engineering or electrical engineering. This expert knowledge is embodied in domain-specific standards, terminologies, people, processes, methods, models, and software. The weak semantic integration of the expert knowledge across domain boundaries of engineering aspects makes changes late in the engineering process risky (Biffl et al. 2009a). Thus, the traditional ASE process follows a strongly sequential procedure and suffers from a lack of systematic feedback loops, low engineering process automation, and weak quality management across domain boundaries, which leads to development delays and risks for system operation (Lüder 2000; Medeia-Consortium 2008).

One of the most prominent problems in current industrial development and research approaches is the lack of data integration between the engineering disciplines (Biffl et al. 2009b; Schäfer and Wehrheim 2007). Different and partly overlapping terminologies are used in the different disciplines, which often hampers understanding between engineering disciplines.

Biffl and Schatten proposed a platform called Engineering Service Bus (EngSB) which integrates not only different tools and systems but also different steps in the software development lifecycle (Biffl et al. 2009a; Biffl and Schatten 2009)—the Automation Service Bus (ASB) is a modified version of the EngSB for the ASE domain. The EngSB addresses requirements such as the capability to integrate a mix of user-centered tools and backend systems and flexible and efficient configuration of new project environments and SE processes. The EngSB platform introduces the concept of tool domains that provide interfaces for solving a common problem, independent of the specific tool instance used. The abstraction of tool instances by tool domains seems possible since different tools, developed to solve the same problem have, more or less, similar interfaces (Biffl and Schatten 2009). On the one hand, a tool domain consists of a concrete engineering tool-specific interface part used by so called connector components which establish communication connections between a specific engineering tool and the tool domain it belongs to. On the other hand, it consists of a general engineering tool independent interface part which enables communication to any other tool domain. This concept allows the EngSB to interact with a tool domain without knowing which specific tool instances are actually present. In heterogeneous ASE environments, this capability enables flexible engineering process automation and advanced quality management.

4.2.2 Semantic Integration of Tool Data Models

The problem of reconciling (data) schema heterogeneity has been a subject of research for decades, but solutions are few. The fundamental reason that makes semantic heterogeneity so hard to address is the independent origin of datasets using varying structures to represent the same (or overlapping) concepts (Bergamaschi et al. 1999; Doan and Halevy 2005). From a practical perspective, one of the reasons why schema heterogeneity is difficult and time consuming is that the solution requires both domain and technical expertise: a domain expert who understands the domain meaning of all schemas being reconciled and technical experts for writing transformations between the schemas to be integrated. Resolving schema heterogeneity is inherently a heuristic, human-assisted process. Unless there are very strong formal constraints on allowed schema differences, one should not hope for a completely automated solution. Therefore, the goal is to reduce the time it takes human experts to create a mapping between a pair of schemas, and enable them to focus on the hardest and most ambiguous parts of the mapping (Rahm and Bernstein 2001).

Semantic integration is defined as solving of problems originating from the intent to share data across disparate and semantically heterogeneous data (Halevy 2005). These problems include the matching of ontologies or schemas, the detection of duplicate entries, the reconciliation of inconsistencies, and the modeling of complex relations in different data sources (Noy et al. 2005). One of the most important and most actively studied problems in semantic integration is establishing semantic correspondences (also called mappings) between vocabularies of different data sources (Doan et al. 2004). In literature (Noy 2004), three major dimensions of the application of ontologies for supporting semantic integration are identified: the task of finding mappings (semi-)automatically, the declarative formal representation of these mappings, and reasoning using these mappings. Ontologies allow modeling the common domain-specific view, tool-specific views, and the mappings between these views in one data model.

A good starting point for semantic integration is both the analysis of the interfaces or export artifacts of the involved engineering tools, as well as the identification and analysis of available standards in the problem domain. AutomationML is a neutral data format based on XML for the storage and exchange of plant engineering information, which is provided as an open standard (Drath et al. 2008). The goal of AutomationML is to interconnect the heterogeneous tool landscape of modern engineering tools in their different disciplines, e.g., mechanical plant engineering, electrical design. AutomationML describes real plant components as objects encapsulating different aspects. An object can consist of sub-objects, and can itself be part of a bigger composition. An object can describe a screw, a claw, a robot, or a complete manufacturing cell in different levels of detail. The result of this analysis typically is a set of overlapping engineering concepts used in the engineering process. While this has proven to be true for well-established and long-running projects, the overlapping engineering concepts may not yet be known or elicited for projects in novel domains. In order to support future engineering projects, the knowledge regarding existing engineering projects should be used to derive guidelines to support similar future engineering projects.

4.2.3 Research Challenges

The scope of the EKB is an engineering team consisting of experts from several engineering disciplines, who work on engineering process tasks with role-specific tools and systems that encapsulate engineering models and project data. As the engineers work together to deliver a product to the end user, they inevitably have to form common concepts on deliverables at interfaces between their work tasks. Such common concepts can be found in elements of requirements, design, and defect descriptions, which concern more than one role.

As shown in Fig. 4.1, each engineering role has a tool set that works on data relevant to the engineer's tasks. In a typical step in the engineering process, an engineer exports data from his tool to a transfer document and integrates this document in a common repository accessible by a set of partner engineering tools. The major challenges here are on the one hand in the identification and description of tool data that should be extracted from tools and made available to other tools. On the other hand, data integration itself poses another challenge, since it is often not possible to determine a common data schema agreed on by all tools. Finally, the reuse of at least parts of integration solutions for other projects with different project partners is mostly not possible. In order to support data exchange between these sets of partner engineering tools, transformations between the different notations and formats of the particular partner engineering tools is needed.

Based on these tasks, the following research challenges for the novel engineering knowledge base (EKB) framework are derived. The EKB framework should be able to store explicit engineering knowledge to support access to and management of engineering models across tools and disciplines. Therefore, a special research focus is required for (1) **data integration** by exploiting mappings between local and

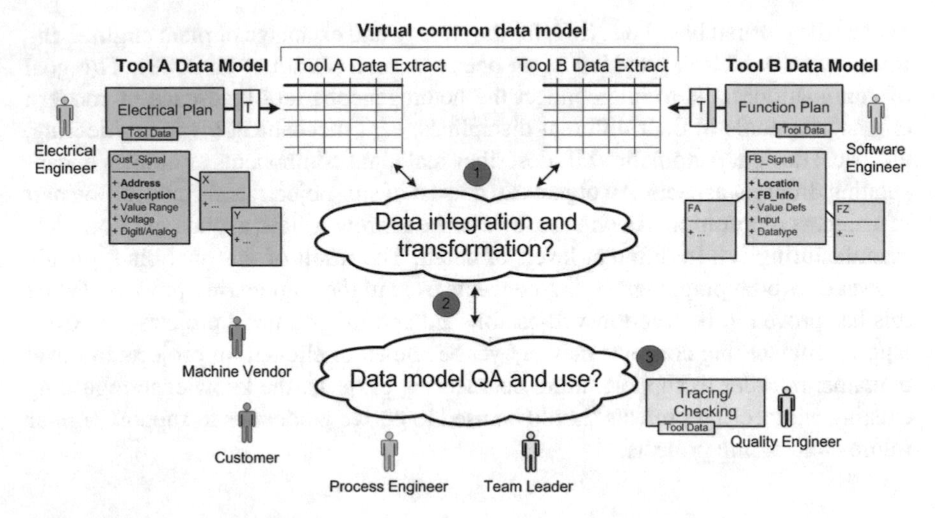

Fig. 4.1 Research challenges

common engineering concepts; (2) **transformations** between local, tool-specific and domain-specific engineering concepts; and (3) **advanced applications** using these foundations, e.g., end-to-end analyses.

4.3 Related Work

Modern industrial automation systems have to be designed for better interoperability and flexibility to satisfy increasing customer needs for product variety, manufacturing agility, and low cost. The strategic goal of making the systems engineering process more flexible without delivering significantly more risky end products translates into the capability to efficiently reconfigure the engineering process and tool instances of a project environment. While there are approaches based on a common repository that holds all relevant project data (Schäfer and Wehrheim 2007), experience has shown that such a repository tends to become large, inflexible, and hard to maintain surprisingly fast, which makes the knowledge in the repository hard to reuse in new projects. Thus, a key goal is to allow all participants to continue using their own data models and provide a mechanism for translation between these data models. In the past several approaches, providing engineering knowledge in machine-understandable syntax have been investigated (Liao 2005; Lovett et al. 2000; McGuire et al. 1993). However, these approaches focus primarily on storing existing homogeneous knowledge rather than providing support for managing and accessing heterogeneous knowledge.

This section discusses the proposed EKB framework by providing the one hand differentiations to other technologies and approaches for engineering tool and process data integration, and on the other hand by mentioning limitations of related approaches which hinder an efficient and effective realization.

4.3.1 Usage of Standards in Development Processes

A possible solution approach is the usage of standards (e.g., the Rational Unified Process (RUP)[1], SysML[2]) for platforms, data models, modeling languages, and tools in the development process. This works well, if the standard is defined in an early phase of the project and if all project partners adhere to the standard, however, it is hard to define and maintain standards for cross-domain engineering, and even harder or nearly impossible for a larger number of project partners to agree on a standard, which usually takes longer than the time horizon of the project (Kruchten 2000; Weilkiens 2008).

The advantages of SysML over UML for systems engineering become obvious when considering a concrete example, such as modeling an automotive system. However, many projects do not yet use a common data schema which could be represented using SysML; therefore, it is important to provide mechanisms allowing also participants with different heterogeneous data models to cooperate.

4.3.2 Usage of Common Project Repositories

The usage of common project repositories ("data dumps") is a typical solution for modern data-driven tool integration which sufficiently solves the challenges of persistency and versioning of data, but poses new challenges since the stored data often is hard to access and query. Databases that are widely used in an engineering context do not allow an intuitive and out-of-the-box storage of heterogeneous concepts that do not follow a homogeneous schema (e.g., use different granularity levels or terminologies for similar concepts) and also do not store metainformation.

Bernstein and Dayal (Bernstein and Dayal 1994) define a repository as "shared database of information about engineered artifacts produced or used by an enterprise". Examples of such artifacts include software, documents, maps, information systems, and discrete manufactured components and systems (e.g., electronic circuits, airplanes, automobiles, industrial plants). Storing this information in a common repository has several benefits. First, since the repository provides storage services, tool developers do not need to create tool-specific databases. Second, a common repository allows tools to share information so they can work together.

[1]http://www.ibm.com/software/rational.

[2]http://www.omg.org/spec/SysML/.

Without a common repository, special protocols would be needed for exchanging information between tools. By conforming to a common data model (i.e., allowable data formats) and information model (i.e., schema expressed in the data model), tools can share data and metadata without being knowledgeable about the internals of other tools. Third, the information in the repository is subject to common control services, which makes sets of tools easier to use. Since a repository is a database, it is subject to database controls, such as integrity, concurrency, and access control. However, in addition, a repository system provides checkout/check-in, version and configuration control, notification, context management, and workflow (Heiler 1995).

However, Bernstein and Dayal (Bernstein and Dayal 1994) state that it is unavoidable that many tools will, for the foreseeable future, have replicated heterogeneous repositories, for the following reasons: (a) Many existing tools are already committed to a private repository implementation, e.g., database systems. These repositories are already well-tuned to the tool's performance requirements. (b) Many tools need to be portable across operating systems. Therefore, they can only depend on a repository manager that runs on those operating systems and there are few such products on the market. (c) In an object-oriented world, some objects will be designed to maintain some state that describes the object. It will be some time before repository technology is so mature that all objects will entrust all their state to a shared repository manager. Thus, the problem of maintaining consistent heterogeneous repositories must be faced.

4.3.3 Complete Transformation Between Project Data Models

The ultimate alternative solution is the complete transformation between data models of tools, i.e., the translation of engineering model parts from one tool for work in another tool. While the vision of this solution is the seamless cooperation between project partners using well-known and established tools and notations, the feasibility of this approach is hard to verify and the effort required for establishing the needed transformations is considerable.

In the Modale project, Assmann et al. (2005) developed an ontology-based data integration approach in order to realize a seamless integration between cooperating partners in the field of digital production engineering. The major obstacle was the syntactic, structural, and semantic heterogeneity of the internally used tools in the digital production engineering domain. As proof of concept, the researchers have also provided a web-service-based prototypic implementation of their approach.

However, many questions still remain open, requiring more research effort to be invested. The main directions that concern in the short to medium term revolve around the following: (a) a methodology that allows an efficient (even automatic) construction of the necessary models and semantic bridges; (b) standardized, domain-specific extensions which would allow a very short start-up time for projects;

and (c) issues concerning the integration with each partner's internal processes, in order to achieve minimal disturbances in their existing workflows.

4.4 Engineering Knowledge Base Framework

This section gives an abstract overview of the engineering knowledge base (EKB) framework. In addition, this section presents the approach using extended mechatronic objects, so-called Engineering Objects (EOs). The following subsection describes the data structure in the EKB framework.

4.4.1 Engineering Knowledge Base (EKB) Overview

In this subsection, we describe a generic approach for semantic integration in systems engineering (see Fig. 4.2) with a focus on providing links between data structures of engineering tools and systems to support the exchange of information between these engineering tools and thus making systems engineering more efficient and flexible. The EKB framework aims at making tasks, which depend on linking information across expert domain boundaries, more efficient. A fundamental example for such engineering tasks is checking the consistency and integrity of design-time and run-time models across tool boundaries (Moser et al. 2010).

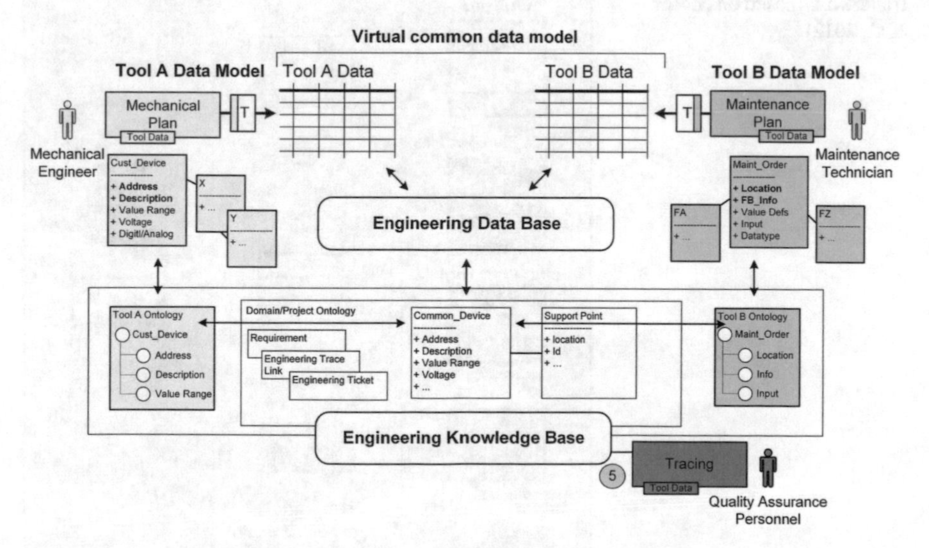

Fig. 4.2 Overview semantic integration approach (based on Moser et al. 2010)

The top part of Fig. 4.2 illustrates the use case scenario with the EKB framework for engineering a production automation system. In this example, there are two types of engineers (electrical engineer, software engineer) who come from two different engineering domains respectively. These roles use specialized engineering tools for their tasks. These tools contain local data sources, which produce and/or consume data with heterogeneous data structures. The EKB is used to facilitate the efficient data exchange between these engineering tools and data sources by providing a "virtual common data model". Based on this data exchange, more complex tasks such as model checking across tools are supported.

4.4.2 Data Structuring in the EKB Framework

The data of two or more engineering tool types is combined using Engineering Objects (EO) contained in a Virtual Common Data Model (VCDM) (Waltersdorfer et al. 2010). The presented concept *device* is an example of an EO. Figure 4.3 illustrates the usage mechanism of EOs. In the EKB framework, there exist three data modeling levels, namely the tool instance level (represented by the mechanical planning tool CAD pro, the two maintenance planning tools *#1* and *#2*, the PLC

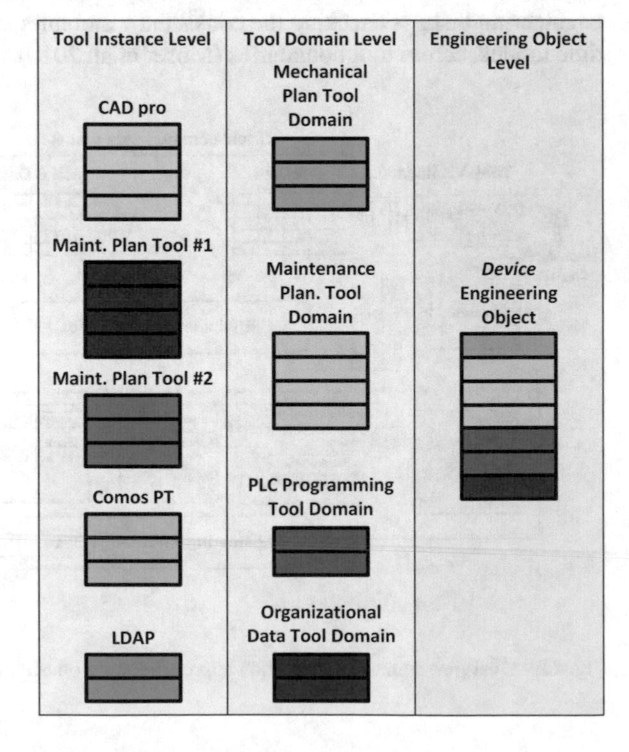

Fig. 4.3 Devices as example for Engineering Objects in the EKB framework (based on Moser et al. 2012)

programming tool *Comos PT*, and *LDAP* for the management of organizational and project-specific data), the tool domain level (represented by the *Mechanical Plan* tool domain, the *Maintenance Planning* tool domain, the *PLC Programming* tool domain and the *Organizational Data* tool domain), as well as the newly introduced third layer specifying virtual data model elements, so-called EOs. In the example used in Fig. 4.3, the EO *device* presents aggregated information of the four tool domains. Since there is no real physical representation of the data aggregated in the EO *device*, the underlying data model is called VCDM.

Similar to a database view (Ullman 2000), the information is queried and aggregated on-the-fly, i.e., if some project manager wants to get an overview of all *devices* with a specific *device* state. Such tool domain-specific information is created during engineering workflow execution and does not have an explicit representation in a particular tool instance. In the EKB framework, such information is called tool domain metainformation. This tool domain metainformation can be included in an EO-specific view, as shown in the example regarding the *device* status.

A typical workflow for querying tool domain data could be defined as follows: (1) the tools can check-in ("push") their data into the EKB framework through tool connectors (Biffl et al. 2009a), which on the one hand provide the technical connection to the EKB framework and on the other hand perform two types of conversions. (2) The parsing of proprietary tool data into EKB framework compliant messages and (3) the model transformation, which transforms data satisfying the data models of the tool instances described in the particular tool ontologies into data satisfying the data model described in the explicit tool domain specification in the form of tool domain ontologies which are stored in the engineering knowledge base (EKB) (Moser 2009). (4) This transformed data is then stored in the Engineering Data Base (EDB) (Waltersdorfer et al. 2010), a versioned database storing data on the level of each tool domain. (5) Queries can now be defined against the data model described in the particular tool domain ontology. (6) these queries then get translated to queries of the data stored in the EDB, (7) and finally the results of these queries are then returned to the user, e.g., in order to be displayed in the Engineering Cockpit (as shown in Chap. 11).

For more details on the usage of ontologies for modeling different types of engineering tool data, please refer to Chap. 5.

As already mentioned, the EKB framework supports engineering processes which are not yet fully automated and include media breaks requiring tedious and error-prone manual human work tasks, e.g., reentering information in other tools because there is no connection between these tools possible yet. Another prerequisite is the solved technical integration, which means that the communication between the engineering tools is established and working correctly, e.g., using an Engineering Service Bus (Biffl et al. 2009a). After all these technical heterogeneities have been addressed and solved, there still exist semantic heterogeneities between the involved tools respectively between their underlying data models. That means that either two or more models use different terminologies for the same concepts or that the concepts are defined and used using different levels of granularity. Goh (1996) classified semantic heterogeneities into three main categories: confounding conflicts (e.g., equating concepts are actually different), scaling conflicts (e.g., using different units for the same concept), and naming conflicts (e.g., synonyms and homonyms). The EKB's focus of addressed semantic heterogeneities are confounding conflicts and naming conflicts.

4.5 Case Study and Evaluation

This section reports on the application of the proposed EKB framework in the industrial domain of software-intensive production automation systems. In addition, we evaluate the EKB approach using two typical usage scenarios, namely data exchange between tools and model consistency checking across tool boundaries in detail.

4.5.1 Case Study Description

This section presents a multidisciplinary engineering use case from an industrial partner of hydropower plants. From discussions with the industry partner it is known that the concept of signals is used for collaboration between heterogeneous engineering tools. In a typical engineering project for developing power plants, there are about 40.000 to 80.000 devices to be administered and managed from different tools of different engineering fields. Devices help link individual engineering tool models and thus represent objects used to transmit or convey information. The application field "Device engineering" deals with managing devices from different engineering disciplines and is facing some important challenges, e.g., (1) to make device handling consistent, (2) to integrate devices from heterogeneous data models and tools, and (3) to manage versions of device changes across engineering disciplines.

Figure 4.4 shows industry partner relevant engineering tools managing device information from the tool's perspective. As shown in the figure, device information is updated by several tools. Devices are not limited to electrical devices in electrical engineering only, but also include mechanical interfaces in mechanical engineering and software I/O variables in software engineering, and thus may be seen by definition as a mechanical object. The process shown in Fig. 4.4 points out the requirements for knowledge sharing between engineering tools. As in the traditional way of plant engineering where information is passed manually between engineers, the process is performed automatically and is called "check-in". However, as several engineers may work at the same time on the same set of devices it is necessary that each check-in is approved by any other engineer. Whenever the engineer checks-in data, the EKB framework provides information about the type of change to the other engineers by sending an email or creating new issues assigned to them for approval. Automated creation of notifications is necessary in order to minimize surprises in the engineering team due to little communication between project members. Notified engineers receive information about changed devices and review the correctness

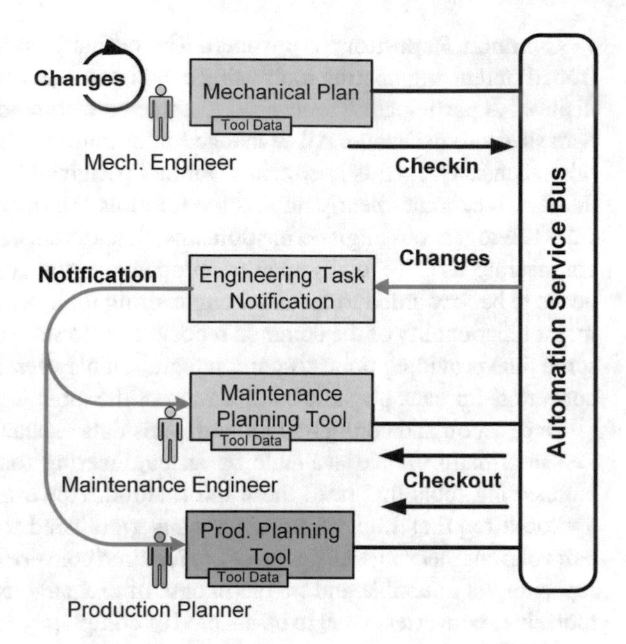

Fig. 4.4 Change management based on Engineering Objects (based on Moser et al. 2012)

of the change from their perspective. By accepting the issue, the change on a device is accepted as well.

4.5.2 *Scenario-Based Evaluation of the EKB*

In this subsection, we describe two typical EKB usage scenarios, Data Exchange Between Tools and Model Consistency Checking Across Tool Boundaries, and compare the usage of both a common repository approach and an EKB-based approach for each of the scenarios.

4.5.2.1 Data Exchange Between Tools

To cooperate, the engineers have to exchange relevant parts of the data structures (i.e., information required in another tool should become available as soon as it has been saved in the original tool) in their tools with each other with the goal of a consistent overall view on certain aspects in the project, e.g., when producing a specification for a subcontractor. Currently, every role uses organization-, domain-, and tool-specific data formats and terms, thus the data exchange takes considerable expert knowledge on the receiving end to make sense of the incoming data, typically as large PDF document or tool-specific import file (Moser et al. 2010).

Common Repository Approach. The exchange of data structures originating from different engineering tools using a common repository requires a set of prerequisites. All participating tools need to agree on a common data schema used for the data structure exchange. All exchanged information is then structured according to this schema. While this is even hard for tools originating from the same engineering domain, it becomes nearly impossible for tools originating from different and typically heterogeneous engineering domains. In addition, changes to one or more of the engineering tools regularly require an update of the common schema, which then needs to be forwarded to the other engineering tools which use this schema. So, the major functionality of the common repository is to store all information, while at the same time providing point-to-point integration between all participating tools using converters for each possible combination of the tools.

Once set up and configured properly, this data exchange method has a low delay, i.e., information made available by an engineering tool is available for all other engineering tools that need these information. However, the configuration of this approach requires high effort, since converters need to be written for all needed pairs of n engineering tools, with $O(n^2)$ required converters. In addition, the common repository is inflexible and fragile in case of exchange of even a single engineering tool, since converters need to be adapted or completely rewritten in this case (Moser et al. 2010).

Engineering Knowledge Base (EKB) Approach. A first step in using the EKB framework is the identification of common concepts used in the participating engineering tools. As a next step, the proprietary tool-specific knowledge is mapped to the more general common concepts. Based on these mappings, the EKB framework semi-automatically generates transformation instructions for transforming data structures between tool-specific formats. This semi-automated generation exploits the mappings stored in the EKB and makes suggestions for possible transformations to be reviewed by a human expert. The human expert can then revise the suggested transformation, change them or add new or more complex transformation rules manually. Since for each of the n participating engineering tools a single transformation instruction is required, the number of overall needed transformation instructions is $O(n)$ (Moser et al. 2010).

While the EKB framework requires similar or at most slightly higher effort for setup and configuration compared to the common repository approach, new benefits come from using ontologies for storing the engineering knowledge. The ontologies enable the semi-automated generation of the required converters, both initially and when engineering tools evolve. The number of required converters is also smaller with $O(n)$ converters for n engineering tools. Further, once set up, the delay of the data exchange method is similar to the delay using the traditional common repository-based approach (Moser et al. 2010).

Figure 4.5 illustrates two different engineering-tool-specific terminologies of the a typical production automation system regarding orders. The business knowledge uses the concept *ClientPurchase* as a local terminology, while in the workshop configuration knowledge the concept *WorkTask* is used as a local terminology. As shown in Fig. 4.5, both concepts are mapped to the corresponding domain concepts, *Cus-*

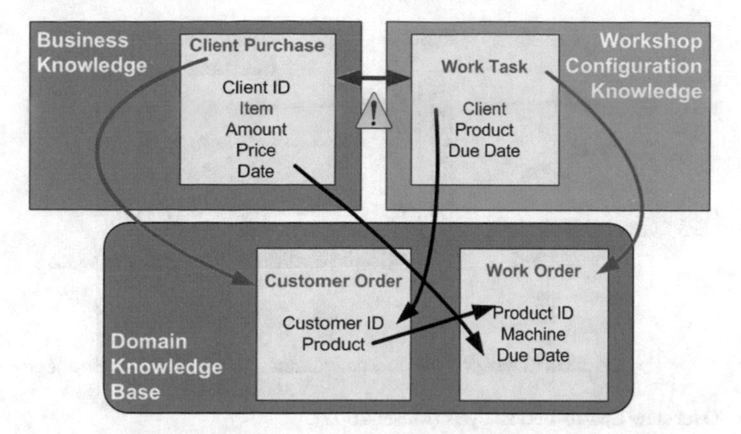

Fig. 4.5 Translation between business and workshop configuration knowledge

tomerOrder and *WorkOrder* respectively. In addition, the attribute *Date* of *ClientPurchase* is mapped to the attribute *DueDate* of *WorkOrder*, while the attribute *Client* of *WorkTask* is mapped to the attribute *CustomerID* of *CustomerOrder*. Further, in the Domain Knowledge Base the concepts *CustomerOrder* and *WorkOrder* are linked by their common attribute, *Product* and *ProductID* respectively. Using these mappings, we can identify work tasks which belong to a specific client purchase and identify the corresponding client purchase for a specific work task, without the need to establish a direct link or mapping between the two local terminologies.

4.5.2.2 Model Consistency Checking Across Tool Boundaries

Model checking, i.e., the validation of model data elements regarding their integrity and consistency, typically is performed at project milestones before the model elements can be used in the next stage of engineering. Currently, model checking is limited to single engineering tools or engineering domains. In addition to syntactical checks, plausibility checks of model elements regarding their usage in other engineering domains are needed. For more details on defect detection in multidisciplinary engineering, please refer to Chap. 11.

In distributed engineering in heterogeneous environments, typically a set of different models is used along the engineering chain. In order to ensure validity and consistency of the overall engineering process, it is important to ensure that required data fields can be enforced during the whole life cycle of the engineering chain (Moser 2009).

In an automation systems context, this may be defined as a list of hardware sensors and software variables (as shown in Fig. 4.6), which are connected to a system interface by virtual links in models or by wiring in the real world. Internally, the devices are mapped from the system interface to a software interface, where these devices

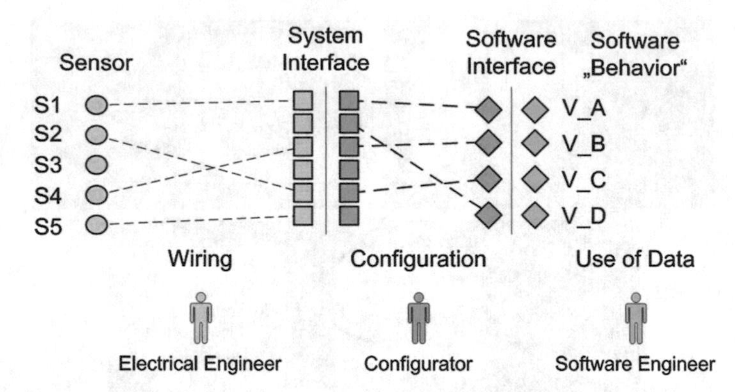

Fig. 4.6 Overview End-to-End analysis (Moser 2009)

are represented as variables. Typical consistency or validity checks may be used to check whether there exist any incomplete chains between variables and sensors.

Another example for consistency and integrity checks of model changes is a hardware pump which supports a certain number of input/output devices (I/Os), and which is controlled by a pump control software using either analog or digital devices. Analog devices can handle 8 I/Os, digital 64 I/Os. If the device type is changed in the pump control software model, it needs to be validated whether the new device type can handle all available I/Os of the hardware pump. Respectively, if the I/Os are changed (e.g., new I/Os added) it has to be checked whether they all can be controlled using the chosen device type of the pump control software. Another example for the derivation of run-time functionality for automated testing and monitoring is again a hardware pump which can handle approximately 1000 liters per hour. A time-based analysis of the events originating from the reservoir located behind the hardware pump could show impossible conditions or sensor states, e.g., if the reservoir capacity of 10000 liters is reached within 5 hours starting from an empty condition (Moser et al. 2010).

Common Repository Approach. Using a common repository enables to perform advanced checks regarding the data structures of more than one engineering tool. The major drawback of this approach of performing model checks is the need for manual involvement of human experts. The experts need to explicitly define the checks and select the data they want to include in these checks. This definition needs to be updated after every change to the involved data elements. Additionally, the nature of the common repository allows only for syntactical checks (e.g., the availability of all obligatory data fields or the validity of data types regarding a certain data schema) of the data, but not for other checks such as regarding the semantic correctness or plausibility of data structures. Other checks, such as checks regarding logical connections of data elements, are not supported out of the box using a common repository, since the data elements in the repository are stored unaltered and without metainformation. However external analysis tools can use the data elements stored in the common repository for performing such model checks (Moser 2009).

EKB Approach. The EKB framework enables automated checks regarding both syntactical issues as well as plausibility checks regarding semantic correctness of data structures. The EKB framework exploits the querying capabilities of ontologies to allow even more advanced checks, such as checks regarding completeness of available information. Human experts define checks regarding specific domain or tool-specific concepts, which are then on-the-fly transformed into checks regarding tool-specific data structures accordingly. The results are then collected and again transformed into the domain concept level, allowing experts both to define checks as well as to view the results of the defined checks in their well-known syntax, terminologies, and notations (Moser 2009).

4.6 Conclusion

Software-intensive systems in business IT and industrial automation and software engineering projects bring together experts from several engineering domains and organizations, who work in a heterogeneous engineering environment with a wide range of models, processes, and tools that were originally not designed to cooperate seamlessly (Schäfer and Wehrheim 2007). A core question is how to integrate data models across tools and domain boundaries. Current semantic engineering environment integration is often ad hoc and fragile, making the evolution of tools and reuse of integration solutions across projects risky (Halevy 2005; Noy et al. 2005).

In order to reach the common goal of developing software products in the engineering team, it is important to share the necessary knowledge for common work processes between engineering domain experts (Schäfer and Wehrheim 2007). However, this knowledge is often only implicitly available and therefore inefficient to share, resulting in time-consuming repetitive tasks; often it is hard or even impossible to create and maintain common shared knowledge repositories. A method and platform for making expert knowledge explicit and efficiently shareable is needed in order to support quality and project managers in their data analyses based on engineering knowledge and concrete data in the engineering tool models, which is currently achieved using inefficient or fragile approaches.

In this chapter, we described the novel Engineering Knowledge Base (EKB) framework for supporting engineering environment integration in multidisciplinary engineering projects with a focus on providing links between data structures of engineering tools and systems to support the exchange of information between these tools and thus making software and systems engineering more efficient and flexible. Since standards are hard to apply in projects where experts from different organizations participate, who have invested into different kinds of local standards or approaches, these experts may use their well-known local tools and data model, and additionally can access data from other tools in their local syntax.

Since the engineering project participants by now already work together, they already use common knowledge for their project tasks. Using the EKB framework, we make this existing knowledge explicit and machine-understandable, and therefore can automate on project-level tasks that build on this explicit and machine-understandable knowledge. Furthermore, using the EKB framework allows a more generic definition and execution of model checks on an application domain level, and additionally enables more advanced checks regarding the plausibility and semantic correctness of data structures by exploiting the querying capabilities of ontologies.

The following Chaps. 5–7 present individual aspects of the EKB framework. Typical applications enabled by implementations of this framework are discussed in Chaps. 9 and 10.

Acknowledgments This work has been supported by St. Pölten University of Applied Sciences, by the Christian Doppler Forschungsgesellschaft, the Federal istry of Economy, Family and Youth, and the National Foundation for Research, Technology and Development in Austria. The author wants to thank Stefan Biffl, Franz Fidler, and Richard Mordinyi for their valuable inputs.

References

Assmann, D., Dörr, J., Eisenbarth, M., Hefke, M., Soto, M., Szulman, P., Trifu, A.: Using ontology-based reference models in digital production engineering integration. In: 16th IFAC World Congress (2005)

Bergamaschi, S., Castano, S., Vincini, M.: Semantic integration of semistructured and structured data sources. SIGMOD Rec. **28**(1), 54–59 (1999)

Bernstein, P.A., Dayal, U.: An overview of repository technology. In: 20th International Conference on Very Large Data Bases, pp. 705–713. Morgan Kaufmann Publishers Inc. (1994)

Biffl, S., Schatten, A.: A platform for service-oriented integration of software engineering environments. In: New Trends in Software Methodologies, Tools and Techniques—Proceedings of the Eighth SoMeT 2009, 23–25 Sept 2009, Prague, Czech Republic, pp. 75–92. doi:10.3233/978-1-60750-049-0-75

Biffl, S., Schatten, A., Zoitl, A.: Integration of heterogeneous engineering environments for the automation systems lifecycle. In: IEEE Industrial Informatics (IndIn) Conference, pp. 576–581 (2009a)

Biffl, S., Sunindyo, W.D., Moser, T.: Bridging semantic gaps between stakeholders in the production automation domain with ontology areas. In: 21st International Conference on Software Engineering and Knowledge Engineering (SEKE 2009), pp. 233–239 (2009b)

Doan, A., Halevy, A.: Semantic integration research in the database community: a brief survey. AI Mag. **26**(1), 83–94 (2005)

Doan, A., Noy, N.F., Halevy, A.Y.: Introduction to the special issue on semantic integration. SIGMOD Rec. **33**(4), 11–13, 1041412 (2004)

Drath, R., Lüder, A., Peschke, J., Hundt, L.: AutomationML—the glue for seamless automation engineering. In: IEEE International Conference on Emerging Technologies and Factory Automation (ETFA 2008), pp. 616–623. IEEE (2008)

Goh, C.H.: Representing and reasoning about semantic conflicts in heterogeneous information systems. Ph.D. thesis, MIT (1996)

Halevy, A.: Why your data won't mix. Queue **3**(8), 50–58 (2005)

Heiler, S.: Semantic interoperability. ACM Comput. Surv. **27**(2), 271–273 (1995)

Kruchten, P.: The Rational Unified Process: An Introduction. Addison-Wesley, Boston (2000)

Lüder, A.: Formaler steuerungsentwurf mit modularen diskreten verhaltensmodellen. Ph.D. thesis, Martin-Luther-Universität (2000)

Liao, S.: Technology management methodologies and applications: a literature review from 1995 to 2003. Technovation **25**(4), 381–393 (2005)

Lovett, P.J., Ingram, A., Bancroft, C.N.: Knowledge-based engineering for SMEs—a methodology. J. Mater. Process. Technol. **107**(1–3), 384–389 (2000)

McGuire, J., Kuokka, D.R., Weber, J.C., Tenenbaum, J.M., Gruber, T.R., Olsen, G.R.: Shade: technology for knowledge-based collaborative engineering. Concurrent Eng. **1**, 137–146 (1993)

Medeia-Consortium: Medeia: Requirements analysis and technology review. Medeia consortium (2008)

Moser, T.: Semantic integration of engineering environments using an engineering knowledge base. Ph.D. thesis, Vienna University of Technology (2009)

Moser, T., Biffl, S., Sunindyo, W.D., Winkler, D.: Integrating production automation expert knowledge across engineering stakeholder domains. In: International Conference on Complex, Intelligent and Software Intensive Systems (CISIS 2010), pp. 352–359. IEEE (2010)

Moser, T., Mordinyi, R., Winkler, D.: Extending mechatronic objects for automation systems engineering in heterogeneous engineering environments. In: 2012 IEEE 17th Conference on Emerging Technologies Factory Automation (ETFA), pp. 1–8 (2012). doi:10.1109/ETFA.2012.6489664

Noy, N.F.: Semantic integration: a survey of ontology-based approaches. SIGMOD Rec. **33**(4), 65–70 (2004)

Noy, N.F., Doan, A.H., Halevy, A.Y.: Semantic integration. AI Mag. **26**(1), 7–10 (2005)

Rahm, E., Bernstein, P.A.: A survey of approaches to automatic schema matching. VLDB J. **10**(4), 334–350 (2001)

Schäfer, W., Wehrheim, H.: The challenges of building advanced mechatronic systems. In: 2007 Future of Software Engineering—International Conference on Software Engineering, pp. 72–84. IEEE Computer Society (2007)

Ullman, J.D.: Information integration using logical views. Theor. Comput. Sci. **239**(2), 189–210, 339543 (2000)

Waltersdorfer, F., Moser, T., Zoitl, A., Biffl, S.: Version management and conflict detection across heterogeneous engineering data models. In: 2010 8th IEEE International Conference on Industrial Informatics (INDIN), pp. 928–935 (2010). doi:10.1109/INDIN.2010.5549617

Weilkiens, T.: Systems Engineering with SysML/UML: Modeling, Analysis, Design. Morgan Kaufmann (2008)

Chapter 5
Semantic Modelling and Acquisition of Engineering Knowledge

Marta Sabou, Olga Kovalenko and Petr Novák

Abstract Ontologies are key Semantic Web technologies (SWTs) that provide means to formally and explicitly represent domain knowledge in terms of key domain concepts and their relations. Therefore, the creation of intelligent engineering applications (IEAs) that rely on SWTs depends on the creation of a suitable ontology that semantically models engineering knowledge and the representation of engineering data in terms of this ontology (i.e., through a knowledge acquisition process). The tasks of semantic modelling and acquisition of engineering knowledge are, however, complex tasks that rely on specialized skills provided by a knowledge engineer and can therefore be daunting for those SWT adopters that do not possess this skill set. This chapter aims to support these SWT adopters by summing up essential knowledge for creating and populating ontologies including: ontology engineering methodologies and methods for assessing the quality of the created ontologies. The chapter provides examples of concrete engineering ontologies, and classifies these engineering ontologies in a framework based on the *Product-Process-Resource* abstraction. The chapter also contains examples of best practices for modelling common situations in the engineering domain using ontology design patterns, and gives an overview of the current tools that engineers ca use to lift engineering data stored in legacy formats (such as, spreadsheets, XML files, and databases, etc.) to a semantic representation.

Keywords Ontology modelling · Ontology engineering methodologies · Ontology evaluation · Classification of engineering ontologies · Ontology design patterns · Ontology population

M. Sabou (✉) · O. Kovalenko · P. Novák
Institute of Software Technology and Interactive Systems, CDL-Flex,
Vienna University of Technology, Vienna, Austria
e-mail: Marta.Sabou@ifs.tuwien.ac.at

O. Kovalenko
e-mail: Olga.Kovalenko@tuwien.ac.at

P. Novák
e-mail: Petr.Novak@tuwien.ac.at

S. Biffl and M. Sabou (eds.), *Semantic Web Technologies for Intelligent Engineering Applications*, DOI 10.1007/978-3-319-41490-4_5

5.1 Introduction

Engineering knowledge is a specific kind of knowledge oriented towards the production of artifacts, and, as such, requires knowledge modelling and representation approaches that differ from other types of knowledge, such as, for example, taxonomical knowledge that is characteristic for the life sciences domain (Sicilia et al. 2009). Ontologies are information artefacts that can be used to explicitly represent such engineering knowledge and as such, they play an important role when creating intelligent engineering applications (IEAs) relying on Semantic Web technologies (SWTs). Chapter 3 provides a detailed insight into ontologies and their role in SWT based solutions.

Yet, the creation of semantic structures (which we refer to as *semantic modelling*) in general, and of ontologies in particular, is a complex process because of a set of factors (Gomez-Perez et al. 2004). First, the creation process needs to lead to a complete and correct representation of the subject domain to the extent required by the intended use of the ontology. This step requires a deep knowledge of the underlying subject domain as well as a good understanding of the way the ontology will be used. Multi-disciplinary engineering settings such as those specific for creating *cyber-physical production systems* (CPPSs), pose a challenge from this perspective because (1) they require knowledge of several engineering disciplines and (2) they are built for only a partially specified set of applications with new applications being added over time. Second, an optimal machine-understandable representation of the engineering knowledge must be chosen that enables taking full advantage of advanced ontology exploitation mechanisms such as querying and reasoning. To support ontology engineers in finding the best conceptualizations for a variety of modelling requirements, the ontology engineering community has distilled a set of *ontology design patterns* (ODP)—modelling best practices applicable to typical conceptualization scenarios (Gangemi and Presutti 2009).

The complexity of the semantic modelling process often hinders the adoption of ontology-based technologies. Adopters of ontology-based solutions from industry are confronted with answering questions which naturally arise while following ontology engineering methodology steps discussed in Sect. 5.2: *What is the process of building an ontology? What kind of knowledge should the ontology contain? What other relevant ontologies exist and can these be reused for the project at hand? How can the quality of an ontology be assessed? If a new ontology must be designed, what are typical modelling issues to be considered and what typical solutions are applied? How can semantic data be created from legacy data sources?* In this chapter, we aim to answer such questions. Concretely, we:

- Provide a brief introduction to *ontology engineering methodologies* (Sect. 5.2) and the main techniques for *ontology evaluation* (Sect. 5.3).
- Synthesize the types of engineering knowledge that are most often captured by ontologies (Sect. 5.4) and propose an *ontology classification framework* based on ontology content.

- Provide an overview of a set of *engineering ontologies* and classify those according to their content (Sect. 5.5).
- Exemplify a set of of *ontology design patterns* and their use for the semantic modelling of engineering knowledge (Sect. 5.6).
- Provide an overview of techniques and tools for transforming engineering data from legacy data formats (spreadsheets, *Extensible Markup Language*[1] (XML), databases) into Semantic Web representations (Sect. 5.7).

Section 5.8 concludes the chapter with a summary of its content and future work.

5.2 Ontology Engineering Methodologies

A wealth of methodologies exists for guiding the creation of ontologies some dating back to as early as 1990 (Lenat and Guha 1990). Overviews of these methodologies are available in (Corcho et al. 2003; Suárez-Figueroa 2012; Poveda-Villalón 2016). In this section, we summarize the most representative ontology engineering methodologies for supporting ontology adopters in choosing the best-suited methodology for their context. Readers not familiar with the notion of ontologies might wish to consult Chap. 3 for foundations about ontologies.

From the proposed methodologies, the ontology creation guideline of Noy and McGuinness (2001) captures the essential steps of any ontology building process and, thanks to its many examples, it offers an excellent resource for developing one's first ontologies. Hereby, we summarize the main steps indicated in this guideline to give an intuitive overview of a typical ontology creation activity:

1. **Determine the domain and scope of the ontology**. This step defines the domain that the ontology will cover (e.g., mechanical engineering, functional testing), the expected usage of the ontology in terms of the questions that the ontology should be able to answer (also known as, *competency questions*), and the expected stakeholders involved with the ontology (maintainers, users). All these aspects are crucial for focusing the development of the ontology. The competency questions provide a means to define the scope of the ontology, and also, to validate the ontology during or at the end of the development cycle.
2. **Consider reusing existing ontologies**. One of the recommended practices in ontology development is to reuse (parts of) existing ontologies. The promise of reuse is that the developers will be able to build ontologies faster and with less effort, and that the applications that make use of these ontologies will interoperate easier. The practice of reuse is particularly useful in enterprises thus ensuring enterprise-level knowledge reuse, especially if the reused parts fit already well for the new purpose.

---

[1]XML: https://www.w3.org/TR/REC-xml/.

3. **Enumerate important terms in the ontology**. After defining the scope of the ontology (Step 1), the development of the ontology starts with collecting the most important terms by using the competency questions as input. The competency questions help identify key concepts, their properties and relationships that hold in the domain. This term list acts as an input to steps 4 and 5 for creating the ontology's class hierarchy and properties.
4. **Define the classes and the class hierarchy**. To create a class hierarchy, ontology engineers can choose one of three approaches. (1) In a *top-down approach*, the hierarchy is constructed starting from the most generic concepts (top concepts) and then creating the more specialized concepts. For example, in the mechanical engineering domain, the concept Device is created first, and then sub-concepts are created for the various types of devices (see for example, the ontology snippet in Fig. 3.3). (2) The opposite of the top-down approach consists in starting with the most specific classes of a domain (e.g., the various types of devices), and then grouping these into more generic classes. This approach is known as the *bottom-up approach*. (3) A *combination* approach is taken that combines the top-down and bottom-up approaches, and intertwines these, as most suitable for the ontology expert.
5. **Define the properties of classes**. Another important step is describing the declared classes in terms of their properties, including both their characteristics or relations to other classes. For example, the class `Device` has characteristics such as *weight* or *maximum allowed working hours*. It is connected to the class `Supplier` with the `hasSupplier` property.
6. **Define the property constraints**. This step envisions further enriching the ontology model by defining characteristics of its properties in more detail. These include the domain and range of properties, as well as any cardinality and value-based constraints, which specify the number and type of values that the property takes. Chapter 3 exemplifies how such constraints can be represented using primitives of the *Web Ontology Language* (OWL).
7. **Create instances**. In this step, users populate the classes in the ontology with concrete entities, that exist in the domain, a.k.a, instances or individuals. For example, a concrete device is represented as an instance of the class `Device`, and a user fills its properties with concrete values (e.g., `weight = 100 kg`). In many cases, the actual instances of an ontology are not part of the ontology itself, but are rather acquired using automatic data transformation processes, such as those described in Sect. 5.7.

The **On-to-Knowledge methodology** (Staab et al. 2001) takes a broader view of the ontology life cycle than (Noy and McGuinness 2001) and focuses on an *application-oriented* development of ontologies in the context of creating Knowledge Management applications. Therefore, its steps consider not only the development of the ontology per se, but also that of the application that will make use of the ontology. The methodology consists of five stages:

1. **Feasibility study** identifies problems and opportunity areas from an organizational perspective, and selects the most promising focus areas and suitable solutions.
2. **Ontology kickoff** produces an ontology requirements specification document, which identifies the ontology's goal, domain and scope. This document also identifies the applications that will make use of the ontology, the knowledge sources from which the ontology could be built (e.g., domain experts, relevant documents) and the competency questions it should provide answers to. This phase also covers the identification of ontologies that could be reused and therefore spans steps 1 and 2 in the Noy and McGuinness' guidelines (2001).
3. **Refinement** refers to the development of a mature and application-oriented ontology, and therefore covers activities such as described by Noy and McGuinness' (2001) steps 3–7. Similarly, it envisions stages for eliciting knowledge from domain experts at an epistemological level (i.e., enumerating important terms and their relations) and then formalizing this knowledge in terms of a formal knowledge-representation language such as OWL.
4. **Evaluation** focuses on assessing the quality of the output ontology both in terms of (1) satisfying the *ontology requirements document,* as well as (2) providing the desired functionality as part of the applications that make use of the ontology. Feedback from the requirements and application-oriented evaluations are used to modify the ontology accordingly. Therefore, *On-to-Knowledge* envisions a feedback loop between the *Evaluation* and *Refinement* stages.
5. **Maintenance** covers activities of updating and versioning the ontology to reflect changes in its environment (e.g., new user requirements, supporting new applications, changes in the engineering knowledge).

The **METHONTOLOGY** methodology (Fernández-López et al. 1997), (Blázquez et al. 1998) goes beyond the *On-to-Knowledge* methodology in terms of breadth and distinguishes between management, development, and support activities. Management activities include scheduling, control, and quality assurance. Development covers the entire life cycle of an ontology, and similarly to *On-to-Knowledge* ranges from feasibility study, through development (specification, conceptualization, formalization, and implementation) and post-developmental use to maintenance. Support activities include knowledge acquisition, evaluation, documentation, merging and alignment.

Proposed in 2010, the **NeOn Methodology** (Suárez-Figueroa 2012; 2015) recognized the need of catering for diverse ontology-engineering scenarios, as opposed to prescribing a generic process for building ontologies, as was the case for the previously-mentioned methodologies. The NeOn methodology identifies nine different scenarios that can be followed when building ontologies. The base scenario refers to those cases when ontologies are created from scratch without reusing existing resources. This scenario typically includes stages for requirements specification (identifying the goal, scope, and relevant competency questions for the desired ontologies), scheduling of the ontology creation task, conceptualizing the knowledge in relevant models, and then formalizing and implementing this knowledge in terms of formal representation languages such as OWL, similarly to the steps envisioned by the guidelines of Noy and McGuinness (2001).

Ideally, the effort needed for the base scenario could be reduced by bootstrapping the ontology-creation activity through the reuse of ontological or non-ontological resources. Scenario 2 of this methodology envisions creating an ontology by re-engineering non-ontological resources such as classification schemes, thesauri, lexicons, and folksonomies. Scenario 3 covers cases when an ontology can be re-used instead of building a new ontology. In some cases, it might be necessary to re-engineer an existing ontology (Scenario 4), to merge several suitable ontologies into one (Scenario 5), to reuse a set of ontologies that are merged and also re-engineered (Scenario 6) or to restructure an existing ontology (Scenario 8). Scenario 7 envisions creating ontologies by reusing *Ontology Design Patterns*, which are high-quality ontology modelling solutions. Scenario 9 refers to creating a new ontology by localizing an ontology to better fit other language or culture communities.

Several ontology engineering methodologies considered scenarios of collaborative ontology development. Representative for this set of methodologies is **DILIGENT** (Pinto et al. 2004), which has a strong focus on managing the collaborative evolution of an ontology. In a first stage, a small and consensual ontology is built by the key stakeholders in the ontology creation process (*Build* phase) by following one of the classical ontology development methodologies described above. This version of the ontology is distributed among its users, who can modify their local ontology copy according to their needs (*Local adaptation* phase). In a follow-up *Analysis* phase, an *ontology control board* discusses the changes made locally by users as well as their arguments for introducing their changes. Based on these inputs and their analysis, in a *Revision* phase, the shared ontology is revised and extended to cover an agreed set of changes. During *Local update*, the users update their local ontology to the new version of the shared ontology.

In summary, ontology-engineering methodologies exist for covering a wide range of settings: focusing on ontology creation per se; considering the lifecycle of the ontology-based applications as well as the organizational context, ontology creation supported by various reuse activities, and collaborative ontology development. These methodologies can provide ample guidelines to practitioners, who wish to create ontologies and ontology-based applications. An important step in ontology engineering is ontology evaluation and, therefore, we focus on this specific aspect in Sect. 5.3.

5.3 Ontology Evaluation

Ontology evaluation is *"the activity of checking the technical quality of an ontology against a frame of reference"* (Sabou and Fernandez 2012). Ontology evaluation is also important during other ontology-related activities, such as: the evolution of an ontology; the selection of an ontology for reuse, or during the process of modularizing an ontology (for assuring the high quality of the resulting modules).

The ontology evaluation goal determines the key aspects of the ontology to be assessed and determines the evaluation approaches and measures to be applied. According to (Sabou and Fernandez 2012), the most frequently evaluated ontology aspects are:

1. *Domain coverage—Does the ontology cover a topic domain?* The goal is to assess the extent to which an ontology contains the knowledge necessary to describe a given (aspect of a) domain. For example, one could assess how well an ontology covers terms and their relations when describing *mechanical engineering* aspects of certain types of production systems. This assessment is important to be made both during the development of an ontology and during the selection of an already built ontology. Typically, evaluations with this goal involve the comparison of the ontology to frames of references such as a reference (i.e., gold standard) ontology (Maedche and Staab 2002), or various data sets that are representative for the domain. These datasets can be user-defined terms (Alani et al. 2006; Fernandez et al. 2006), folksonomy tag sets (Cantador et al. 2007), or representative document corpora (Brewster et al. 2004). Typical evaluation measures are similarity measures that compare two ontologies at lexical (i.e., average string matches between the set of gold standard terms and the set of ontology terms), taxonomic, or relation level (Maedche and Staab 2002).
2. *Quality of the modelling.* The evaluation can either focus on the quality of the design and development process (*Does the ontology development process comply with ontology modelling best practices?*) or on the quality of the resulting ontology (*Is the ontology model correct?*). The quality of the ontology model can be assessed using a wide range of approaches focusing on logical correctness or syntactic, structural, and semantic quality. Logical correctness (e.g., logical consistency) is automatically assessed by reasoners. Other aspects, such as syntactic quality, require human judgment and therefore rely on approaches involving human assessment (Burton-Jones et al. 2005). Last but not least, semantic quality can be assessed with metrics such as *essence* (assess if an entity is true in every possible world) or *unity* (recognizes all the parts that form an individual entity) (Guarino and Welty 2004). The OOPS! (OntOlogy Pitfall Scanner!) tool[2] provides online support for verifying ontology modelling quality (Poveda-Villalón et al. 2014).
3. *Suitability for an application/task—Is the ontology suitable for a specific application/task*? While the previous evaluation goals focus on assessing the ontology quality in general, here the aim is to assess to what extent an ontology is suitable for use within a concrete application or for a certain task (e.g., semantic search, question answering). Different applications/tasks might require ontologies with diverse characteristics. For example, for applications that use

[2]OOPS! Tool: http://oops.linkeddata.es/.

ontologies to support natural language processing tasks such as semantic search, domain coverage is often more important than logical correctness. Therefore, measuring ontology quality generically is not enough to predict how well the ontology (developed or reused) will support an application or a task. Task-centric evaluations help assessing suitability for a task or application (Porzel and Malaka 2004; Fernandez et al. 2009). The typical approach here is to measure ontology quality indirectly and as an approximation of an expected application/task performance. For example, for a semantic-search system the precision and/or recall of the system obtained when using different ontologies will be an indication of the ontology's suitability for that task. The best ontology is the one that leads to the best task performance.

4. *Adoption and use—Has the ontology been reused (imported) as part of other ontologies? How did others rate the ontology?* (Cantador et al. 2007) When selecting an ontology for reuse, the extent of its adoption is of particular interest. The assumption is that there is a direct correlation between the quality of the ontology and the level of adoption by the community. One approach to assess adoption is analyzing the degree of interlinking between an ontology and other ontologies (e.g., in terms of reused terms or ontology imports). Ontology libraries often offer such statistics (d'Aquin and Noy 2012). Social rating systems have also been used to reflect community-level reuse and evaluation of ontologies (Cantador et al. 2007).

The evaluation process is guided by the evaluator's understanding of what is better and what is worse. In some cases, these boundaries (which we refer to as frame of reference) are clearly defined and tangible (e.g., a reference ontology, a reference alignment), but in other cases, they are weakly defined and may be different from one person to another, or even across evaluation sessions. This often renders ontology evaluation a non-trivial task.

In summary, several approaches and measures exist for evaluating ontologies, and their selection should be derived based on the goal of the evaluation.

5.4 Classification of Engineering Ontologies

The domain of engineering cyber-physical production systems (CPPS) is a broad and complex domain. Building IEAs for this domain will therefore require creating a diverse variety of ontologies covering the many aspects of the domain. When creating such ontologies, two important aspects to consider are:

- *What kinds of knowledge should be covered by the engineering ontology?* This is an important question for determining the scope of the ontology, but that generally, ontology engineers (i.e., Semantic Web practitioners), who have little knowledge of this complex domain, find difficult to answer.

- *How to find suitable engineering ontologies for reuse?* Ontology reuse is an important step envisioned by all ontology engineering methodologies. Yet, Legat et al. (2014) observe that the reuse of ontologies in the automation domain is difficult, in part because of a lack of a principled way to create and classify ontologies in an use-case agnostic way.

One approach to alleviate these difficulties in scoping, classifying, and reusing ontologies is the availability of meaningful schemes for classifying the type of knowledge relevant for engineering complex CPPS. Such a scheme could support better scoping ontologies in terms of use-case-independent topics, and could greatly support meaningful selection of relevant ontologies.

In this section, we propose such a classification scheme, which we derived by combining two orthogonal views on the types of engineering knowledge in CPPS. First, the *Product-Process-Resource (PPR)* abstraction (Sect. 5.4.1) provides an intuitive view of the domain of production systems (Schleipen and Drath 2009). Second, we made use of an initial ontology-classification scheme (Sect. 5.4.2), proposed by Legat et al. (2014). This scheme is much more oriented on structure than the PPR view, because it considers the different physical and logical views on the elements of a production system. The proposed categorization framework for engineering ontologies, which combines these two views, will allow practitioners to find easier the ontologies that cover the relevant engineering content for their project. The categorization framework will also be useful to Semantic Web practitioners, who require a better understanding of knowledge needs, when building IEAs and scoping their ontologies.

5.4.1 The Product-Process-Resource Abstraction

To better understand the types of engineering data needed to describe a complex manufacturing plant, we start with the *Product-Process-Resource* abstraction explained in (Schleipen and Drath 2009). The three views of this abstraction are dominant, and of key interest for industry, as described in more detail next.

Product. The intention of any production system is to produce products (e.g., cars, bread). The term "product" refers to the final produced goods as well as to the produced artefact in one of its intermediary stages. A product designer describes a product by its geometric, functional, material, and other relevant characteristics. Depending on the kind of product, the *product structure* can be described with mechanical information, e.g., 3D CAD and kinematics.

Process. To create a product, a set of *production process* steps is applied to manufacture the product from raw materials or semi-finished products. Processes modify intermediate products and create a final product. Example processes are welding, transporting, and filing. The process view is concerned with the set of processes (their parameters, and their chains) needed to create a product from input materials. The production process corresponds to the function description of the production resources, and to the control, e.g., a PLC program. However, the production process can also be described more explicitly, e.g., with GANTT charts or some general form of behaviour description, e.g., state machines.

Resource. Production resources are entities involved in production that provide functions to production processes. Resources include both hardware (e.g., robots, conveyors, machines) and software entities (SCADA systems), and are typically organized hierarchically into a *plant topology*. The properties of production resources are mostly their function capabilities (e.g., welding, transporting, and filing), mechanical information, electrical information, and control-related information; in addition, further technical information can be specified.

There is a strong connection between these three views: a *product* is manipulated by a *resource* as part of a *process*. For example, in the case of a tetra-pack production system described in (Schleipen and Drath 2009), the process *transport* might be linked to a resource *conveyor* and a product *tetra pack*.

Although it is useful to consider these three views in separation, they all contribute to the definition of a *production system*. The intention of any *production system* is to execute *production processes* using *production resources* in order to create *products*. Example production systems are: car body welding shops, combustion-turbine-based power plants or process plants for polyvinyl chloride generation. Therefore, a production system contains a set of interlinked production resources each of them able to execute a certain production process step (drilling, welding, etc.). The production system designer describes a production system by the information given in Fig. 2.6 and provided by the different engineers named in Chap. 2. This information includes:

- topology information describing the hierarchy of production resources;
- geometry and kinematics information;
- network information (electrical, pneumatic, communication, …); and
- control information.

The aim of each production system is the creation of products. Therefore, the *PPR abstraction* follows the Input-Processing-Output principle. Input materials are processed in processes to lead to output products (and waste) by exploiting resources.

5.4.2 A Classification Scheme for Engineering Ontologies

While the *PPR* view provides a concise description of the main concepts in the production systems engineering domain, which could be covered by ontologies, Legat et al. (2014) propose a complementary view on typical ontology content which was developed in a bottom-up fashion during a survey and classification of already available automation ontologies. Concretely, Legat et al. (2014) observe that the reuse of ontologies in the automation domain is difficult. One solution that they propose is that ontology reuse could be improved by modularizing ontologies along topics that are independent of a specific use case. The module structure they propose, provides good insight into the types of information that are relevant when describing CPPS.

In the following, we discuss the main categories of information identified by Legat et al. (2014) and how these relate to the PPR abstraction. By combining these two orthogonal views we obtain a broader classification scheme of engineering knowledge that connects the well-accepted topics from industry (PPR) with use-case-agnostic topics typically covered by ontologies. This framework is depicted in Table 5.1, where columns correspond to PPR concepts and rows represent to the main categories of information according to Legat et al. (2014), namely:

- **Physical objects** description. This description category refers to the categorization of available objects, such as equipment in a plant (e.g., sensors, actuators), product parts that need to be assembled. The properties, functionalities, and structures of these objects are covered by other ontologies. As such, this category in Legat et al.'s classification covers both products and production

Table 5.1 Ontology classification framework combining PPR concepts (columns) and Legat et al.'s modular ontology view (rows)

	Product	Production process	Production resource
Physical objects	Ont. of product types, *OntoCAPE, eClassOWL*	*OntoCAPE, ISO 15926*	Ont. of resource types, *OntoCAPE, CCO, AMLO, ManufOnto, eClassOWL, ISO 15926, AutomOnto*
Structure	Ont. of product structure, *OntoCAPE*	NF	Ont. of resource structures, *OntoCAPE, ManufOnto, EquipOnt, CCO, AMLO, ISO 15926, AutomOnto*
Functionality	NF	Ont. of production process types, *OntoCAPE, ISO 15926, ManufOnto*	Ont. of production resource capabilities (skills), *AMLO, ManufOnto, EquipOnt, ISO 15926*
Process	NF	Ont. of production process structures, *OntoCAPE, ISO 15926, ManufOnto*	*ManufOnto*
Materials	Ont. of bills of materials, *eClassOWL*	NF	NF
Observations, measurements	NF	Ont. of process states and its observability, *SSN, OntoCAPE, ISO 15926*	Ont. of resource states, *SSN, AutomOnto*
Quantities, dimensions, units	Ont. of product characteristics, *eClassOWL*	Ont. of production processes characteristics, *OntoCAPE, ISO 15926*	Ont. of production resource characteristics, *ManufOnto, CCO, SSN, AutomOnto*

Example ontologies added in *italics*. NF = not feasible

resources in PPR, meaning that ontologies that describe types of products or types production resources can be considered, from Legat et al.'s perspective, as ontologies describing physical objects (see Table 5.1).

- **Structures** description. Ontologies are also needed to describe various structural information about how the physical objects are composed. For example, one can describe how products need to be assembled, how production resources are structured, or how production systems (plants) are constructed. Structure-related information can be conveyed in different ways, according to Legat et al. (2014): (1) **interface-based composition** describes the capabilities expected from an interface and can enable reasoning tasks about the correctness of a system's structure; (2) **containment** hierarchies are a well accepted and frequently occurring organizational paradigm in mechatronic engineering settings.
- **Functionality** describes the behaviour of devices (or organizational units), i.e., what these elements can do in terms of contributions to a production process. Functionality descriptions are characterized by their properties, parameters and constraints, for example in terms of materials they can be applied on. Functionality descriptions are relevant both for individual production resources (e.g., a description of the function fulfilled by a resources, such as filling or drilling) as well as by production processes (that is, a function achieved by a process).
- **Process** descriptions are closely related to production resource functionality; they enable describing the composition of functionalities into obtaining functionalities that are more complex. Examples here are workflow patterns, such as the sequential or parallel execution of functions. Process ontologies provide useful primitives for describing the structure of production processes, for example, the workflow structure in which its steps are organized.
- **Material** descriptions contain collections of materials used within a production process and their hierarchical organization (e.g., wood, iron). Ontologies describing *bills of materials* needed for creating a product fit this category.
- **Observations and measurements** capture data produced during a system's operation, and as such differ in nature from the previously mentioned knowledge categories. Such information is essential to capture physical changes that happen in the physical world in which a CPPS operates. Relevant types of ontologies would be those describing the state of production processes or production resources.
- **Physical quantities, dimensions, and units** provide auxiliary information for describing aspects of most of the knowledge categories mentioned above, related to temporal aspects, weight, spatial dimensions (e.g., length). Such ontologies can be used to describe the various characteristics of products, production processes, production resources, and production systems.

From Table 5.1 it becomes evident that ontologies in engineering vary widely and span different aspects of different engineering concepts. Beyond the identified types of ontologies, there are several engineering-discipline-specific ontologies in place. These ontologies usually cover only the concepts relevant within a special application case, such as production system simulation at resource level, or, they may cover special engineering discipline knowledge, such as an automation glossary.[3] In Sect. 5.5 we provide examples of engineering ontologies, and classify them according to the classification scheme proposed in this section.

5.5 Examples of Engineering Ontologies

In this section, we provide an overview of engineering ontologies and classify them according to the classification scheme defined in Sect. 5.4.2. We start by describing ontologies published in literature, and conclude with two ontologies created at the *CDL-Flex* Laboratory[4]: the *AutomationML Ontology (AMLO)* and the *Common Concepts* ontology (CCO), in Sects. 5.5.1 and 5.5.2 respectively.

OntoCAPE[5] is an ontology for supporting *computer-aided process engineering* (CAPE). It was designed at the RWTH Aachen University, and it is discussed in numerous publications in details (Morbach et al. 2009). It has a modular structure consisting of 60 + OWL files (with the meta-model, it has more than 80 files).

OntoCAPE has a layered logical design spanning from foundational to domain-specific layers. The lowest level is called the *application-specific layer,* and it contains application-specific knowledge. The next level is the *application-oriented layer*, which describes the plant and process control equipment. This level also includes the view on the particular technology from the process point of view. The third level is the *conceptual layer*, which provides supporting concepts for modelling of processes, materials, and other representations and characteristics needed for modelling of processes. The top level is called the *upper layer,* and provides expressive means for representing networks, coordinate systems, and others. *OntoCAPE* is defined through a meta model, which is also denoted as the *meta layer* of *OntoCAPE*. The meta layer is represented as a stand-alone OWL ontology, which provides foundations for meta-modelling structures and other fundamental concepts.

In terms of the terminology introduced in Chap. 3, *OntoCAPE* combines ontology layers that include domain ontologies, generic ontologies (i.e., the conceptual layer) and foundational ontologies (i.e., the meta-layer). *OntoCAPE* is highly axiomatized. Because of its breadth, *OntoCAPE* can be classified in our classification scheme (Sect. 5.4.2) under several types of ontologies. From the PPR perspective, it entirely covers the production process criterion. *OntoCAPE* also addresses all other criteria, but focuses mainly on physical objects and structure levels, according to the classification by Legat et al. (2014).

[3]Automation glossary: http://ai.ifak.eu/glossar2/.

[4]CDL-Flex Laboratory: http://cdl.ifs.tuwien.ac.at/.

[5]*OntoCAPE* is available at: https://www.avt.rwth-aachen.de/AVT/index.php?id=730.

ISO 15926[6] is a complex standard dealing with industrial automation systems and integration (ISO 15926). Although it has been originally intended for the oil industry, its ideas and approaches are general and usable also in other domains. The main part of the standard is Part 2, dealing with description of objects and activities during various stages of the plant life cycle. The ontology includes diverse views on the process plant depending on the involved engineering disciplines. The original version of the standard (Parts 2–4) use the EXPRESS language, which was difficult to use and has a limited tool support. Hence, the standard was enhanced with Parts 7–10, relying on the OWL language. The OWL representation of Part 2 is available online.

The ISO 15926 defines an ontology that is generic, heavyweight and covers mainly information about production processes. It also includes information about production resources and their evolution, especially in terms of physical objects, structure, functionality, and materials.

ManufOnto. Alsafi and Vyatkin (2010) developed the Manufacturing Ontology to model modular manufacturing systems. The ontology (*ManufOnto* in Table 5.1) describes the machinery and operations in a semantic way, in order to facilitate flexibility and agility of manufacturing. The ontology is not available online for reuse, but nevertheless provides an example of the types of ontologies designed for manufacturing systems. The top (e.g., most generic) part discussed in (Alsafi and Vyatkin 2010) includes 29 concepts and 39 properties. The main idea reflected in the ontology is to separate and interrelate (i) the required operations, (ii) the physical machinery performing the required operations, and (iii) the control of the machinery. Based on the available description, we classify it as a lightweight, domain ontology, which covers aspects such as: resource types, process types, and products from the perspective of processes, functionalities, and structures.

EquipOnt. Lohse et al. (2006) describe the *Equipment Ontology, EquipOnt,* designed to support *Reconfigurable Assembly Systems* (RAS), i.e., systems that allow configuring and reconfiguring assembly systems based on changing requirements of the assembled product. The authors rely on the function-behaviour-structure paradigm in the design of the ontology, because they consider these three aspects of an equipment essential for their use case (i.e., deciding whether a piece of equipment can be reused in an assembly system). In their view, "functions express the capabilities of a module based on the intention of their designer". The behaviour of an equipment defines its reaction to changes in the environment. The physical model of the equipment represents its structure. The authors adopt a modular ontology design, with different modules being dedicated to covering the three aspects of function, behaviour and structure. The class *Equipment* is a key class in the ontology, and a superclass for different types of equipment such as *Device*, *Unit*, *Cell*, *Workstation*. The internal structure of elements is modelled with the *subComponents* relation (partOf) as well as with elements to describe connections between components. Each *Equipment* is also associated to a

[6]ISO15926 is available at: https://www.posccaesar.org/wiki/ISO15926inOWL.

Function and a *Behaviour* object. Similar to *Equipment*, functions and behaviour are also organized in specialization hierarchies.

The E*quipment Ontology* is not available online. Based on the available description, we classify *EquipOnt* as a lightweight, domain ontology, which covers aspects of production resource structures and capabilities.

eClassOWL[7] (Hepp 2006) is an OWL representation of the *eClass* standard,[8] which is a cross-industry catalogue describing the nature and features of various products and services. *eClass* spans more than 30,000 product and service types and includes over 5,000 product or services properties. The parts of *eClassOWL* related to engineering include: machines or devices (for special applications) (17,000,000); electrical engineering, automation, process control engineering (27,000,000); and automotive technology (28,000,000).

eClassOWL aims to provide semantic representation of the types and properties of products and services while preserving the original semantics of *eCl@ss* as much as possible. Since the *eCl@ss* catalogue was designed from a purchasing manager perspective, it does not have a proper subsumption hierarchy. This is the reason why *eClassOWL* creates two classes for each *eCl@ss* category: a "*generic*" class to represent the actual product or service, and a "*taxonomic*" class with a wider semantics (meaning that something can be related to this category). The intention is to use generic concepts for describing the actual products (e.g., a "machine"), and to use taxonomic concepts to describe something that is not a product, but is related to it (e.g., "machine maintenance").

eClassOWL is tightly connected with GoodRelations ontology, developed for the eCommerce domain and describing such product aspects as demand, prices and delivery options. The GoodRelations metamodel is used as a schema skeleton for *eClassOWL*. Therefore, *eClassOWL* inherits the following property types from the GoodRelations meta-model: a) *quantitative* properties (for product features with a numeric range); b) *qualitative* properties (for product features with predefined value instances); and c) *datatype* properties (used only for a few features with the datatypes string, date, time, datetime, or Boolean). The domain of all properties in *eClassOWL* is gr:ProductOrService.

Due to copyright and pricing issues, the current version of the *eClassOWL* ontology represents version 5.1.4 of the standard, and therefore lags behind significantly from the latest version of *eCl@ss* which is 9.0. Nevertheless, the process of extracting an OWL ontology from *eCl@ss* is well documented and available (Hepp 2006). Therefore, it can be potentially reused for extracting the ontology from newer versions of the standard. *eClassOWL* is a large, lightweight, domain ontology which covers various aspects of product types and characteristics.

The **Semantic Sensor Network (SSN) ontology**[9] was developed by W3C's Semantic Sensor Networks Incubator Group (SSN-XG) (Compton et al. 2012).

[7]eCl@ssOWL: http://www.ebusiness-unibw.org/ontologies/eclass/5.1.4/eclass_514en.zip.

[8]eCl@ss standard: http://www.eclass.de/eclasscontent/standard/overview.html.en.

[9]SSN Ontology: www.w3.org/2005/Incubator/ssn/ssnx/ssn.owl.

The main motivation for this group was to develop an ontology, which captures (a) the event based nature of sensors and sensor networks, for the cases in which the temporal and spatial relationships need to be taken into account; and (b) complex physical constraints (e.g., limited power availability, limited memory, variable data quality, and loose connectivity) that need to be considered for effective reasoning and inference.

The SSN ontology describes sensors, their observations, and related concepts in a general way. This means that no domain-specific information is given: domain semantics, specific units of measurement, etc., can be included, if necessary, via OWL imports, while instantiating the SSN ontology for a particular domain. The SSN ontology focuses on the description of the physical and processing structure of sensors. "Sensors" are not limited simply to physical sensing devices. Rather a sensor is anything that can estimate or calculate the value of some phenomenon. Thus, a device or computational process, or combination of those can be considered as a sensor. A "sensor" in the ontology links together *what it measures* (the domain phenomena), the *physical sensor* (the device) and *its functions and processing* (the models). Therefore, depending on the application at hand, the SSN ontology allows focusing on different perspectives: (a) a sensor perspective (what senses, how it senses, and what is sensed); (b) a data or observation perspective; (c) a system perspective; and (d) a feature and property perspective.

The SSN ontology is a generic, heavyweight ontology. It has a careful ontological design. For example, it is aligned with the *DOLCE Ultra Lite* upper ontology to facilitate its usage with other ontologies or linked data resources developed elsewhere. It is also based on the *Stimulus-Sensor-Observation* Ontology Design Pattern[10] introduced by (Janowicz and Compton 2010). The SSN ontology is best suited to describe process states and their observability, as well as resource states.

The **Automation Ontology** captures knowledge about industrial plants and their automation systems to support engineering of simulation models (*AutomOnto* in Table 5.1). It has been presented in (Novák et al. 2015). The automation ontology has a mechatronic nature and provides support for simulation model design and integration.

The automation ontology covers four domains and their mappings: a real plant domain, a variable domain, a parameter domain, and a simulation domain. The real plant domain represents the topology of a real system, i.e., it includes physical devices and their connections. Each real device can have assigned one or more parameters and can have input and output variables. Both parameters and variables are formalized in their respective domains. Parameters are considered as physical properties representing device features, such as size or length and other characteristics. They are constant values (i.e., independent on time). On the contrary, variables annotate process variables, inputs, and outputs. A tag can be assigned to

[10]The Stimulus-Sensor-Observation ODP: http://www.w3.org/2005/Incubator/ssn/XGR-ssn-20110628/#The_Stimulus-Sensor-Observation_Ontology_Design_Pattern.

each variable, which is a unique name for the variable shared among automation system tools and algorithms. Values of variables/tags are considered as time-series of physical quantities that are either measured by sensors in the real plant or exported by software parts of automation systems. Finally, the simulation domain is focused on supporting the engineering process of simulation models, which approximate the behavior of the real plant. This is the reason why the simulation domain and the real plant domain are mapped, which is useful for redesigning and reusing simulation artifacts. The simulation artifacts annotated with the automation ontology are simulation modules, which are parts of specific simulation models, or simulation blocks, which are the smallest artifacts supported in this formalization. In addition, the ontology includes expressiveness for representing how all these entities are hierarchically organized and how they correspond to each other. The simulation domain is mapped to parameters and variables/tags as well to support efficient interface description and configuration. Further details about this ontology from an application perspective can be found in Chap. 10.

5.5.1 The AutomationML Ontology

AutomationML (AML) is an open, XML-based data exchange format developed to support exchange of engineering data within the engineering process in production systems (Drath 2010). AML includes information about system topology, geometry, kinematics, and control behaviour. For more details about AML one can check the specification (IEC 62714). The AML representation is based on the four main concepts of CAEX (Computer Aided Engineering Exchange), a data format for storing hierarchical object information, such as the hierarchical architecture of a plant (IEC 62424 2008):

- The *RoleClassLibrary* allows specifying the vendor-independent requirements (i.e., the required properties) for *production-system-equipment objects*. A *role class* describes a physical or logical object as an abstraction of a concrete technical realization (e.g., a robot or a motor). In this way, a role class specifies the semantics of an object enabling automatic interpretation by a tool. Additionally, a role class allows defining attributes that are common for an object.
- The *SystemUnitClassLibrary* allows specifying the capabilities of solution equipment objects that can be matched with the requirements of objects defined with the role classes. A *system unit* class describes a physical or logical object including the concrete technical realization and internal architecture. For example, a System Unit Class *KR1000* that matches with the role class *robot* may describe attributes of the *KUKA KR 1000,* which has a payload of 1,000 kg. Thereby *system unit* classes form a multi-level hierarchy of vendor-specific objects that can be instantiated within the *InstanceHierachy*.
- The *InterfaceClassLibrary* defines a complete set of interfaces required to describe a plant model. An *interface class* can be used to define two types of

relations. First, it can define relations between the objects of a plant topology (these include all kind of relations, e.g., of mechanical nature or signals and variables related to the PLC code). Second, an interface class can serve as a reference to information stored outside the CAEX file (e.g., a 3D description for a robot).

- The *InstanceHierarchy* contains the plant topology, comprising the definition of a specific equipment for an actual project—the instance data. Therefore, all project participants can refer to the instance hierarchy to define the context for their work tasks and results. The instance hierarchy contains all data including attributes, interfaces, role classes, relations, and references.
- Although facilitating data exchange is already an important improvement, there is still a lack of infrastructure for supporting advanced engineering activities across the disciplines and tools in AutomationML-based projects, e.g., for data linking, change propagation across connected datasets, data analysis and consistency checking. In order to address this need, we developed an AutomationML ontology. We provide a solution for moving from AutomationML XML files to an OWL representation, as a prerequisite for developing an ontology-based integration and cross-disciplinary analytics on the top of data represented in this format.

Ontology creation process. As a starting point for the ontology creation, we took the available *XML Schema Definitions* (XSDs) providing the AutomationML language definitions in machine-readable and structured form. The actual ontology-building process is divided into two main steps. In the first step, we performed an *automatic transformation* from XML Schema to OWL using Apache Jena[11] to obtain an initial ontology draft. Here our goal was to provide an ontology draft, that is equivalent to the available XSDs, and that will allow to build efficient tool support. In the second step, we enhanced the draft ontology with additional axioms that would reflect the domain knowledge, but which was not available in the XSD. We also optimized the draft ontology obtained in the first step to reflect more accurately the AutomationML specification, and to use more efficiently the graph-based nature of ontologies. In particular, we replaced the string values of some properties (storing the path expressions to AutomationML structural elements) with relations, which actually refer to those elements.

Using the automatic conversion of the XML schema document into OWL allowed us to rapidly bootstrap the ontology. The final ontology directly matches the original XSD structure, which makes the data transformation from the original AML format to OWL straight-forward. Figure 5.1 presents the core concepts of the AutomationML ontology[12] derived from the AutomationML XSD schema. The AutomationML ontology depicts the inheritance taxonomy of the main concepts as well as selected containment (part-of) relationships.

[11]Apache Jena: https://jena.apache.org.

[12]AutomationML Ontology: http://data.ifs.tuwien.ac.at/aml/ontology#.

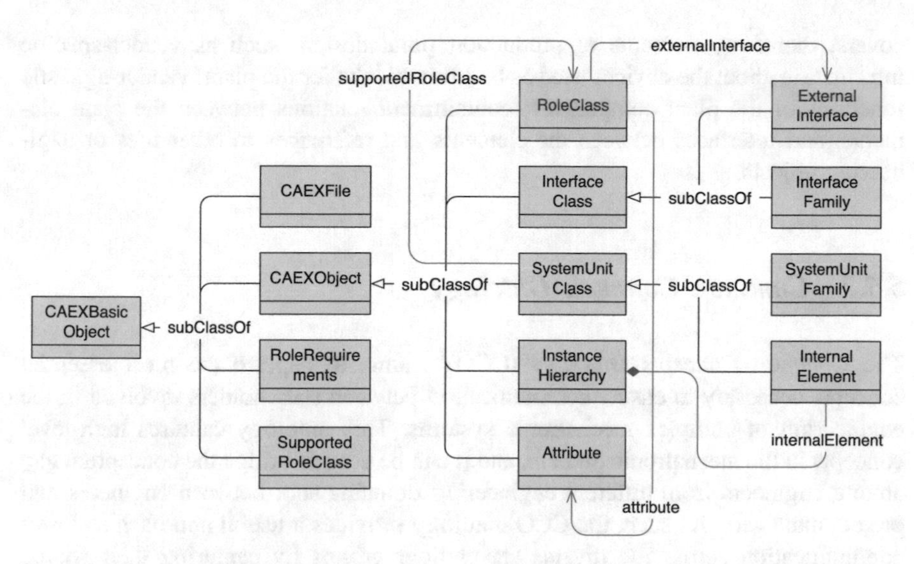

Fig. 5.1 AutomationML ontology: core concepts and their relations

The ontology concepts describe various aspects of production plant design. *RoleClass* represents production plant elements by describing vendor-agnostic functionality of the plant components, e.g., a robot or a motor. *SystemUnitClass* is used for storing vendor-specific plant elements, i.e., physical plant "building blocks" that can be ordered from a vendor and used for plant implementation. The instantiation of system units takes place in the *InstanceHierarchy*. *InstanceHierarchy* stores current plant system, putting the defined *InternalElements* in a hierarchical structure. *InternalElement* describes a specific element of the actual designed plant, including attributes, interfaces, relations, and references. *InterfaceClass* encodes the syntactical and semantic specification of user specific interfaces. Detailed description of the engineering elements can be given by defining various *Attributes*, e.g., material, weight, maximum power consumption. Attributes can be assigned to interfaces, roles, system units and internal elements. The AutomationML ontology also contains other auxiliary concepts (not shown in Fig. 5.1) covering additional information about the tool that was used to create a specific AutomationML file (such as the name of the exporting software tool, its vendor, version and release information), or a project identifier and title.

The AutomationML ontology supports the full representation of the content of any AutomationML file. As a result, Semantic Web tools and technologies may be applied on AutomationML data as well. For example, a semantic representation of AutomationML files is a pre-requisite for providing methods that semantically integrate those files, and allow advanced analytics on the integrated datasets. An example of such a tool is the AutomationML Analyzer (Sabou et al. 2016).

In summary, the **AutomationML ontology** is a lightweight domain ontology, capturing the structural elements of the AutomationML standard. The ontology covers, therefore, elements of production plant design, such as vendor-specific information about the devices used as building blocks for the plant; vendor-agnostic functions of the plant components; containment relations between the plant elements; and interfaces between the elements and references to other files or engineering objects.

5.5.2 Common Concepts Ontology

The *Common Concepts Ontology* (CCO)[13] aims to capture the most essential concepts necessary to ensure communication between stakeholders involved in the engineering of complex mechatronic systems. This ontology captures high-level concepts in the mechatronic domain, and it can be used to bridge the conceptual gap among engineers from different engineering domains and, between engineers and project managers. As such, the CCO ontology provides a useful approach to foster communication across the diverse stakeholder groups by capturing their shared understanding of the engineering domain.

Besides its value in supporting human communication, the CCO is an information model that acts as a key element of the *Engineering Knowledge Base* (EKB) described in Chap. 4. Namely, it is used as a *global ontology* to which concepts from *local ontologies* are mapped in settings where project-level data integration and access is needed. In such settings, IEAs built to explore the integrated data should answer a variety of competency questions, such as: questions about the project and various engineering roles; questions about engineering objects created by different engineering disciplines and their interconnection. Therefore, the focus of the ontology is primarily on describing the product of the engineering process, without considering in detail information about engineering processes and resources.

The methodology for engineering the CCO followed the guidelines of Noy and McGuinness (2001), where the knowledge elicitation step involved workshops with various domain experts. CCO was built in the CDL-Flex Laboratory and it was re-used and improved in several engineering projects for building diverse mechatronic objects (e.g., steel mills, hydro power plants). CCO represents a set of concepts that can be reused in other engineering projects of similar nature, especially focusing on the creation of production plants. This ontology might have limited use for different engineering projects than the ones that inspired its derivation. Naturally, projects that reuse this ontology might need to adapt it to meet their own needs and characteristics.

The CCO consists of two major parts. First, as depicted on the left hand side of Fig. 5.2, the ontology contains concepts that describe organizational level aspects. These concepts include the *Project*, the *Customer* for whom the project is

[13] CCO Ontology: http://data.ifs.tuwien.ac.at/engineering/cco.

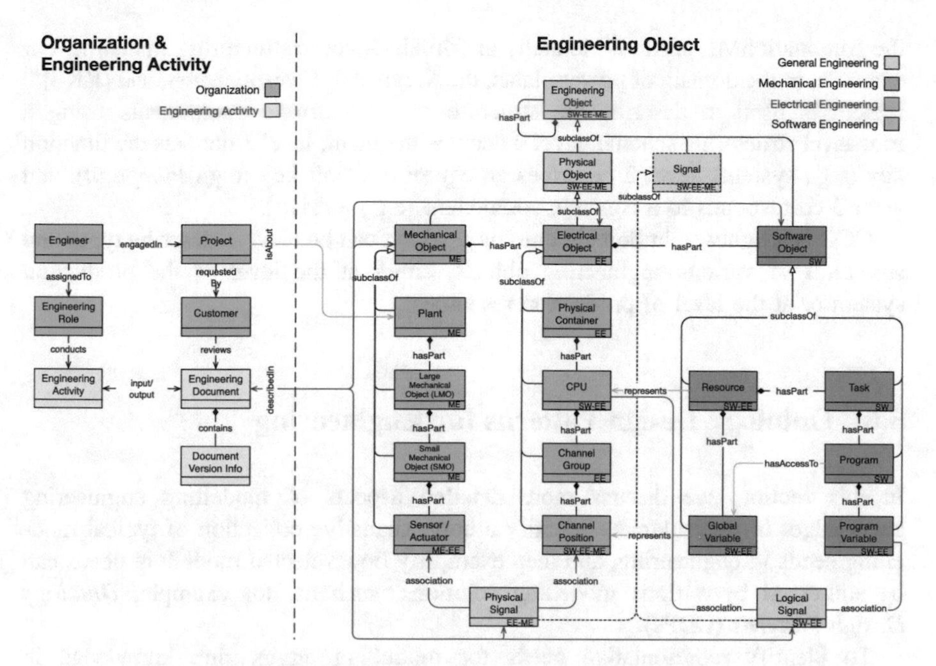

Fig. 5.2 The common concepts ontology

performed, as well as the *Engineers* (and *Engineering Roles*) necessary to realise the project. Engineers conduct *Engineering Activities,* which take as input and create as their output various *Engineering Documents* (e.g., signal lists, design documents). Engineering documents are versioned and reviewed by the customer, thus constituting an important exchange medium between the customer, who requested a project, and the engineering team executing that project.

Second, the CCO describes various *Engineering Objects* created during the engineering project (right-hand side of Fig. 5.2). The ontology identifies different types of engineering objects, such as *Software Objects*, *Mechatronic Objects*, and *Electrical Objects*. Each engineering project leads to the creation of a *Plant*, which is a type of *Mechatronic Object*. The ontology also clarifies the various parts of a mechatronic object, their internal structure and connections among them. To that end, the ontology captures different types of Mechatronic, Electrical and Software objects, and details their internal structure at a high level of abstraction. *Physical Signals* and *Logical Signals* represent the links between engineering objects created by different engineering disciplines and how these diverse components can command or exchange data with each other. In addition to these signals, detailed descriptions of the various mechatronic components are also available as engineering documents (e.g., PLC programs for software objects, or ECAD diagrams for the electrical wiring).

The internal structure of the components captured by the ontology emerged during several projects, and can also be represented using other approaches, such as the AutomationML instance hierarchy or domain-specific structuring standards. For example, in the domain of power plants, the *Kraftwerk-Kennzeichensystem* (KKS)[14] system is used to describe the structure of mechatronic components using a four-level structuring scheme: level 0 denotes the plant, level 1 denotes the function key (e.g., system), level 2 describes an equipment unit key (e.g., *Pumpunit*), and level 3 corresponds to a concrete component (e.g., a drive).

CCO is a lightweight domain ontology, which can be used to describe types and structures of various engineering objects, either at the level of the production system or at the level of production resources.

5.6 Ontology Design Patterns for Engineering

In this section, we discuss more detailed aspects of modelling engineering knowledge. In particular, we identify a non-exhaustive collection of typical modelling needs in engineering, and then exemplify how selected modelling needs can be addressed by various modelling solutions, such as, for example, *Ontology Design Patterns* (ODPs).

To identify representative needs for modelling engineering knowledge in mechatronic systems, we consider elements of *systems engineering* (Stevens et al. 1998), which have been adopted as the basis of mechatronic development process (VDI 2004). More concretely, we consider the elements of the SysML language (OMG 2006), which is a wide-spread modelling language for describing systems. As also discussed in Chap. 12, SysML distinguishes between a component's *definition* and its *usage*. A system's definition consists of the declaration of parts and their connections as well as the declaration of constraints on parts; e.g., *each car has exactly 1 engine*. Usage related aspects refer to the role that a component plays as part of another component. Accordingly, important modelling scenarios in system's engineering include:

- *Modelling Part-whole relations*. Legat et al. (2014) observe that *containment* hierarchies are a well-accepted and frequently occurring organizational paradigm from modelling part-whole relations in mechatronic engineering settings.
- *Modelling connections between components*. Legat et al. (2014) observe that *interface-based composition* describes the capabilities expected from an interface and can enable reasoning tasks about the correctness of a system's structure.
- *Modelling component roles*. Component roles refer to their functions and behaviour that they play in the system.

In what follows, we discuss modelling approaches for part-whole relations and connections between components.

[14]KKS System: https://de.wikipedia.org/wiki/Kraftwerk-Kennzeichensystem.

Modelling Part-Whole Relations. Expressing part-whole relations is an important aspect in several domains, most prominently in engineering and life sciences. Typical use cases involve the creation of inventory applications that, given an engineering object, could report all its parts and the subparts of those; or applications where the composition of an engineering object is explored one level at a time (i.e., only the direct components of the object are shown at any time).

Mereology (or mereotopology) refers to the study of part-whole relations and has been extensively investigated (Artale et al. 1996; Odell 1994; Winston et al. 1987). However, ontology modelling languages, such as RDF(S) and OWL, do not provide built-in primitives for modelling part-whole relations as they do, for example, for modelling subsumption hierarchies (e.g., `rdfs:subClassOf`). There are, however, several approaches to model different types of part-whole relations. We hereby provide an overview of some modelling options.

In Chap. 12, the authors report on the *Components* ontology (depicted in Fig. 12.2), which was designed for capturing part-whole relations in a use case from the car manufacturing industry. The ontology describes `Components` in general and distinguishes between `Composite Components` and `Atomic Components`. A non-transitive `hasPart` relation is introduced to represent *direct* subparts of a component. This modelling caters for use cases, where it is sufficient to retrieve the direct parts of a given component. Inventory-type applications, which should recursively return all the parts of subparts, require the use of a *transitive* relation `hasSubpart`. As discussed in Chap. 3, a transitive property satisfying equation Eq. 5.1 supports the computation of the transitive closure of all parts.

$$hasSubPart(A, B) \wedge hasSubPart(B, C) \vDash hasSubPart(A, C) \qquad (5.1)$$

The *Component* ontology also declares an inverse relation for `hasPart`, namely `isPartOf`. Note that inverse relations can significantly slow description logic reasoners, and therefore they should be used with care (Rector and Welty 2005).

The ontology engineering community has identified generic modelling approaches for common modelling problems, such as part-whole relations. The ontologydesignpatterns.org is a community-curated portal,[15] which contains *Ontology Design Patterns* (ODP) for *meronomy* and other common situations. This portal recommends three ODPs for modelling part-whole relations:

- The PartOf ODP[16] allows modelling of part-whole relations in a transitive fashion. To that end, it introduces the `hasPart` transitive relation between two `Entities`, as well as its inverse, `isPartOf` (also transitive).
- The Componency ODP[17] is a specialization of the *PartOf* ODP, and offers a solution for modelling part-whole relations in such a way that a distinction can be made between direct and non-direct (i.e., transitively-assessed) parts of an

[15] ODP community portal: http://ontologydesignpatterns.org/.

[16] PartOf ODP: http://ontologydesignpatterns.org/wiki/Submissions:PartOf.

[17] Componency ODP: http://ontologydesignpatterns.org/wiki/Submissions:Componency.

Object. The difference from the *PartOf* ODP is the introduction of the non-transitive hasComponent relation as a subproperty of hasPart. Note that sub-properties do not inherit the constraints of their super-properties, and therefore hasComponent is non-transitive. Furthermore, isComponentOf is declared as the inverse of isPartOf. Similar to the modelling described in Chap. 12, this ODP caters for deducing both direct and indirect parts of an object. Since isComponentOf is a sub-Property of part of, it is sufficient to assert it between relevant Object instances, as the part of relation will be automatically deduced (by virtue of the semantics of subPropertyOf construct). Interested readers can further investigate an example of how this pattern is used to model components of a car.[18] This pattern corresponds to the basic modelling pattern[19] recommended by Rector and Welty (2005).
- The TimeIndexedPartOf ODP[20] caters for situations, which require modelling of a situation in which an object is part of another for a specified time interval. The pattern relies on *n-ary* relationships to establish time-based part-whole relations between a part, a whole, and a time interval.

Besides the patterns described above, other modelling approaches have also been put forward, but their discussion exceeds the scope of this chapter. For example, Rector and Welty (2005) provide five different modelling patterns, while the modelling of ISO 15926 uses property chains for modelling part-whole relations.[21]

While several modelling solutions are proposed, it is also important to avoid confusing part-whole relations with other relations. Rector and Welty (2005) mention typical confusions with relations such as containment, membership, connections, constituents and subClassOf. For example, *constituency* refers to a relation without a clear part of relationship (e.g., different types of wood constitute a table) and special ODPs are offered for modelling constituency.[22]

Modelling Connections. As discussed in Chap. 12, the behaviour of a system is determined by interactions between its parts, and such interactions are abstracted to connections representing flows of energy, matter, or signals between components. Therefore, an important aspect in defining a system is declaring connections among its components. Chapter 12 presents an approach for modelling connections as part of the *Connection* ontology depicted in Fig. 12.3. This ontology provides a pattern for modelling connections at a high-level of abstraction for any entity that can have connections (which is conceptualized as TopologicalIndividual). For example, systems and individual system components can be considered as types of TopologicalIndividuals, when applying this pattern to describe connections between them. A Connection is established between two TopologicalIndividuals. The concept Connection represents an *n-ary* relationship

[18] Example use of componency ODP: http://mowl-power.cs.man.ac.uk/2011/07/sssw/lp.html.
[19] https://www.w3.org/2001/sw/BestPractices/OEP/SimplePartWhole/part.owl.
[20] TimeIndexedPartOf: http://ontologydesignpatterns.org/wiki/Submissions:TimeIndexedPartOf.
[21] Part-whole modelling in ISO15926: https://www.posccaesar.org/wiki/ISO15926inOWLPart2.
[22] Constituency ODP: http://ontologydesignpatterns.org/wiki/Submissions:Constituency.

between two components: it points to the two components that are involved in the connection, it specifies the direction of the connection and also specifies the component level `Connectors` that are involved (i.e., a connector corresponds to the notion of a port in UML or SysML).

While in this section the main focus was on how to correctly model engineering knowledge in ontologies, an important next step is the population of engineering ontologies with concrete instances (c.f. guideline Step 7 of Noy and McGuinness 2001). This process requires extracting data from engineering artefacts and converting it into Semantic Web formats, i.e., RDF(S) and OWL. Section 5.7 describes how semantic data may be acquired from engineering artefacts stored in legacy data formats such as spreadsheets, XML files and databases.

5.7 Acquisition of Semantic Knowledge from Engineering Artefacts

Section 5.2 describes in detail different strategies for ontology engineering. In short, two main phases can be distinguished in the ontology development process: (1) constructing the ontology skeleton (i.e., classes and properties) that can be achieved either manually or by using *ontology learning techniques* (Maedche 2012); and (2) creating ontology instances, a process known as *ontology population* (Petasis et al. 2011).

Ontology learning and population for engineering applications often require extracting data from legacy systems and existing proprietary data. Considering the scale of data in such systems, the (automated) tool-support for these processes is of vital importance, as performing them manually is time-consuming and error-prone. Engineering data comes in many formats, most commonly as (semi-) structured data, i.e., spreadsheets, XML, and databases (Villazón-Terrazas et al. 2010).

Spreadsheets as data representation format are often used to share, store, and exchange data in engineering environments despite of a set of important drawbacks, such as: the implicit data schema hampers automatic and effective processing; high-level of freedom and, therefore, high variability in data representation, which often does not adhere to best practices of data representation. These weaknesses are balanced by the positive characteristics of spreadsheets. Indeed, from a user point of view, spreadsheets are easy to understand, they do not require sophisticated skills to create and work with; and have adequate representational power and expressiveness for many common tasks.

XML facilitates the encoding of documents readable for both humans and machines. XML stores data in plain text, wrapping the information in tags. XML supports nested elements, which allows easy capturing of hierarchical structures omnipresent in engineering. The fact that XML does not restrict the use to predefined tags has both advantages and disadvantages. On the positive side, it leads to high flexibility in defining XML document structures according to the user or corporate preferences. However, XML (and its extensions) can become overly verbose and complex. An additional drawback is that XML does not define the semantics of data, but rather its structure, leaving space for ambiguity in data interpretation (Bray et al. 2008).

A large amount of engineering data (similar to other domains) is stored *in relational databases* (RDBs) thanks to their maturity of technology and tools, enabling support for scalability, optimized query execution, efficient storage and security (Sahoo et al. 2009). RDBs represent data in structured ways and, if modeled according with the best practices potentially incorporate significant share of domain knowledge, especially in a large company, where RDBs are typically maintained over a long period. This makes RDBs a valuable source for data extraction as not only the stored data, but also the schemas, defined queries and procedures can be used (with the support of domain experts) for ontology learning and population (Spanos et al. 2012).

Besides the three main data formats described above, a variety of other company- or discipline-specific proprietary formats are used in multi-disciplinary engineering settings. In this section we will focus on acquiring semantic knowledge from spreadsheets, XML-based documents, and databases.

The Semantic Web community and tool vendors have developed numerous tools and mapping techniques to enable (semi-)automated data transformation from legacy data sources into Semantic Web formats. For instance, the following tools can be used for extracting data *from spreadsheets* and converting that data into Semantic Web formats: *Open Refine* (former *Google Refine*),[23] *Anzo* for Excel[24] from Cambridge Semantics, RDF123 (Han et al. 2006), *XLWrap* (Langegger and Wöß 2009), *MappingMaster* plug-in for *Protégé* (O'Connor et al. 2010) and the Cellfie plugin,[25] and to some extent *Populous* (Jupp et al. 2012). For *converting from XML* files one can use Protégé's *XMLTab,*[26] *Rhizomik ReDeFer*[27] or the *xSPARQL* language (Bischof et al. 2012). Data *transformation from an RDB* into OWL and RDF formats is supported by e.g., *Relational.OWL* (De Laborda and Conrad 2005), the *DataMaster* plug-in for Protégé (Nyulas et al. 2007), and D2RQ (Bizer and Cyganiak 2006). Also, much work on this topic is covered within the W3C RDB2RDF Working Group[28] which focuses on the development of R2RML,[29] a language to define customized mappings from relational databases to RDF datasets. Some tools support generating semantic data *from various types of data sources*, e.g., the RML language (Dimou et al. 2014) allows specifying

[23]Open Refine: http://openrefine.org.

[24]Anzo for Excel: http://www.cambridgesemantics.com/solutions/spreadsheet-integration.

[25]Cellfie: https://github.com/protegeproject/cellfie-plugin/wiki.

[26]Protégé XML Tab: http://protegewiki.stanford.edu/wiki/XML_Tab.

[27]Rhizomik ReDeFer: http://rhizomik.net/html/redefer/.

[28]W3C's RDB2RDF Working Group: https://www.w3.org/2001/sw/rdb2rdf/.

[29]R2RML: https://www.w3.org/TR/r2rml/.

mappings from CSV and XML to RDF, and TopBraid Composer[30] from TopQuadrant can manage spreadsheets, XML and databases.

Another approach to manage the legacy engineering data is *ontology-based data access (OBDA)*. According to OBDA, an ontology is used as a mediator to access local data sources. Here data remains stored in the original databases and the ontology defines a global schema that provides a common vocabulary for query formulation, thus separating user from the details of the actual structure of the local data sources. The OBDA-based system rewrites user queries (formulated in terms of global schema) into queries built in terms of local data sources, and then delegates the query execution to data sources (Rodriguez-Muro et al. 2008; Civili et al. 2013). This approach is especially intended for applications that rely on large amounts of data. An example of tool supporting the ODBA is Ontop (Rodriguez-Muro et al. 2013).

As the available tools for data extraction and transformation from legacy data formats vary in many aspects (e.g., input and output formats supported, license and price, or a required level of user expertise) it can be difficult to select the best fitting tool for a specific usage context. Therefore, practitioners need to be supported to select the most suitable tool for their context.

To address the need for supporting practitioners in choosing the most appropriate data transformation tool for their context, Kovalenko et al. (2013) developed a *tool selection framework* as a means to facilitate choosing appropriate tools for ontology population from spreadsheet data. The framework has been applied to a set of transformation tools from spreadsheets. The developed framework considers nine criteria: *general information* (maturity, license and type of tool-plug-in or stand-alone tool); *usability* (availability of GUI and required user knowledge to start working with a tool); supported *input/output formats*; *mapping definition* aspects (i.e., how mappings are represented internally and to the user; and how they are stored); *expressiveness* (what complexity of data can be managed with a tool); *multi-user support*; *required additional software* (e.g., for plug-ins); and *additional features* (any other functionality that is provided by a tool). The proposed framework is used to classify the existing tools, thus, providing a focused summary on selection-critical aspects of tools for non-Semantic Web experts. Depending on a project and available project resources (e.g., budget or availability of knowledge engineer), various weights will be assigned for different criteria. The selection framework is adaptable based on specific engineering project requirements and needs. Similar tool selections frameworks should be developed to facilitate the selection of ontology learning and population tools from XML documents and relational databases for engineering practitioners.

[30]TopBraid Composer: http://www.topquadrant.com/tools/modeling-topbraid-composer-standard-edition/.

5.8 Summary and Future Work

This chapter aimed to provide introductory material on major issues related to the semantic modelling and acquisition of engineering knowledge that potential adopters of SWTs should consider. Concretely, the chapter introduced and discussed the following key topics:

- *Ontology engineering methodologies* prescribe concrete steps for creating and maintaining ontologies (Sect. 5.2). A wealth of methodologies exists focusing on different settings and contexts, starting from generic ontology creation guidelines suitable for beginners (the guidelines of Noy and McGuinness), to methodologies that cover a broader view of ontology engineering including their use in applications and organizations (On-to-Knowledge, METHONTOLOGY), scenario-based methodologies (NeOn Methodology) as well as methodologies suitable in collaborative ontology engineering settings (DILLIGENT).
- *Ontology evaluation* is an important issue that needs to be considered when building, reusing or modularizing ontologies (Sect. 5.3). There is an abundance of techniques and metrics to perform ontology evaluation, but their choice depends on the goal of the evaluation, which, most often focuses on one of the following issues: domain coverage, quality of modelling, suitability for an application or task and adoption by the community.
- *Classification of engineering ontologies* is currently an open research topic, with the exception of an initial classification scheme, which was derived based on a review of the content of existing ontologies by Legat et al. (2014). In Sect. 5.4, we report on aligning this classification scheme with the *PPR* view in order to create a classification scheme that is meaningful to ontology experts and industry practitioners alike.
- *Examples of engineering ontologies* in Sect. 5.5 give an insight into the variety of the existing ontologies and demonstrate the usefulness of the previously proposed scheme for ontology classification.
- *Ontology modelling with ontology patterns described in* Sect. 5.6, approaches the issue of semantic modelling in more depth. It provides examples of recurring, engineering-specific modelling needs, and shows how these can be addressed, for example, by modelling best practices, such as ontology design patterns.
- *Ontology learning and population from legacy data formats* are important tasks as they facilitate the transition from legacy information systems to Semantic Web-based solutions. As discussed in Sect. 5.7, engineering data is most often stored in databases, XML files or spreadsheets. Several tools are available for transforming data from these formats into ontologies and ontology instances. A major issue for practitioners is the selection of the most appropriate tool for their context. This issue is alleviated through tool selection frameworks, such as the one developed for evaluating and selecting tools for translating spreadsheet data.

The following topics emerged as interesting for future investigations. There is a clear need for supporting both industry adopters and Semantic Web experts with tools for identifying existing ontologies. A prerequisite is the availability of ontology classification schemes, which can be easily understood by both stakeholder groups, as well as the availability of surveys that would provide a comprehensive view of engineering ontologies. We expect that ontology reuse would be highly facilitated, if these elements were in place. For supporting the actual modelling of ontologies ODPs are useful. However, these are currently presented in a domain-agnostic manner, which hampers their adoption. Future work could therefore also investigate how to bring ODPs closer to creators of engineering ontologies. This step could involve, for example, a catalogue of frequently emerging modelling needs and guidelines of solving these with ODPs adapted to the engineering domain. Finally, practitioners would highly benefit from the availability of tool selection frameworks that support them in evaluating and selecting the most suitable tools for their context of data transformation from legacy data sources.

Acknowledgments This work was supported by the Christian Doppler Forschungsgesellschaft, the Federal Ministry of Economy, Family and Youth, and the National Foundation for Research, Technology and Development in Austria.

References

Alani, H., Brewster, C., Shadbolt, N.: Ranking ontologies with AKTiveRank. In: 5th International Semantic Web Conference (ISWC), Athens, GA, USA, pp. 1–15 (2006)

Alsafi, Y., Vyatkin, V.: Ontology-based reconfiguration agent for intelligent mechatronic systems in flexible manufacturing. J. Robot. Comput. Integr. Manuf. **26**(4), 381–391 (2010)

d'Aquin, M., Noy, N.F.: Where to publish and find ontologies? A survey of ontology libraries. J. Web Semant. **11**, 96–111 (2012)

Artale, A., Franconi, E., Guarino, N., Pazzi, L.: Part-whole relations in object-centered systems: an overview. Data Knowl. Eng. **20**(3), 347–383 (1996)

Bischof, S., Decker, S., Krennwallner, T., Lopes, N., Polleres, A.: Mapping between RDF and XML with XSPARQL. J. Data Semant. **1**(3), 147–185 (2012)

Bizer, C., Cyganiak, R.: D2R server-publishing relational databases on the semantic web. In: Poster at the 5th International Semantic Web Conference, pp. 294–309 (2006)

Bray, T., Paoli, J., Sperberg-McQueen, C.M., Maler, E., Yergeau, F.: Extensible markup language (XML) 1.0 (2008)

Brewster, C., Alani, H., Dasmahapatra, S., Wilks, Y.: Data driven ontology evaluation. In: 4th International Conference on Language Resources and Evaluation (LREC), Lisbon, Portugal, pp. 164–169 (2004)

Burton-Jones, A., Storey, V.C., Sugumaran, V., Ahluwalia, P.: A semiotic metrics suite for assessing the quality of ontologies. Data Knowl. Eng. 84–102 (2005)

Blázquez, M., Fernández-López, M., García-Pinar, J.M., Gómez-Pérez, A.: Building ontologies at the knowledge level using the ontology design environment. In: Gaines, B.R., Musen, M.A. (eds.) 11th International Workshop on Knowledge Acquisition, Modeling and Management (KAW), Banff, Canada, SHARE4:1–15 (1998)

Cantador, I., Fernandez, M., Castells, P.: Improving ontology recommendation and reuse in WebCORE by collaborative assessments. In: Workshop on Social and Collaborative Construction of Structured Knowledge, 16th International World Wide Web Conference (WWW) (2007)

Civili, C., Console, M., De Giacomo, G., Lembo, D., Lenzerini, M., Lepore, L., Mancini, R., et al.: MASTRO STUDIO: managing ontology-based data access applications. Proc. VLDB Endow. **6**(12), 1314–1317 (2013)

Compton, M., Barnaghi, P., Bermudez, L., Garcia-Castro, R., Corcho, O., Cox, S., Graybeal, J., et al.: The SSN ontology of the W3C semantic sensor network incubator group. J. Web Semant. **17**, 25–32 (2012)

Corcho, O., Fernández-López, M., Gómez-Pérez, A.: Methodologies, tools and languages for building ontologies: Where is their meeting point? Data Knowl. Eng. **46**(1), 41–64 (2003)

De Laborda, C.P., Conrad, S.: Relational.OWL: a data and schema representation format based on OWL. In: Proceedings of the 2nd Asia-Pacific Conference on Conceptual Modelling, vol. 43, pp. 89–96. Australian Computer Society (2005)

Dimou, A., Vander Sande, M., Colpaert, P., Verborgh, R., Mannens, E., Van de Walle, R.: RML: a generic language for integrated RDF mappings of heterogeneous data. In: Proceedings of the 7th Workshop on Linked Data on the Web (LDOW) (2014)

Drath, R. (ed.): Datenaustausch in der Anlagenplanung mit AutomationML: Integration von CAEX, PLCopen XML und COLLADA. Springer DE (2010)

Fernandez, M., Cantador, I., Castells, P.: CORE: a tool for collaborative ontology reuse and evaluation. In: 4th International Workshop on Evaluation of Ontologies for the Web at the 15th International World Wide Web Conference (WWW 2006), Edinburgh, Scotland (2006)

Fernandez, M., Overbeeke, C., Sabou, M., Motta, E.: What makes a good ontology? A case-study in fine-grained knowledge reuse. In: 4th Asian Semantic Web Conference (ASWC), Shanghai, China, pp. 61–75 (2009)

Fernández-López, M., Gómez-Pérez, A., Juristo, N.: METHONTOLOGY: from ontological art towards ontological engineering. In: Spring Symposium on Ontological Engineering of AAAI, Stanford University, California, pp. 33–40 (1997)

Gangemi, A., Presutti, V.: Ontology design patterns. In: Staab, S. et al. (eds.) Handbook of Ontologies, 2nd edn., pp. 221–244. Springer (2009)

Gomez-Perez, A., Corcho, O., Fernandez-Lopez, M.: Ontological Engineering: With Examples from the Areas of Knowledge Management, 404 p. Springer (2004)

Guarino, N., Welty, C.: An overview of OntoClean. In: Handbook on Ontologies, pp. 151–172. Springer, Berlin (2004)

Han, L., Finin, T., Parr, C., Sachs, J., Anupam, J.: RDF123: a mechanism to transform spreadsheets to RDF. In: Proceedings of the Twenty-First National Conference on Artificial Intelligence (AAAI). AAAI Press (2006)

Hepp, M.: Products and services ontologies: a methodology for deriving OWL ontologies from industrial categorization standards. Int. J. Semant. Web Inf. Syst. (IJSWIS) **2**(1), 72–99 (2006)

IEC 62424: Representation of process control engineering—Requests in P&I diagrams and data exchange between P&ID tools and PCE-CAE tools (2008)

IEC 62714 (all parts): Engineering data exchange format for use in industrial systems engineering—Automation Markup Language

Industrial automation systems and integration—Integration of life-cycle data for process plants including oil and gas production facilities. http://www.iso.org/

Janowicz, K., Compton, M.: The stimulus-sensor-observation ontology design pattern and its integration into the semantic sensor network ontology. In: Taylor, K., Ayyagari, A., Roure, D. (eds.) The 3rd International Workshop on Semantic Sensor Networks (SSN10) at the 9th International Semantic Web Conference (ISWC) (2010)

Jupp, S., Horridge, M., Iannone, L., Klein, J., Owen, S., Schanstra, J., Wolstencroft, K., Stevens, R.: Populous: a tool for building OWL ontologies from templates. BMC Bioinform. **13**(1) (2012)

Kovalenko, O., Serral, E., Biffl, S.: Towards evaluation and comparison of tools for ontology population from spreadsheet data. In: Proceedings of the 9th International Conference on Semantic Systems, pp. 57–64. ACM (2013)

Langegger, A., Wöß, W.: XLWrap—Querying and Integrating Arbitrary Spreadsheets with SPARQL. Springer, Berlin (2009)

Legat, C., Seitz, C., Lamparter, S., Feldmann, S.: Semantics to the shop floor: towards ontology modularization and reuse in the automation domain. In: 19th IFAC World Congress (2014)

Lenat, D.B., Guha, R.V.: Building Large Knowledge-Based Systems: Representation and Inference in the CycProject. Addison-Wesley, Boston (1990)

Lohse, N., Hirani, H., Ratchev, S.: Equipment ontology for modular reconfigurable assembly systems. Int. J. Flex. Manuf. Syst. **17**(4), 301–314 (2006)

Maedche, M., Staab, S.: Measuring similarity between ontologies. In: 13th International Conference on Knowledge Engineering and Knowledge Management (EKAW), pp. 251–263 (2002)

Maedche, A.: Ontology Learning for the Semantic Web, vol. 665. Springer Science & Business Media (2012)

Morbach, J., Wiesner, A., Marquardt, W.: OntoCAPE—A (re)usable ontology for computer-aided process engineering. Comput. Chem. Eng. **33**(10), 1546–1556 (2009)

Novák, P., Serral, E., Mordinyi, R., Šindelář, R.: Integrating heterogeneous engineering knowledge and tools for efficient industrial simulation model support. Adv. Eng. Inform. **29**, 575–590 (2015)

Noy, N.F., McGuinness, D.L.: Ontology Development 101: A Guide to Creating Your First Ontology, Stanford University Knowledge Systems Laboratory Technical Report KSL-01-05 (2001)

Nyulas, C., O'Connor, M., Tu, S.: DataMaster—a plug-in for importing schemas and data from relational databases into Protege. In: Proceedings of the 10th International Protege Conference (2007)

O'Connor, M.J., Halaschek-Wiener, C., Musen, M.A.: Mapping Master: a flexible approach for mapping spreadsheets to OWL. In: The Semantic Web–ISWC, pp. 194–208. Springer, Berlin (2010)

Odell, J.J.: Six different kinds of composition. J. Object Oriented Program. **5**(8), 10–15 (1994)

Object Management Group (OMG): OMG Systems Modeling Language Specification. http://www.sysml.org/docs/specs/OMGSysML-FAS-06-05-04.pdf (2006)

Petasis, G., Karkaletsis, V., Paliouras, G., Krithara, A., Zavitsanos, E.: Ontology population and enrichment: state of the art. In: Knowledge-Driven Multimedia Information Extraction and Ontology Evolution, pp. 134–166. Springer (2011)

Pinto, H.S., Tempich, C., Staab, S.: DILIGENT: towards a fine-grained methodology for DIstributed, Loosely-controlled and evolvInG Engineering of oNTologies. In: Proceedings of the 16th European Conference on Artificial Intelligence (ECAI), pp. 393–397. IOS Press (2004)

Porzel, R., Malaka, R.: A task-based approach for ontology evaluation. In: Proceeding of the ECAI Workshop on Ontology Learning and Population (2004)

Poveda-Villalón, M., Gómez-Pérez, A., Suárez-Figueroa, M.C.: OOPS! (OntOlogy Pitfall Scanner!): an on-line tool for ontology evaluation. Int. J. Semant. Web Inf. Syst. (IJSWIS) **10**(2) (2014)

Poveda-Villalón, M.P.: Ontology Evaluation: a pitfall-based approach to ontology diagnosis. Ph.D. Thesis, UPM Madrid (2016)

Rector, A., Welty, C.: Simple part-whole relations in OWL Ontologies. W3C Editor's Draft 11 (2005)

Rodriguez-Muro, M., Lubyte, L., Calvanese, D.: Realizing ontology based data access: a plug-in for Protégé. In: IEEE 24th International Conference on Data Engineering Workshop (ICDEW), pp. 286–289. IEEE (2008)

Rodriguez-Muro, M., Kontchakov, R., Zakharyaschev, M.: Ontology-based data access: ontop of databases. In: The Semantic Web, ISWC, pp. 558–573. Springer, Berlin (2013)

Sabou, M., Fernandez, M.: Ontology (network) evaluation. In: Suarez-Figueroa, M.C., et al. (eds.) Ontology Engineering in a Networked World. Springer, Berlin (2012)

Sabou, M., Ekaputra, F.J., Kovalenko, O.: Supporting the engineering of cyber-physical production systems with the AutomationML analyzer. In: Proceedings of the CPPS Workshop, at the Cyber-Physical Systems Week. Vienna (2016)

Sahoo, S.S., Halb, W., Hellmann, S., Idehen, K., Thibodeau Jr, T., Auer, S., Sequeda, J., Ezzat, A.: A survey of current approaches for mapping of relational databases to RDF. W3C RDB2RDF Incubator Group Report (2009)

Schleipen, M., Drath, R.: Three-View-Concept for modeling process or manufacturing plants with AutomationML. In: IEEE Conference on Emerging Technologies & Factory Automation (ETFA) (2009)

Sicilia, M., Garcia-Barriocanal, E., Sanchez-Alonso, S., Rodriguez-Garcia, D.: Ontologies of engineering knowledge: general structure and the case of software engineering. Knowl. Eng. Rev. **24**(3), 309–326 (2009)

Spanos, D.E., Stavrou, P., Mitrou, N.: Bringing relational databases into the semantic web: a survey. Semant. Web J. **3**(2), 169–209 (2012)

Staab, S., Schnurr, H.P., Studer, R., Sure, Y.: Knowledge processes and ontologies. IEEE Intell. Syst. **16**(1), 26–34 (2001)

Stevens, R., Brook, P., Jackson, K., Arnold, S.: Systems Engineering: Coping with Complexity. Prentice Hall PTR (1998)

Suárez-Figueroa, M.C.: NeOn Methodology for Building Ontology Networks: Specification, Scheduling and Reuse. Dissertations in Artificial Intelligence, vol. 338. IOS Press (2012)

Suárez-Figueroa, M.C., Gómez-Pérez, A., Fernández-López, M.: The NeOn methodology framework: a scenario-based methodology for ontology development. Appl. Ontol. **10**(2), 107–145 (2015)

Verein Deutscher Ingenieure (VDI): Design methodology for mechatronic systems. VDI-Richtlinie 2206. Beuth Verlag, Berlin (2004)

Villazón-Terrazas, B., Suárez-Figueroa, M.C., Gómez-Pérez, A.: A pattern-based method for re-engineering non-ontological resources into ontologies. Int. J. Semant. Web Inf. Syst. (IJSWIS) **6**(4), 27–63 (2010)

Winston, M., Chaffin, R., Hermann, D.: A taxonomy of part-whole relations. Cogn. Sci. **11**(4), 417–444 (1987)

Chapter 6
Semantic Matching of Engineering Data Structures

Olga Kovalenko and Jérôme Euzenat

Abstract An important element of implementing a data integration solution in multi-disciplinary engineering settings, consists in identifying and defining relations between the different engineering data models and data sets that need to be integrated. The ontology matching field investigates methods and tools for discovering relations between semantic data sources and representing them. In this chapter, we look at ontology matching issues in the context of integrating engineering knowledge. We first discuss what types of relations typically occur between engineering objects in multi-disciplinary engineering environments taking a use case in the power plant engineering domain as a running example. We then overview available technologies for mappings definition between ontologies, focusing on those currently most widely used in practice and briefly discuss their capabilities for mapping representation and potential processing. Finally, we illustrate how mappings in the sample project in power plant engineering domain can be generated from the definitions in the Expressive and Declarative Ontology Alignment Language (EDOAL).

Keywords Ontology matching · Correspondence · Alignment · Mapping · Ontology integration · Data transformation · Complex correspondences · Ontology mapping languages · Procedural and declarative languages · EDOAL

6.1 Introduction

Ontology and data matching tasks are key steps for semantic data integration (Breslin et al. 2010) and therefore they often surface in real-world applications that have the need to exploit relations between the concepts and instances of two data sources.

O. Kovalenko (✉)
Institute of Software Technology and Interactive Systems, CDL-Flex,
Vienna University of Technology, Vienna, Austria
e-mail: olga.kovalenko@tuwien.ac.at

J. Euzenat
INRIA & Univ. Grenoble Alpes, Grenoble, France
e-mail: jerome.euzenat@inria.fr

S. Biffl and M. Sabou (eds.), *Semantic Web Technologies for Intelligent Engineering Applications*, DOI 10.1007/978-3-319-41490-4_6

Ontology alignments express these relations explicitly in order to be processed as mappings for various applications, e.g., for data transformation or query rewriting (Shvaiko and Euzenat 2013).

Although much work has been done on ontology schema matching and defining data-level mappings, it is not obvious, especially for non Semantic Web experts, which of the provided tools and technologies will be well-suited in a specific real-life application case and how to apply them (Vyatkin 2013). For engineering practitioners, it might be challenging to understand (a) which technologies can be used to represent the relations between their data models and data sets; and (b) how to use these for a specific application. On the other hand, Semantic Web researchers might lack insight on (1) what are the needs (in terms of mappings) of real-world applications in engineering; and (2) how well existing mapping technologies can support defining such mappings. In this chapter, we aim to address these gaps for the benefit of practitioners and Semantic Web researchers alike. Therefore, we formulated the following *research questions* for the chapter: (Q1) What kind of relations are expected to occur often between the engineering data models and data sets? (Q2) What are the available mapping languages developed by Semantic Web community and what are their characteristics? (Q3) How well the existing mapping languages can support defining and representing the relations identified for Q1?

In order to answer Q1 we synthesized our experiences from (a) implementing semantic integration in the real-life application scenario from our industry partner, a power plant systems integrator; and (b) literature analysis in the fields of ontology matching and schema matching. As a result, we derive a catalog of schema and data correspondences that are expected to arise often in ontology mediation applications (see Sect. 6.4 for details). We focus on complex correspondence types that are not obvious with respect to their representation, i.e., those that go beyond simple one-to-one mapping (like `owl:sameAs`) between the entities. These are already well-explored and supported by existing ontology matching solutions. For more information about existing matching techniques, systems, and tools please see a comprehensive review by (Otero-Cerdeira et al. 2015) or (Euzenat and Shvaiko 2013).

Our application scenario, a power plant engineering project, belongs to the automation system engineering domain where mappings between ontologies are used to support data integration. The intended application in this case is enabling data transformation from the local ontologies of the different engineering disciplines involved in a project into a common ontology in order to allow the project-level integration (see Sect. 6.3 for the detailed description). As in the presented use case ontology mapping is used for data transformation, we will use it as default application for the sake of coherence across the chapter. Therefore, when speaking about mappings in this chapter, it can be assumed that they are done for data transformation.

In Sect. 6.5 we overview available technologies for mappings definition, representation and processing and briefly discuss their main usages in current practice and potential strengths and limitations w.r.t. representing correspondences identified in Sect. 6.4 (therefore, addressing Q2 and Q3). Finally, Sect. 6.6 introduces the EDOAL language, gives the examples of how the identified correspondence types can be implemented in this language and shows how they can be used for data transformation. We chose EDOAL because, contrary to other languages, it combines the features of both declarative and procedural languages: (a) it has been designed for expressing correspondences (conforms to declarative languages); (b) but at the same time those correspondences can be interpreted to perform linking, i.e., primitive Silk scripts (Volz et al. 2009), SPARQL[1] queries, SPARQL+SPIN[2] can be generated to perform data transformation for instance.

Our main contributions therefore are the following:

1. A catalog of complex correspondence types with the examples of those in real-world application scenario in the automation systems engineering domain;
2. A succinct overview of existing mapping languages, their principal features and capabilities;
3. A detailed analysis of EDOAL capabilities to support identified correspondences types.

6.2 Ontology Matching: Background Information and Definitions

In this section we provide background information on ontology matching and give the definitions of the important terms, which will be used across the chapter. We will follow the terminology from the "Ontology Matching" book (Euzenat and Shvaiko 2013) and from (Scharffe et al. 2014):

Ontology matching is the process of finding relations between the entities of different ontologies, e.g., identifying the entities that represent the same (or similar) semantics in these ontologies.

Correspondence is the expression of a relation holding between entities (classes, properties, or individuals) of different ontologies. Correspondences express the essence of how the entities are related independently from any application or implementation details in a specific mapping language.

Alignment is a set of correspondences between two ontologies. The alignment is the output of the ontology matching process.

Mapping specifies the connection between ontology entities in enough details to process or execute them for a certain task. Execution in this case means using a specific engine/tool that can read the mapping, understand its semantics and run it according to the intended application. The output of the execution will vary depending on specific application. For instance, for data transformation it will be data set conforming to the target ontology; and for query rewriting the initial query (formulated in the source ontology vocabulary) will be transformed into the query formulated according to the target ontology vocabulary. A mapping, therefore, is a correspondence expressed in a specific mapping language with a certain exploitation in mind.

---

[1] https://www.w3.org/TR/sparql11-query/.

[2] http://www.w3.org/Submission/spin-overview/.

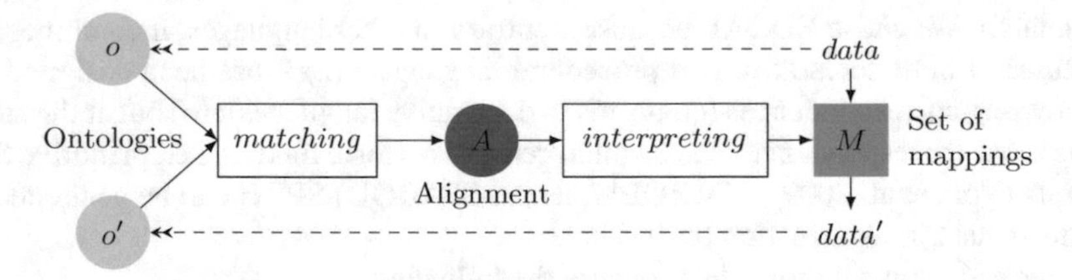

Fig. 6.1 Ontology matching for data transformation. Ontology matching generates an alignment A between ontology o and o'. This alignment can be interpreted as a set of mappings M, which can be used for transforming data expressed in ontology o into ontology o'

Both correspondences and mappings can be specified on the data (ontology instances) or schema (ontology classes and properties) level.

The main terms defined above as well as possible articulation between these terms (for the data transformation scenario) are illustrated in Fig. 6.1.

Three main dimensions may be distinguished in the ontology matching area (Noy 2004): (a) alignments discovery, i.e., ontology matching, (b) alignments representation, and (c) applying them as mappings.

The majority of works in the area is dedicated to *ontology matching*, which is a challenging and complex problem. Although many tools have been developed using different approaches, e.g., language-based, structure-based, or context-based, discovery processes tend to be highly iterative and cannot be fully automated in the majority of cases (Bernstein and Melnik 2007). Interested reader can find the detailed description of various alignments discovery approaches in (Euzenat and Shvaiko 2013).

Another important issue in the ontology matching area is the *application of mappings*, i.e., to be applied for a specific task alignments should be implemented in some mapping language; and then the obtained mappings can be processed for an application at hand. The precise type of application will for a large part determine the choice of a language to express mappings. Typically applications aim to reduce semantic heterogeneity. The often reported applications are ontology merging, data translation, and ontology integration.

Finally, the third dimension of the ontology matching area is dedicated to *defining and representing* alignments and mappings. Different languages may be used for that purpose with various characteristics (see Sect. 6.5 for more details). At the same time, there is a lack of guidelines (especially for non Semantic Web experts) on how to select a language for specific application and what are the important features of these languages that will influence the choice.

In this chapter we focus on the representation problem, in particular when this involves complex relations between entities. As we see in next sections, this is the case in the engineering domain.

6.3 Running Example: The Power Plant Engineering Project

Power plant engineering projects represent complex environments where participants from different disciplines (e.g., mechanical, electrical, and software engineering), have to collaborate to deliver high quality end products while satisfying tight timeframes (Mordinyi et al. 2011).

This application scenario is based on a case study implemented in a power plant engineering project of our industry partner, an automation system solution provider. Initially, different disciplines within the project applied isolated engineering processes and workflows: (a) isolated data sets stored in various proprietary data formats and discipline-specific relational databases; and (b) tools were loosely coupled with limited capabilities for cross-disciplinary data exchange or/and data analysis. The goal of the case study was to implement an ontology-based integration solution in order to allow cross-disciplinary data exchange and data analysis for advanced applications, such as change propagation and consistency checking across the discipline boundaries.

After analyzing the project requirements the hybrid ontology approach for integration according to (Wache et al. 2001) (see Fig. 6.2) has been chosen for the final system. According to this approach, a local ontology is built for each engineering discipline, gathering discipline-specific data models, knowledge, and constraints, and finally, data. Then an additional integrating layer is introduced—a common knowledge base—that contains only concepts, which are important on the project level. These concepts are called *common concepts* (Biffl et al. 2011), because they are typically relevant within at least two engineering disciplines. For more detailed description of common concepts please see Chap. 4 *"The Engineering Knowledge Base Approach."* The final system can be implemented following one of the two basic approaches: (a) the CC ontology defines the structure of the common knowledge base and data is stored in database(s); or (b) the CC ontology is populated with

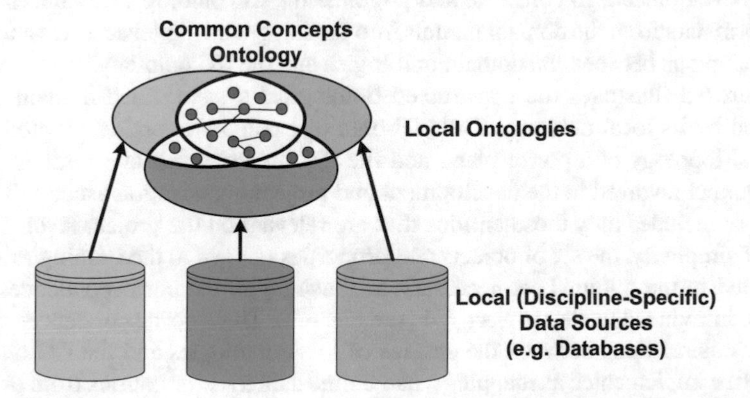

Fig. 6.2 Hybrid ontology integration (adapted from (Wache et al. 2001))

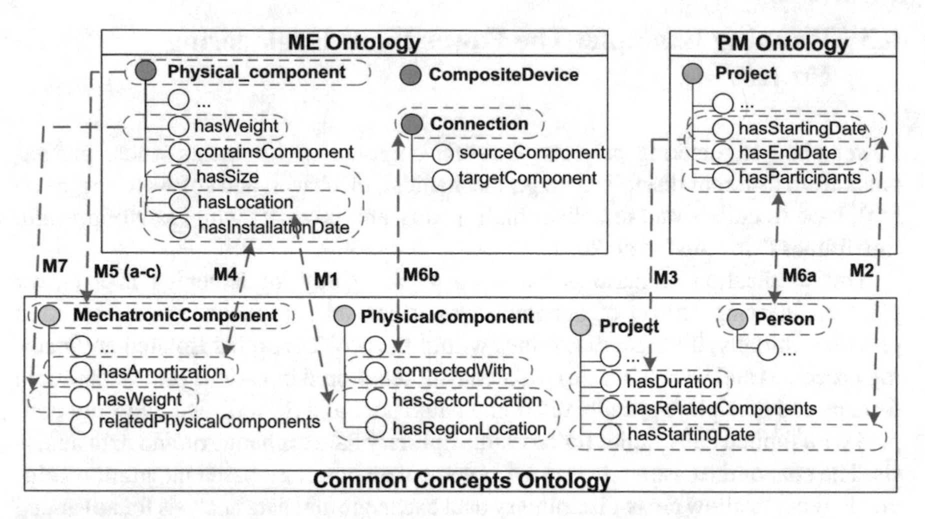

Fig. 6.3 Ontology-based integration in the power plant engineering project

instances and serves itself as a common knowledge base for a project. In both cases the advantage is that the CC ontology defines a common vocabulary on the top of the engineering project and provides a single access point for data analysis across the disciplines. Within the case study the second approach was chosen. In the application scenario, mappings are defined between the local discipline-specific ontologies and the CC ontology to serve as a basis for data transformation from the local level into the CC ontology.

We present the part of the ontological system developed for the case study as the running example in this chapter. In this example, the two domains are integrated: (a) *mechanical engineering* (ME), responsible for designing the physical structure of devices and connections between them, and (b) *project management* (PM), responsible for managing the information about past and current projects, and people involved in the development. To construct and populate the CC ontology, it is necessary to transform data from the domain models into the integrated model according to specified mappings between the domain ontologies and the CC ontology.

Figure 6.3 illustrates the constructed ontological system. Each domain is represented by its local ontology. The ME ontology comprises entities related to the physical topology of a power plant, and the PM ontology includes entities related to personnel involved in the development and project organization aspects. The CC ontology includes only those entities that are relevant on the project level. For the sake of simplicity, the set of objects and properties (shown in the running example) is limited to the minimal set necessary to illustrate all the correspondences types that are introduced further in Sect. 6.4 (see Fig. 6.3). These correspondences specify the various relations between the entities of local ontologies and the CC ontology and, when implemented as mappings, can define data transformations from the local storages to the integrated storage.

6.4 Representing Relations Between Engineering Objects

In this section, we overview what kind of correspondences and mappings between ontology entities may occur while capturing relations between the engineering data models and data (though the described ones are not strictly specific to the engineering domain and may arise in other domains as well).

In order to identify the presented correspondences we followed a bottom-up approach. First, we analyzed what types of relations occur often in the engineering data, summarizing our experiences with implementing semantic integration in the multi-disciplinary engineering projects of the industry partner. Second, we performed a literature analysis in the ontology mediation, ontology matching, and schema matching fields, in order to verify that identified correspondences are not specific to our scenarios, but are indeed widespread and recognized by researchers and practitioners from different domains and can be found in a wide range of application scenarios. Besides describing the correspondences in general, we provide concrete examples of those from the application scenario presented in Sect. 6.3. To better position the identified correspondences in the ontology matching field we link them to the ontology alignment design patterns[3] and the work of F. Scharffe on correspondence patterns (Scharffe 2009).

The presented correspondences can be expected to occur frequently in various ontology mediation tasks. At the same time, they require the use of an expressive formalism in order to be properly defined. We proceed with the detailed description of these correspondences in general and examples for those from the real-life application scenario.

Value Processing (M1–M4)

Often, the relation between values of two entities can be represented by some function that takes a value on one side as an input and returns a value on another side as an output, i.e., some processing is needed to map the entities. The complexity of this processing varies from simple string operations to sophisticated mathematical functions. This type of correspondences is considered in (Scharffe 2009) as the "Property value transformation" pattern, where the author distinguishes between the string-based, operations on numbers and unit conversion transformations. In the following, several types of such correspondences are described in detail.

String Processing (M1)

Description. As string values are widely used for data representation, the processing of string values is often needed while transforming data from one ontology into another. Expressing such correspondences requires using special functions on string values, e.g., "concat," "substring," or "regex." An example of this correspondence for the "concat" function can be found in (Scharffe 2009) under the name of "Class Attribute to Attribute Concatenation Pattern."

Example. In the ME ontology each physical component's location is defined via the `hasLocation` property, whose value is a string combining sector and region

[3]http://ontologydesignpatterns.org/wiki/Category:AlignmentOP.

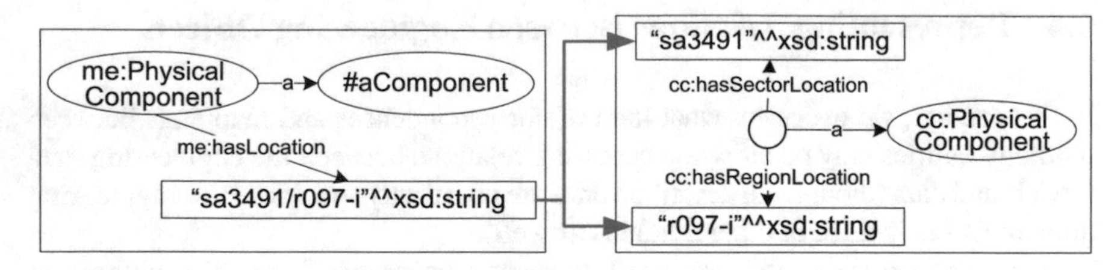

Fig. 6.4 M1: String processing correspondence example

(defines location within a specific sector) information for a specific component. In the CC ontology the location information of a physical component is explicitly divided into two separate properties. The correspondence specifies that the initial string must be split into two parts, which then will be used to construct values for the `hasSectorLocation` and the `hasRegionLocation` properties in the CC ontology (see Fig. 6.4).

Data Type Transformation (M2)
Description. It can happen that in a proprietary data model the data type of a certain property was not modeled in an optimal way, e.g., the date can be represented in the format of "string," instead of "Date" or "DateTime." This type of correspondence encodes, therefore, how the value of one type in source ontology can be transformed into value of a different type in a target ontology.

Generally, the data types can be compatible, partially compatible or incompatible (Legler and Naumann 2007). For instance, "integer" and "string" are compatible data types (although only uni-directionally); "date" and "string" are partially compatible (the compatibility will depend on a specific value); and "DateTime" and "integer" are incompatible data types. For the compatible data types there is also a possibility to specify correspondence in a more general way, i.e., specifying how these data types can be transformed into each other. For the example above it could be defining how any "Date" value should be transformed into a "string" value. In this case, the inverse mapping cannot be defined in a general way.

Example. All values of the `hasStartingDate` property in the PM ontology are strings in the following format "DD/MM/YYYY." But because the data type of the corresponding property in the CC ontology is "Date," a data type transformation correspondence takes place between these two properties (see Fig. 6.5).

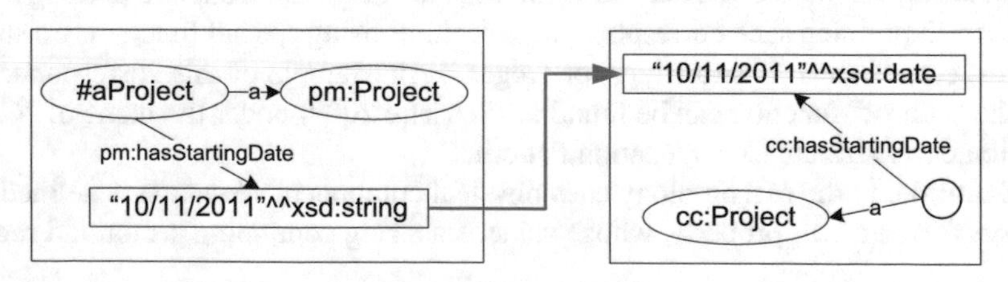

Fig. 6.5 M2: Transforming a string into xsd:date

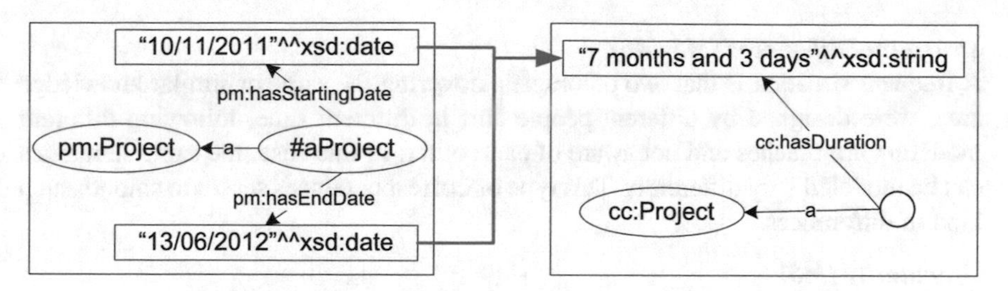

Fig. 6.6 M3: Computing the duration of a project

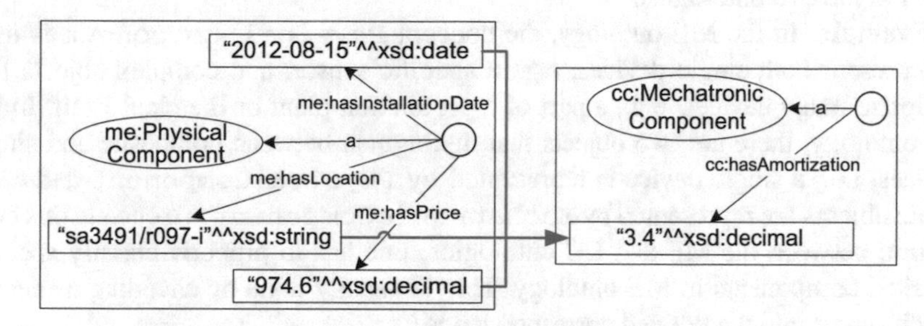

Fig. 6.7 M4: Computing the amortization of a component based on its properties

Math Functions (M3)
Description. In this case value processing involves some mathematical operations or is specified by a formula representing mathematical, physical, or other natural laws. This, for an instance, can be such simple mathematical operations as addition or multiplication, or more complex functions such as finding an integral or logarithm. However, relation capturing in this case is done by the means of the used mapping language.

Example. The value of the `hasDuration` property in the CC ontology is equal to substracting `hasStartingDate` from `hasEndingDate` in the PM ontology (see Fig. 6.6).

User-defined Functions (M4)
Description. For this type of correspondences, the relation is expressed by functions that are not supported by the used mapping language/technology, but must be additionally implemented. Therefore, it must be possible to call an external function, e.g., implemented in Java, that will generate a property value or an entity in a target ontology.

Example. The concept `MechatronicComponent` in the CC ontology captures information regarding complex composite components, which consist of many physical components and basically can represent a plant itself. The anticipated amortization value can be an important characteristic for such objects. The exact value will depend on the location, price and installation date of a specific mechatronic component (see Fig. 6.7).

Structural Differences (M5–M6)
A frequent situation is that two ontologies, covering the same or similar knowledge area, were designed by different people and in different time, following different modeling approaches and not aware of each other. In this case, the same semantics can be modeled very differently. This type of correspondences serves to smooth such kind of differences.

Granularity (M5)
Description. In this case, the same real-life object is modeled at a different level of detail in the two ontologies.

Example. In the ME ontology, the concept `Physical_component` is used to represent both single devices, e.g., a specific sensor, and complex objects that comprise many devices, e.g., a part of a production plant or the plant itself. In the CC ontology, there are two objects that distinguish between composite and single devices, i.e., a single device is represented by `PhysicalComponent` and composite objects are represented by `MechatronicComponent`. To encode this connection between the ME and CC ontologies, one has to properly classify specific physical components in ME ontology. This is usually done by encoding a specific conditioning into the defined correspondence.

For instance, for the presented example one can perform filtering of the mechatronic components based on property value, i.e., saying that those physical components, which weight more than a specific threshold, are mechatronic components (see Fig. 6.8).

Another option could be to filter based on property occurence, i.e., saying that mechatronic components are those physical components that contain other devices. To check that one can use the existense of `containsComponent` property for a specific physical component in the ME ontology (see Fig. 6.9).

One more example of filtering could be checking whether a physical component also belongs to a specific type, e.g., saying that mechatronic components are those physical components that are of type `compositeDevice` in the ME ontology (see Fig. 6.10).

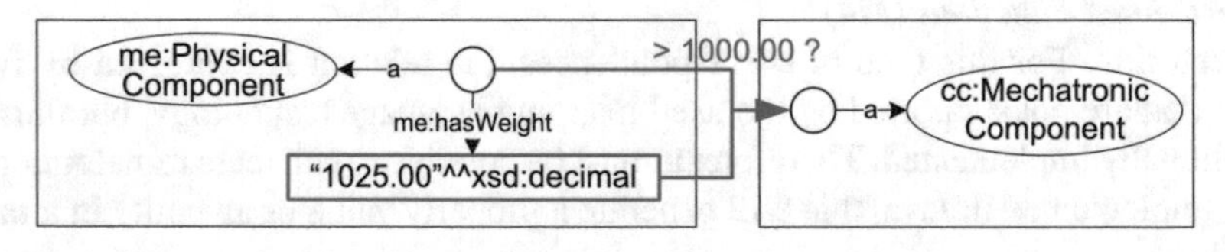

Fig. 6.8 M5a: Defining mechatronic components by property value

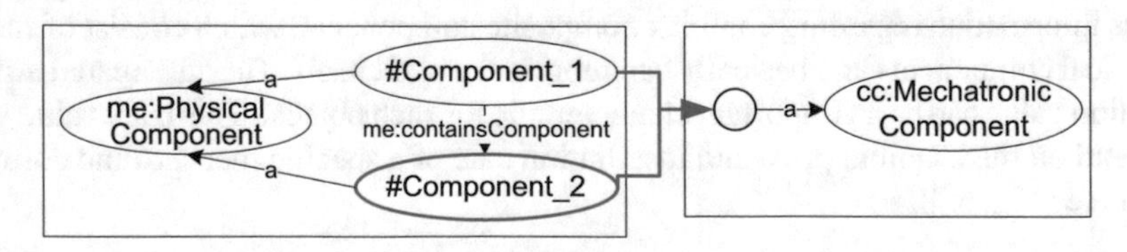

Fig. 6.9 M5b: Defining mechatronic component by property occurence

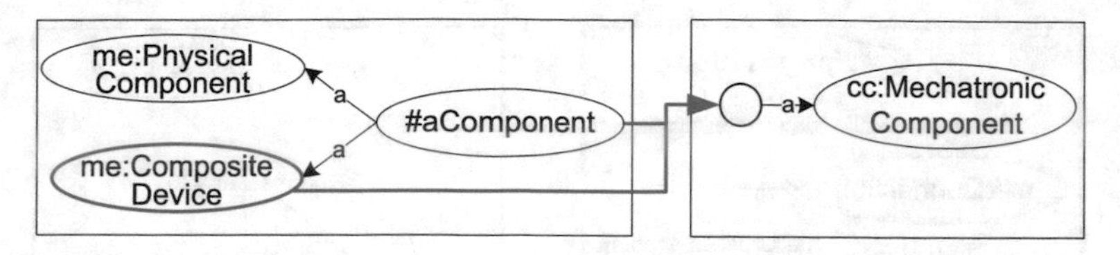

Fig. 6.10 M5c: Defining mechatronic component by instance type

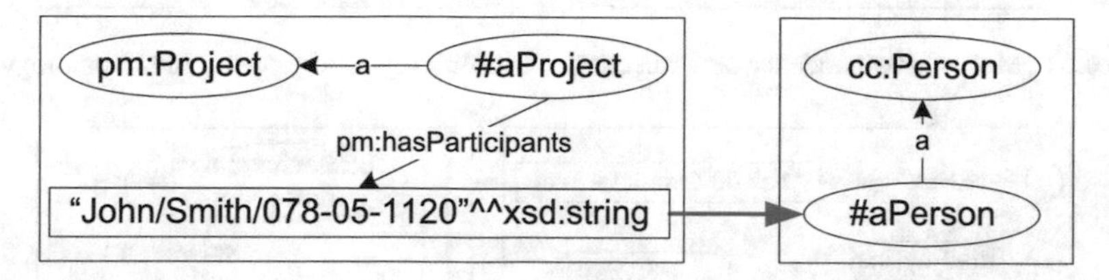

Fig. 6.11 M6a: Correspondence between the property value in ME ontology and instance in CC ontology

Similar types of correspondences and examples for them are reffered in (Scharffe 2009) as the "Class Correspondence by Path attribute Value," "Property Correspondence by Value," "Class by Attribute Occurence Correspondence' patterns. Also, similar patterns are described by the "Class_by_attribute_occurence," "Class_by_attribute_type," "Class_by_attribute_value," and "Class_by_path_attribute_value" ontology design paterns.[4]

Schematic Differences (M6)

Description. In this case, there are substantial differences in the way the same semantics is represented in the two ontologies.

Example1. Each employee in the PM ontology is represented by a string value of the `hasParticipants` property, while in the CC ontology the concept `Person` serves the same purpose. The correspondence captures this relation between a property value and a class instance (see Fig. 6.11).

Example2. A connection between physical devices in the ME ontology is represented by the `Connection` concept with the `sourceComponent` and `target Component` properties, while in the CC ontology the same semantics is expressed with the `connectedWith` property of the `PhysicalComponent` concept (see Fig. 6.12).

The correspondences with similar semantics and corresponding examples can be found in (Scharffe 2009) denoted as the "Class Relation Correspondence," "Property–Relation Correspondence," and "Class Instance Correspondence" and also within the ontology design patterns ("Class correspondence defined by relation domain").

---

[4]Mentioned design patterns can be found under http://ontologydesignpatterns.org/wiki/Submissions:#pattern_name.

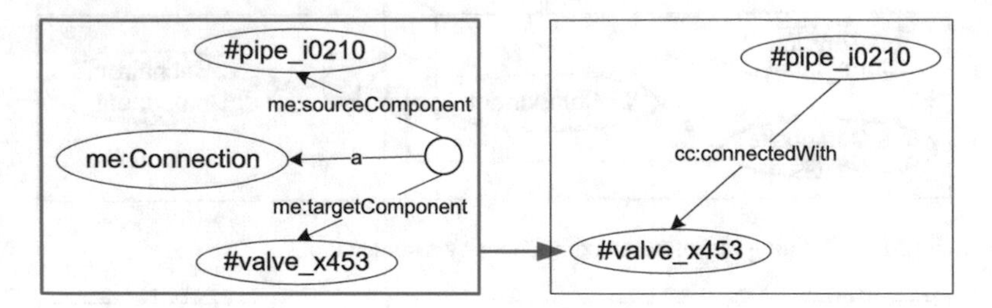

Fig. 6.12 M6b: Correspondence between class in ME ontology and property in CC ontology

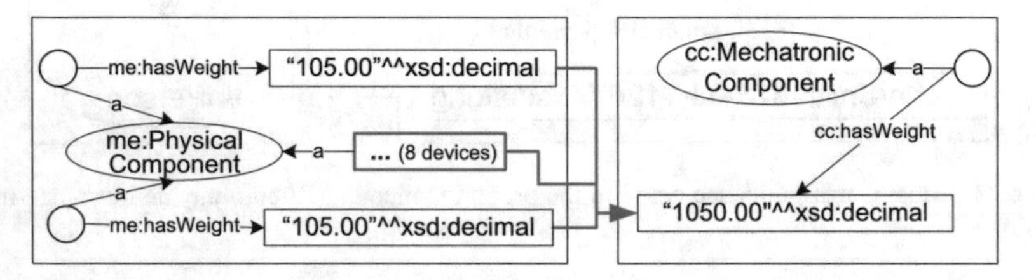

Fig. 6.13 M7: Aggregation of property values to get the weight of a mechatronic device

Grouping and Aggregation (M7)
Description. In some cases it is important to use grouping or/and aggregation of entities in one or several ontologies in order to set the relation to another ontology. This type of correspondence is also presented in (Scharffe 2009) as the "Aggregation" pattern.

Example. In order to calculate a value of the property `hasWeight` for a specific `MechatronicComponent` in the CC ontology, values of the `hasWeight` property of all devices from the ME ontology, which are contained in this component, should be summed up (see Fig. 6.13).

Mapping Directionality
When speaking about mappings, an important characteristic is that they can be directional (Ghidini et al. 2007), i.e., can be specified in a direction from source to target and the data flow cannot occur in the opposite direction. However, for some applications, such as for a data transformation, it could be beneficial to define bidirectional mappings between the engineering objects. It would help to reduce the total amount of mappings, thus facilitating their maintenance. However, in some cases, it may be impossible to specify a bidirectional mapping—e.g., in the example for mapping type M3 it will not be possible to specify the specific values for start and end dates based only on the duration of a specific project.

Example. Examples for M1 and M2 (if specified in a specific mapping language) can also serve as examples of bidirectional mappings.

6.5 Languages and Technologies for Mapping Definition and Representation

This section provides a description of languages and technologies that can be applied for ontology mapping. Even though many initiatives exist to map heterogeneous sources of data to RDF such as XSPARQL (Akhtar et al. 2008) to transform XML into RDF and RML (Dimou et al. 2014) to map CSV, spread sheets and XML to RDF, we will only examine those languages that allow expressing alignments and mappings between different ontologies.

Although the languages described below are of very different nature for the sake of uniformity hereafter we will call them “mapping languages.” For this chapter, we focus on those languages which are already well known and/or widely used. This means that these languages already have implementations and tool support and, therefore, would be the most probable and convenient choice for practitioners.

All mapping languages can be divided into the two categories: declarative and procedural languges. A language is *declarative*, if it expresses something independently from the way it is processed. Therefore, one should use external tools to process the defined correspondences for an application at hand. A *procedural* language, on the other hand, expresses how mappings are processed (for a specific or various applications).

Another important characteristic of a mapping language is whether it is suited to express correspondences at *schema* (classes and properties) or *data* (ontology instances) level. Below we provide a brief description of existing mapping languages. Table 6.1 position them with respect to these categories.

The **Web Ontology Language (OWL)** is an ontology language where one can declare relations between concepts such as equivalence, subsumption, etc., and allows one to infer additional information about instances by reasoning over the properties of classes and relations. Although one can define one-to-one correspondence

Table 6.1 Mapping languages and their characteristics: ● = compliant; ○ = non compliant

	Declarative	Procedural	Schema	Data
OWL	●	○	●	●
SWRL	●	○	○	●
SPARQL CONSTRUCT	○	●	○	●
Jena rules	○	●	○	●
SPIN	○	●	○	●
SILK	○	●	○	●
SKOS	●	○	●	●
SEKT	●	●	●	●
Alignment format	●	○	●	●
EDOAL	●	●	●	●

between the ontology entities using `owl:equivalentClass`, `owl:equivalentProperty` and `owl:sameAs` for classes, properties and individuals correspondingly, OWL itself has no means to define more complex correspondences as those described in Sect. 6.4. Also, as OWL is a knowledge representation language, it by itself possesses no means for data transformation between ontologies. One will need to use additional tools for that, such as OWL reasoners to infer additional triples[5] from an OWL file. Thus, reasoners could be used in combination with SPARQL CONSTRUCT queries to create a "reasoner-enabled" mapping.

The **Semantic Web Rule Language, (SWRL)**[6] is a W3C submission for a Semantic Web rule language, combining OWL DL—a decidable fragment of OWL—with the Rule Markup Language.[7] Rules are thus expressed in terms of OWL concepts. Rules are of the form of an implication between an antecedent (body) and consequent (head). The intended meaning is: whenever the conditions specified in the antecedent hold, then the conditions specified in the consequent must also hold. Note that SWRL rules are not intended for defining mappings, but to infer additional information from an ontological system, i.e., if the intended application is instance translating, they should be used in combination with SPARQL CONSTRUCT queries for instance to create a target RDF graph out of a source graph.

One way to define mappings is to use a **SPARQL CONSTRUCT**[8] query, which returns an RDF graph created with a template for generating RDF triples based on the results of matching the graph pattern of the query. To use this construct, one needs to specify how patterns in one RDF graph are translated into another graph. The outcome of a CONSTRUCT query depends on the reasoner and rule engine used. A SPARQL endpoint not backed by an OWL reasoner will only do simple graph matching for returning triples. A software agent that needs to compute these inferences will therefore have to consume all the necessary triples and perform this computation itself. The same holds for inferring additional information via business rules. SPARQL CONSTRUCT, however, is not a rule language and "merely" allows one to make a transformation from one graph match to another graph, i.e., a one-step transformation.

Another option to define mappings is using rules that can be declared on top of OWL ontologies. Apache Jena[9] includes a rule-based inference engine called **Jena Rules**[10] for reasoning with RDF and OWL data sources based on a Datalog implementation. Datalog is a declarative logic programming language that is popular for data integration tasks (Huang et al. 2011).

SPARQL Inference Notation (SPIN)[11] is currently submitted to W3C and provides—amongst others—means to link class definitions with SPARQL queries

[5]RDF triple: https://www.w3.org/TR/2004/REC-rdf-concepts-20040210/#section-triples.

[6]http://www.w3.org/Submission/SWRL/.

[7]http://wiki.ruleml.org/index.php/RuleML_Home.

[8]http://www.w3.org/TR/rdf-sparql-query/#construct.

[9]Apache Jena: http://jena.apache.org/.

[10]https://jena.apache.org/documentation/inference/#rules.

[11]http://www.w3.org/Submission/spin-overview/.

(ASK and CONSTRUCT) to infer triples. An implementation of SPIN is available from TopBraid.[12] It is built on top of the Apache Jena framework, and therefore inherits its properties.

Silk (Volz et al. 2009) is a link discovery framework for RDF data sets available as a file or via SPARQL endpoints. It allows one to declare how different RDF data sets relate to each other by specifying so-called linkage rules. These linkage rules are used to identify which resources are related to generate, for instance, owl:sameAs predicates. One is also able to define correspondences using other predicates, which depend on the use case. These linkage rules can make use of aggregates, string metrics, etc. SILK allows describing how resources in two existing data sets relate to each other, but does not possess means to process them for a certain application, e.g., to perform the transformation.

Two systems for expressing relations between entities worth mentioning are **SKOS** (Miles et al. 2005) and the **Alignment format** (David et al. 2011). However, they can only express correspondences between pairs of named entities of two ontologies, so they are not suited to address the requirements in Sect. 6.4.

Another language to define mappings was developed by the **SEKT** project. This language is designed to be independent from any ontology language, thus, it can be used for ontologies written in different languages. Several syntaxes are available—verbose human readable syntax and RDF and XML syntaxes. A Java API is also available allowing parsing and serializing to and from the object model of the mapping document. This mapping language is quite expressive—it allows specifying mappings between classes, properties, and instances of an ontology (also across) using a set of operators, which have a cardinality, an effect and some related semantics. One can also specify conditions, annotations, direction info (bidirectional or unidirectional mapping) and extend the defined constructs with arbitrary logical expressions (Scharffe et al. 2006).

The **EDOAL**[13] **(Expressive and Declarative Ontology Alignment Language)** (David et al. 2011) is a language for expressing alignments which is supported by the Alignment API. The Alignment API allows to generate and parse alignments, to manipulate them and to render these alignments in different languages, eventually executable. EDOAL can express correspondences between more precise and complex terms than the named entities. EDOAL also supports expressing transformations on property values, which is of particular interest in our context.

Table 6.2 summarizes the level of support of the mapping languages listed above for defining and representing complex relations between the ontologies described in Sect. 6.4. The evaluation was done based on (a) checking the specification documents for each language; (b) authors' practical experiences with implementing ontology-based integration solutions; and c) knowledge obtained during authors' involvement in the language development (for some languages).

Due to space limits we cannot provide a detailed analysis for each of the described mapping languages. From Table 6.2, it is clear that everything can be written directly

[12]http://www.topquadrant.com/.

[13]http://alignapi.gforge.inria.fr/edoal.html.

Table 6.2 Support for the complex relations definition and representation in various mapping languages: ●–supported; ◐–partially supported; ○– no support; *–vendor dependent

	M1	M2	M3	M4	M5a	M5b	M5c	M6a	M6b	M7
SWRL	◐	●	●	◐	●	●	●	●	●	●
SPARQL CONSTRUCT	●	●	●	●	●	●	●	●	●	●
Jena rules	◐	◐	◐	●*	●	●	●	●	●	●
SPIN	●	●	●	●*	●	●	●	●	●	●
SILK	●	●	●	●	●	●	●	○	●	◐
SKOS	○	○	○	○	○	○	○	○	○	○
SEKT	●	●	●	●	●	●	●	○	●	○
Alignment Format	○	○	○	○	○	○	○	○	○	○
EDOAL	●	●	●	●	●	●	●	○	●	●

with SWRL or SPARQL CONSTRUCT. Such languages have enough expressivity and can be considered, at least for SPARQL, to have efficient implementations. However, they lack declarativity. For instance, they define oriented rules and changing the orientation requires rewriting the rules. A language like EDOAL allows to express declaratively the relations between two ontologies and can generate SPARQL CONSTRUCT (or eventually SWRL in simple cases) to implement the transformations in one way or the other. Therefore, we decided to focus on EDOAL. We continue with detailed analysis of the EDOAL's capabilities for representing complex correspondences and the examples of those identified in the use case scenario (see Sect. 6.3).

6.6 Representing Complex Relations with EDOAL

Expressive and Declarative Ontology Alignment Language (EDOAL) is an extension of the Alignment format supported by the Alignment API (David et al. 2011). It offers the capability to express correspondences that go beyound putting in relation named entities. In EDOAL correspondences may be defined between compound descriptions, which allow to further constrain those entities that are put in correspondences. Compound descriptions may be combination of concepts, e.g., a physical component that is also a composite device, or restriction of concepts, e.g., a physical component whose weight is over 125 pounds. Compound descriptions are defined through a description language similar to that of description logics.

This is possible through:

Construction that constrains the object put in correspondence with classical Boolean operators (disjunction, conjunction, complement) or property construction operators (inverse, composition, reflexive, transitive, and symmetric closures);

Restriction that restrains the values of objects put in correspondence (through domain, range, cardinality and value restrictions). This is typically achieved by requesting all or some values of a property to be in some domain (`weight` is above 1025) or class (some member being a senior manager).

These constraints can be composed together to obtain more elaborate correspondences. For instance, the meaning of some of the correspondences in Sect. 6.4 can be simply expressed as EDOAL correspondences.

The snippet in Listing 6.1 combines M5a and M5c examples in one correspondence (M5b can be treated in the exact same way). The first part (within the "`and`") is a class construction, while the `AttributeValueRestriction` part is a class restriction.

Listing 6.1 EDOAL correspondence capturing M5a and M5c examples

```
<align:Cell rdf:about="#M5">
  <align:entity1>
    <Class>
      <and rdf:parseType="Collection">
        <Class rdf:about="&me;PhysicalComponent"/>
        <Class rdf:about="&me;CompositeDevice"/>
         <AttributeValueRestriction>
           <onAttribute><Relation rdf:about="&me;hasWeight"/></
               onAttribute>
           <comparator rdf:resource="&edoal;greater-than"/>
           <value><Literal edoal:type="&xsd;decimal"edoal:string
               ="1025.00"/></value>
         </AttributeValueRestriction>
      </and>
    </Class>
  </align:entity1>
  <align:entity2>
    <Class rdf:about="&cc;MechatronicComponent"/>
  </align:entity2>
  <align:relation>=</align:relation>
  <align:measure rdf:datatype="&xsd;float">1.0</align:measure>
</align:Cell>
```

The Alignment API offers EDOAL alignment manipulation and rendering. So, the correspondence presented in Listing 6.1 can be automatically transformed into the SPARQL CONSTRUCT query shown in Listing 6.2, which can be used for transforming data.

Listing 6.2 SPARQL CONSTRUCT query generated from the EDOAL correspondence in Listing 6.1

```
CONSTRUCT { ?s rdf:type cc:MechatronicComponent . }
WHERE {
  ?s rdf:type me:PhysicalComponent .
  ?s rdf:type me:CompositeDevice .
  ?s me:hasWeight ?w .
  FILTER( ?w >="1025.00"^^xsd:decimal ).
}
```

The same can be achieved with the M6b correspondence example that connects an object and a relation between objects (see Listing 6.3).

Listing 6.3 EDOAL correspondence capturing M6b example

```
<align:Cell rdf:about="#M6b">
  <align:entity1>
    <Relation>
      <compose rdf:parseType="Collection">
        <Relation>
          <inverse><Relation rdf:about="&me;sourceComponent"/></
              inverse>
        </Relation>
        <Relation>
          <and rdf:parseType="Collection">
            <Relation rdf:about="&me;targetComponent"/>
            <RelationDomainRestriction>
              <class><Class rdf:about="&me;Connection"/></class>
            </RelationDomainRestriction>
          </and>
        </Relation>
      </compose>
    </Relation>
  </align:entity1>
  <align:entity2><Relation rdf:about="&cc;connectedWith"/></
      align:entity2>
  <align:relation>=</align:relation>
  <align:measure rdf:datatype="&xsd;float">1.0</align:measure>
</align:Cell>
```

In addition to expressing the usual correspondences between two entities, EDOAL correspondences are extended to support two types of information:

Transformations that specify how to transform an instance of the related entities into the other;

Link keys that define under which conditions two individuals must be considered the same (Atencia et al. 2014).

For this chapter we are mostly concerned with transformations. They allow expressing how property values of two equivalent relations can be transformed into one another. For that purpose EDOAL usually applies transformation functions on the values.

This is, in general, sufficient to express the examples for M1–M4 and M7. In Listing 6.4 the correspondence covering M2 and M3 (where M2 is a simple conversion from string to date; and M3 is a function call though the call of substract-dates) is presented. In general, all M1–M4 examples would follow the same pattern.

Listing 6.4 EDOAL correspondence capturing M2 and M3 examples

```
<align:Cell rdf:about="#M1extended">
  <align:entity1><Class rdf:about="&pm;Project"/></align:entity1
      >
  <align:entity2><Class rdf:about="&cc;Project"/></align:entity2
      >
  <align:relation>=</align:relation>
  <align:measure rdf:datatype="&xsd;float">1.0</align:measure>
  <transformation>
    <Transformation edoal:direction="o-">
      <entity1>
            <Apply edoal:operator="xsd:date">
          <arguments rdf:parseType="Collection">
            <Property rdf:about="&pm;hasStartingDate"/>
```

```
        </arguments>
      </Apply>
    </entity1>
    <entity2><Property rdf:about="&cc;hasStartingDate"/></
        entity2>
  </Transformation>
  <Transformation edoal:direction="o-">
    <entity1>
      <Apply edoal:operator="op:subtract-dates">
        <arguments rdf:parseType="Collection">
          <Property rdf:about="&pm;hasStartingDate"/>
          <Property rdf:about="&pm;hasEndDate"/>
        </arguments>
      </Apply>
    </entity1>
    <entity2><Property rdf:about="&cc;hasDuration"/></entity2>
  </Transformation>
 </transformation>
</align:Cell>
```

As it can be seen in the snippet in Listing 6.4, the same EDOAL correspondence can support several transformations as, for two objects in the correspondence, there may be several properties to be transformed.

Listing 6.5 EDOAL correspondence capturing M7 example

```
<align:Cell rdf:about="#M1extended">
  <align:entity1><Class rdf:about="&me;PhysicalComponent"/></align
     :entity1>
  <align:entity2><Class rdf:about="&cc;MechatronicComponent"/></
     align:entity2>
  <align:relation>=</align:relation>
  <align:measure rdf:datatype="&xsd;float">1.0</align:measure>
   <transformation>
    <Transformation edoal:direction="-o">
      <entity1>
            <Aggregate edoal:operator="&fn;sum">
              <arguments rdf:parseType="Collection">
            <Property>
              <compose rdf:parseType="Collection">
                <Relation rdf:about="&me;component"/>
                <Property rdf:about="&me;hasWeight"/>
              </compose>
            </Property>
            </arguments>
          </Aggregate>
      </entity1>
      <entity2><Property rdf:about="&cc;hasWeight" /></entity2>
    </Transformation>
  </transformation>
</align:Cell>
```

The example for M7 can be expressed in a similar way (see Listing 6.5) Alternatively, the transformation here could have been added to the previous M5 example.

With regard to directionality, EDOAL correspondences are in general not oriented: they express a relation between two terms and not a function from one term to another. However, when generating a mapping from a correspondence, this may direct the use of the correspondence, e.g., the SPARQL query above. Moreover, transformations may be oriented. This is specified by the direction attribute: indeed, it is possible to compute a duration from the dates or a total weight of the compound component from its component part weights but not the other way around.

Although EDOAL can express such transformations, it cannot process them. These transformations are rendered in other languages that can be processed, such as SPARQL. Transformations often rely on external functions, most of the time identified by XPath URIs. Hence, a renderer using them should be able to interpret such operations as SPIN functions or SPARQL built-in functions, for instance.

One type of correspondences that is difficult to represent in EDOAL, are those, like M6a, which introduce some new resources that have no counterpart in the initial data set. In M6a, the `Person` in the CC ontology does correspond to `John/Smith/078-05-1120`, which is a simple string. Sometimes this can be easily expressed, like for M6b, because the `connectedWith` relation relates two existing resources. But in general, this requires more information.

6.7 Conclusion

In this chapter, we described what types of complex relations between engineering objects may need to be captured, while implementing ontology-based integration solutions. We presented a catalogue of complex correspondences between the ontologies and illustrated each correspondence type with an example from the real-life power-plant engineering project. We also provided an overview on available mapping languages and technologies that can be used to define, represent, and sometimes also process the correspondences between the different ontologies and briefly explained their key characteristics and capabilities w.r.t. the complex relations capturing. As space limitations do not allow us to perform the detailed analysis of each mapping language, we decided to focus on EDOAL, because of its interesting quality to combine the features of both declarative and procedural languages (contrary to other mapping languages). We therefore explained in detail what are the EDOAL's capabilities for representing complex relations and gave the examples of how the correspondences identified for the power-plant engineering project can be implemented with EDOAL.

As future work we would like to explore what languages and techniques are currently applied by engineers to link different models and data structures (especially across the engineering disciplines and tools) during the engineering process and to compare their representational capabilities with those provided by the Semantic Web community. Another interesting direction for the future work will be analyzing languages to define relations and constraints on different data models and data sets developed within the Model-Driven engineering field and compare their application aspects with the languages discussed in this chapter.

Acknowledgments This work was supported by the Christian Doppler Forschungsgesellschaft, the Federal Ministry of Economy, Family and Youth, and the National Foundation for Research, Technology and Development in Austria.

References

Akhtar, W., Kopecký, J., Krennwallner, T., Polleres, A.: XSPARQL: Traveling Between the XML and RDF Worlds—and Avoiding the XSLT Pilgrimage. Springer (2008)

Atencia, M., David, J., Euzenat, J.: Data interlinking through robust linkkey extraction. In: Proceeding 21st European Conference on Artificial Intelligence (ECAI), Praha (CZ), pp. 15–20 (2014)

Bernstein, P.A., Melnik, S.: Model management 2.0: manipulating richer mappings. In: Proceedings of the 2007 ACM SIGMOD International Conference on Management of Data, pp. 1–12. ACM (2007)

Biffl, S., Moser, T., Winkler, D.: Risk assessment in multi-disciplinary (software+) engineering projects. Int. J. Softw. Eng. Knowl. Eng. **21**(02), 211–236 (2011)

Breslin, J.G., O'Sullivan, D., Passant, A., Vasiliu, L.: Semantic web computing in industry. Comput. Ind. **61**(8), 729–741 (2010)

David, J., Euzenat, J., Scharffe, F., Trojahn Dos Santos, C.: The Alignment API 4.0. Semant. Web J. **2**(1), 3–10 (2011)

Dimou, A., Vander Sande, M., Colpaert, P., Verborgh, R., Mannens, E., Van de Walle, R.: RML: a generic language for integrated RDF mappings of heterogeneous data. In: Proceedings of the 7th Workshop on Linked Data on the Web (LDOW2014), Seoul, Korea (2014)

Euzenat, J., Shvaiko, P.: Ontology Matching, 2nd edn. Springer, Heidelberg (DE) (2013)

Ghidini, C., Serafini, L., Tessaris, S.: On relating heterogeneous elements from different ontologies. In: Modeling and Using Context, pp. 234–247. Springer (2007)

Huang, S.S., Green, T.J., Loo, B.T.: Datalog and emerging applications: an interactive tutorial. In: Proceedings of the 2011 ACM SIGMOD International Conference on Management of Data, pp. 1213–1216. ACM (2011)

Legler, F., Naumann, F.: A classification of schema mappings and analysis of mapping tools. BTW, Citeseer **103**, 449–464 (2007)

Miles, A., Matthews, B., Wilson, M., Brickley, D.: SKOS core: simple knowledge organisation for the web. In: International Conference on Dublin Core and Metadata Applications, p. 3 (2005)

Mordinyi, R., Winkler, D., Moser, T., Biffl, S., Sunindyo, W.D.: Engineering object change management process observation in distributed automation systems projects. In: Proceedings of the 18th EuroSPI Conference, Roskilde, Denmark (2011)

Noy, N.F.: Semantic integration: a survey of ontology-based approaches. ACM SIGMOD Rec. **33**(4), 65–70 (2004)

Otero-Cerdeira, L., Rodríguez-Martínez, F.J., Gómez-Rodríguez, A.: Ontology matching: a literature review. Exp. Syst. Appl. **42**(2), 949–971 (2015)

Scharffe, F.: Correspondence patterns representation. Ph.D. thesis, University of Innsbruck (2009)

Scharffe, F., de Bruijn, J., Foxvog, D.: Ontology mediation patterns library v2. Deliverable **D4**, 3 (2006)

Scharffe, F., Zamazal, O., Fensel, D.: Ontology alignment design patterns. Knowl. Inf. Syst. **40**(1), 1–28 (2014)

Shvaiko, P., Euzenat, J.: Ontology matching: state of the art and future challenges. IEEE Trans. Knowl. Data Eng. **25**(1), 158–176 (2013)

Volz, J., Bizer, C., Gaedke, M., Kobilarov, G.: Silk: a link discovery framework for the web of data. LDOW 538 (2009)

Vyatkin, V.: Software engineering in industrial automation: state-of-the-art review. IEEE Trans. Ind. Inf. **9**(3), 1234–1249 (2013)

Wache, H., Voegele, T., Visser, U., Stuckenschmidt, H., Schuster, G., Neumann, H., Hübner, S.: Ontology-based integration of information—A survey of existing approaches. In: IJCAI-01 Workshop: Ontologies and Information Sharing, Citeseer, vol. 2001, pp. 108–117 (2001)

Chapter 7
Knowledge Change Management and Analysis in Engineering

Fajar Juang Ekaputra

Abstract Knowledge is changing rapidly within the engineering process of Cyber-Physical Production Systems (CPPS) characterized by the collaborative work of engineers from diverse engineering disciplines. Such rapid changes lead to the need for management and analysis of knowledge changes in order to preserve knowledge consistency. Knowledge change management and analysis (KCMA) in Multidisciplinary Engineering (MDEng) environments is a challenging task since it involves heterogeneous, versioned, and linked data in a mission-critical fashion, where failure to provide correct data could be costly. Although, there are several available solutions for addressing general issues of KCMA, from fields as diverse as Model-Based Engineering (model co-evolution), Databases (database schema evolution), and Semantic Web Technology (ontology versioning), solving KCMA in engineering remains a challenging task. In this chapter, we investigate issues related to KCMA in MDEng environments. We provide a definition of this task and some of its challenges and we overview technologies that can be potentially used for solving KCMA tasks from the three research fields mentioned above. We then define a technology agnostic solution approach inspired by the Ontology-Based Information Integration approach from Semantic Web research as a first step toward a complete KCMA solution and provide an indication of how this solution concept could be implemented using state of the art Semantic Web technologies.

Keywords Knowledge change management and analysis • Ontology evolution • Ontology versioning • Ontology-based information integration • Change detection • Change validation • Change propagation • Multidisciplinary engineering

F.J. Ekaputra (✉)
Institute of Software Technology and Interactive Systems, CDL-Flex, Vienna University of Technology, Vienna, Austria
e-mail: fajar.ekaputra@tuwien.ac.at

S. Biffl and M. Sabou (eds.), *Semantic Web Technologies for Intelligent Engineering Applications*, DOI 10.1007/978-3-319-41490-4_7

7.1 Introduction

The process of designing a Cyber-Physical Production System (e.g., modern power plants or steel mills) often requires teams of engineers from diverse engineering domains (e.g., mechanical, electrical and software engineering) to work together. As a result, this design process typically takes place in a multidisciplinary engineering (MDEng) environment, in which experts from various engineering domains and organizations work together toward creating complex engineering artifacts (Serral et al. 2013). Figure 7.1 depicts such a typical MDEng setting. Domain specific engineers (shown on the right hand side) use their own tools to create models that represent parts of the final system. Therefore, the MDEng environment is highly heterogeneous, as it involves a wide range of data models, processes, and tools that were originally not designed to cooperate seamlessly. Despite this situation, as shown on the left-hand side of Fig. 7.1, other engineers and project managers need to perform tasks that require access to project-level data as opposed to domain specific data alone. For these actors, there is a need for accessing integrated data at project level. In response to this need, knowledge engineers aim to integrate models and data from different engineering domains based on the requirements and feedback of engineers and project managers.

In addition to the characteristics described above, the process of designing complex mechatronic objects, such as CPPS, requires iterations and redesign phases, which lead to continuous changes of the data and knowledge within the MDEng environment. To deal with these changes, industrial partners need to keep data versions, move backwards to previous versions, and query different versions of large data (schema and instances) from heterogeneous local data sources. Furthermore, the effective and considerate propagation of changes is essential to ensure

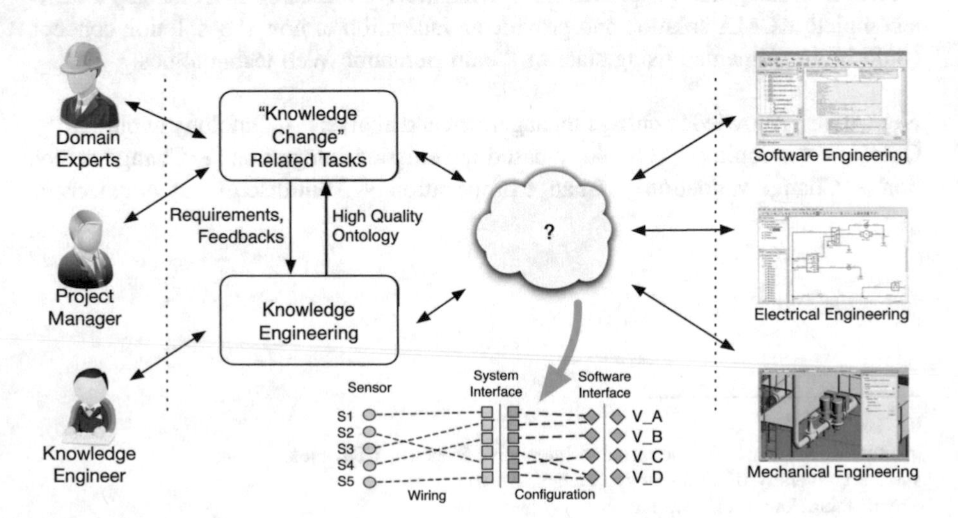

Fig. 7.1 Problem setting for KCMA in MDEng environments

a consistent view of the project, to minimize defects and risks, as well as facilitate acceptance of new solutions by domain experts. To achieve this, changes originating from one engineering discipline need to be communicated and coordinated with participants of other disciplines, where those changes are relevant. Ideally, this communication should focus on high-level changes (e.g., defined in terms of domain concepts such as "Motor X updated to new version") as opposed to low-level individual changes (i.e., change operations on versioned files) to ease the data analysis process. To cater for all these needs, a process and tool support should be available for knowledge change management and analysis (KCMA) within MDEng environments.

Technology solutions for data integration and knowledge change management in general, and to some extent for multidisciplinary engineering in particular, have been proposed by research fields as diverse as Database Systems, Model-Based Engineering (MBE), and the Semantic Web. Although the general strengths and weaknesses of these solution approaches are somewhat known, a precise comparison of how well they could support KCMA in the MDEng environments is hampered by two major factors. First, there is a lack of understanding of requirements for KCMA in MDEng, i.e., a characterization of the problem that needs to be solved. Second, a baseline setting that would allow an objective comparison of these technologies is also missing.

To overcome these shortcomings, this chapter makes two important contributions. First, it provides a characterization of KCMA by means of key requirements that should be fulfilled (Sect. 7.2). These requirements were derived from concrete industry specific projects where the author investigated the need for knowledge change management. Second, this chapter provides a brief introduction in typical solutions to knowledge change management from the areas of Database Systems, Model-Based Engineering (MBE), and the Semantic Web (Sects. 7.3 and 7.4). Based on this analysis spanning the three major research fields of interests, the chapter provides a technology agnostic reference process for solving KCMA (Sect. 7.5). This reference process is suitable to play the role of a baseline for comparing the strengths and weaknesses of implementations relying on either of the three major families of technologies described in this chapter. Additionally, a proposal of implementation based on Semantic Web technologies is presented in Sect. 7.6. We sum up the chapter and discuss future work in Sect. 7.7.

7.2 KCMA in Engineering

In this section, we take a detailed look to the needs for KCMA in engineering settings. To that end, we provide an illustrative real-world example (Sect. 7.2.1) and then discuss a set of requirements for KCMA in Sect. 7.2.2.

7.2.1 KCMA Example

An illustrative MDEng setting is the engineering of a modern power plant. Similar to any other large-scale project, the development of a power plant requires coordinated work of engineers from multiple disciplines that needs to converge into a high-quality product. Such a heterogeneous team of experts should be coordinated in a way that fulfills important technical and management project-level constraints (e.g., the mass and dimension constraints of the base plate are not exceeded by individual equipment). Such coordination requires aggregating relevant data across teams from various disciplines, but it is hampered by the semantic heterogeneity of the data, with different disciplines possibly using diverse terms to refer to the same entities.

For illustration purposes, we provide an excerpt of the data (usually called a "signal list") exchanged between the engineers participating the engineering process of a hydropower plant (Table 7.1). A signal list is typically serialized and used by engineers as spreadsheet files. The header of Table 7.1 represents the data schema used within the signal list, while its body represents data instances. The combination of the first four columns (ANR, L1, L2, and SIG) identifies the engineering signals/objects, while the later three (DIFF MAX, GRADIENT MAX, and RESISTANCE VALUE) represent signal/object properties.

To add more complexity, the MDEng project models and the project data change over time due to:

1. **Changes in the represented domains**, such as the introduction/removal of domain concepts (e.g., the removal of RESISTANCE VALUE column from Table 7.1 since it is not relevant anymore in the domain) or granularity changes (e.g., adding more detailed information than at signal level);
2. **Changes in the underlying data sources**, such as when new data elements become available and old data elements become obsolete (e.g., a new spreadsheet file is produced to replace the old file without changing the data schema); or

Table 7.1 Excerpt of the engineering data of a hydropower plant engineering process

ANR	L1	L2	SIG	S3 DIFF MAX	S3 GRADIENT MAX	RESISTANCE VALUE
0	BAA30	GS100	XB01	0.025	0.8	500
0	BAA30	GS191	YB01	0.025	0.8	500
0	BAA30	GS191	YB02	0.025	0.8	500
0	BFB10	GS100	XB01	0.025	0.8	500
0	BFB10	GS100	XM01	0.025	0.8	500
0	BFB10	GS100	YB01	0.025	0.8	500

3. **Changes in the intended use of the models and data**, such as by changing requirements of the currently supported tools or the design of new tools (e.g., a new data schema is introduced and it has to be mapped into the old schema in Table 7.1).

To address such changes, an integrated versioning of the MDEng data needs to be prepared for facilitating this evolution and the consequent data transformations and propagation, according to the evolved model.

7.2.2 Requirements for KCMA in Engineering

Dealing with the types of possible changes described above requires both activities for managing and analyzing changes. In terms of change management, it is important to record data versions and to be able to move backwards to previous versions, as well as to query different versions of integrated data (schema and instances) originating from heterogeneous local data sources.

Furthermore, on a more analytics related level, it is essential to enable the effective and considerate propagation of changes across data from different disciplines. This will ensure a consistent view of the project and it will minimize defects and risks. To achieve this, changes originating from one discipline need to be communicated and coordinated with participants of other disciplines, where those changes are relevant, especially in cases when data from different engineering disciplines is closely interlinked. The KCMA approach should also provide high-level change definitions instead of low-level ones to ease the analysis process of data.

Based on our involvement and experiences in several industrial engineering settings where KCMA was required, we have identified a set of requirements and characteristics of KCMA in engineering (specifically in MDEng environments) as follows:

1. **Closely interlinked knowledge**. In the engineering process of a CPPS, engineering models and data created by different engineering disciplines reflect diverse views on the same system and are therefore naturally interlinked (e.g., an electrical signal activates a mechanical component as a result of executing a certain software routine). Therefore, knowledge changes within one discipline may require changes in other disciplines too due to this relation between data in different disciplines. For example, a signal change in electrical engineering area will require the adaptation of the corresponding machinery part (mechanical engineering) or reprogramming of the relevant software components (software engineering).
2. **Large amounts of data**. Engineering projects typically deal with large amounts of data required to describe any complex system. For example, an average size power plant design data contains data about hundreds of thousands to tens of

millions of signals. This already large data size is further multiplied due to many iteration processes during the design time of the system, which should all be versioned and stored (Mordinyi et al. 2014).

3. **Changes in schema and instances**. In MDEng environments, both data models (i.e., schema) and actual data (instances) is likely to change. Indeed, the heterogeneity of data sources within MDEng environments and the environment's dynamism means that additional tools could be added anytime, which may imply changes in the data models of all engineering disciplines involved. At the same time, data instances (e.g., signals with changed characteristics, added, or deleted signals) within MDEng environment will change even more frequently due to revisions and engineering process iterations.
4. **Change validation support**. Given the mission-critical nature of projects in MDEng environments, domain experts and engineers do not want to fully rely on automatic change validation mechanisms (e.g., to decide whether changes initiated by one discipline will break the overall consistency of the project wide data). Therefore, instead of fully automated change validation, the involvement of domain experts in the validation workflow is important for making critical decisions about changes.
5. **High-level change definition and detection**. Typical tools currently used in individual engineering disciplines are able to produce report data that consists of signal lists that represent those parts of a CPPS, which these specific tools handle (Vogel-Heuser et al. 2014). The differences between two versions of signal lists represent changes between them. However, it is challenging for a project manager to grasp the meaning of such low-level changes in data, i.e., signal changes. Instead, they would highly benefit from changes to data being presented in a more meaningful manner as high-level changes. Such presentation could be achieved in terms of domain level common concepts, e.g., relocation of specific engine to different machine rack.
6. **Support for data evolution and versioning**. A KCMA approach should be able to address both data evolution and data versioning. Data evolution focuses on how to modify the data in response to the changes in the surrounding. Data versioning covers the management of different versions of data schema caused by the data evolution (Noy and Klein 2004; Roddick 1995).

7.3 Solutions for KCMA in the Engineering Domain

This section introduces brief summaries of approaches related to the KCMA in the engineering domain from the Database systems (Sect. 7.3.1) and Model-Based Engineering research communities (Sect. 7.3.2).

7.3.1 Database Schema Evolution and Versioning

One of the earliest works concerning knowledge change management is reported in the field of database systems. Roddick summarized the issues of schema evolution and versioning in database systems (Roddick 1995). He explains that change management is closely related to data integration, claiming that both areas are the flavors of a more generic problem: using multiple heterogeneous schemata for various tasks. To solve the issues suggested by Roddick, there were several proposed conceptual approaches. One of them is the Hyper-graph Data Model (HDM), which targets schema evolution for heterogeneous database architectures (McBrien and Poulovassilis 2002). The HDM schema consists of Nodes, Edges, and Constraints. Nodes and Edges in the schema define a labeled (Nodes require unique names), directed (the Edges may link sequences of Nodes or Edges), and nested (Edges could link an unlimited number of other Nodes and Edges) hypergraph, while Constraints define a set of Boolean valued queries over the HDM schema.

In the field of MDEng, there are limited concrete solutions that are utilizing relational databases as the basis for managing and analyzing changes in engineering data from multiple engineering datasets. One exception is the Engineering Database (EDB), a solution based on relational databases that was introduced as an attempt to provide versioning management of engineering data using database technology (Waltersdorfer et al. 2010). The Engineering Database is a concrete implementation of Engineering Knowledge Base (EKB) that is explained in Chap. 4. The EDB stores engineering data as a flat database table, consisting of objects, properties, values, and important metadata information such as change commit information and provenance from the original data sources. The approach is capable of handling closely linked knowledge from different engineering disciplines.

To conclude, the maturity of database approaches in general provides a solid basis for handling change management of a large number of data. Additional KCMA-related solutions from database systems, not covered in this chapter, are also worth further investigations.

7.3.2 Model-Based Engineering (MBE) Co-Evolution

A lot of attention has been given to the comparison and versioning of software models in the Model-Based Engineering (MBE) research community, with more than 450 papers written in this area[1], covering various topics such as change modeling and definition of evolution frameworks. Similar to other research areas, the domain of Engineering is not the main application area of these works. An exception is the work of Göring and Fay, which proposes a metamodel language for modeling temporal changes of automation systems together with their physical and

[1]http://pi.informatik.uni-siegen.de/CVSM.

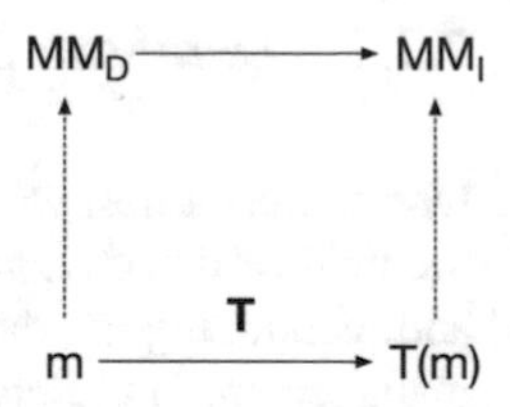

Fig. 7.2 Basic co-evolution schema, adapted from (Meyers and Vangheluwe 2011)

functional structures (Goring and Fay 2012). Their work extends the IEC 81346 standard, which already includes product, function, and location aspects. In the approach, however, they do not explain how to map their metamodel to other metamodels that are potentially used in the same system and therefore it is not clear how data integration is achieved.

Another line of MBE work focuses on change propagations of model variants of a single metamodel, to ensure the consistency of changes as well as the adoption of relevant changes in different model variants (Kehrer et al. 2014). Recently, the authors of (Berardinelli et al. 2015) provide means of a prototype-based model co-evolution, showing the capability of providing various levels of validation configuration to be applied in a top-down co-evolution approach.

Meyers and Vangheluwe propose one of the most recent frameworks for evolution of model and modeling languages, which claim that any possible model co-evolution could be derived as a composite of the basic co-evolution schema shown in Fig. 7.2 (Meyers and Vangheluwe 2011). It consists of a model *m* that conforms to metamodel domain (MM_D). Model *m* needs to be transformed into T (*m*) via transformation T, which again will conform to image metamodel (MM*i*). This co-evolution framework could theoretically address the requirements of knowledge change management within closely linked discipline data.

In conclusion, these works from the MBE research community partially address the requirements of KCMA in MDEng environment as mentioned in Sect. 7.2.2. Benefits of using MBE techniques for knowledge change management in engineering projects include the availability of a good tool support and solid theoretical frameworks (Meyers and Vangheluwe 2011; Taentzer et al. 2014).

7.4 Semantic Web for KCMA in Engineering

Advances in Semantic Web research could also provide partial solutions to the challenges of KCMA in MDEng discussed in Sect. 7.2. For example, the Provenance Ontology (PROV-O[2]), a recent W3C recommendation, is highly suitable to represent and interchange provenance information and could be used to record which changes were performed by which stakeholders.

[2]https://www.w3.org/TR/prov-o/.

An important observation is that, in MDEng environments, the *integration* of heterogeneous data sources is a prerequisite to addressing KCMA challenges and that knowledge change management must be applicable over the totality of integrated data sources that represent the data of an engineering project.

In the Semantic Web area, such integration settings can be addressed by relying on the Ontology-Based Information Integration (OBII) approach (e.g., in Calvanese et al. 2001; Wache et al. 2001), which provides a solution approach for integrating data from heterogeneous data sources using Semantic Web technologies. There are three different ways of applying the OBII approach as described by Wache et al. (2001).

1. **Single Ontology approach**, which relies on a single ontology for integrating all data sources. The approach is typically hard to maintain since it is susceptible to changes in each information sources. In other words, the ontology must be updated anytime a change in a data source occurs and then its compatibility with the other data sources must be ensured.
2. **Multiple Ontology approach** is characterized by defining one ontology per data source and subsequently linking the ontologies with each other independently. The drawback of this approach is that the mappings between different ontologies are challenging to define and maintain, due to the different granularities and aggregation between ontologies.
3. **Hybrid approach**, which combines the benefits of both single and multiple ontology-based approaches thus eliminating most of their drawbacks. An example of a Hybrid-style OBII system used to integrate engineering data of a power plant is shown in Fig. 7.3 and consists of three components. First, local ontologies represent data specific to one engineering discipline, i.e., local data. Second, a common ontology represents the aggregation of relevant and related concepts at the organizational level, e.g., power plant. Third, mappings between local and common ontologies enable linking and integration between the heterogeneous data sources. Similar to (Calvanese et al. 2001), we will use the term OBII to refer to the Hybrid OBII approach, unless stated otherwise.

The OBII approach aims to solve data integration and, as such, by itself it does not provide support for KCMA, which is essential for MDEng environments. Ontology change management has been investigated as a generic problem, not in conjunction with data integration, to address the dynamics of ontology data and its derivative challenges (e.g., Klein 2004; Noy et al. 2006; Stojanovic 2004; Zablith 2011).

However, it is not clear how applicable the currently available generic ontology change management solutions are to improve the KCMA support of an OBII approach in the context of MDEng environment. Therefore, the availability of a generic reference process of KCMA in MDEng environment is crucial to assess whether these ontology change management approaches sufficiently address the

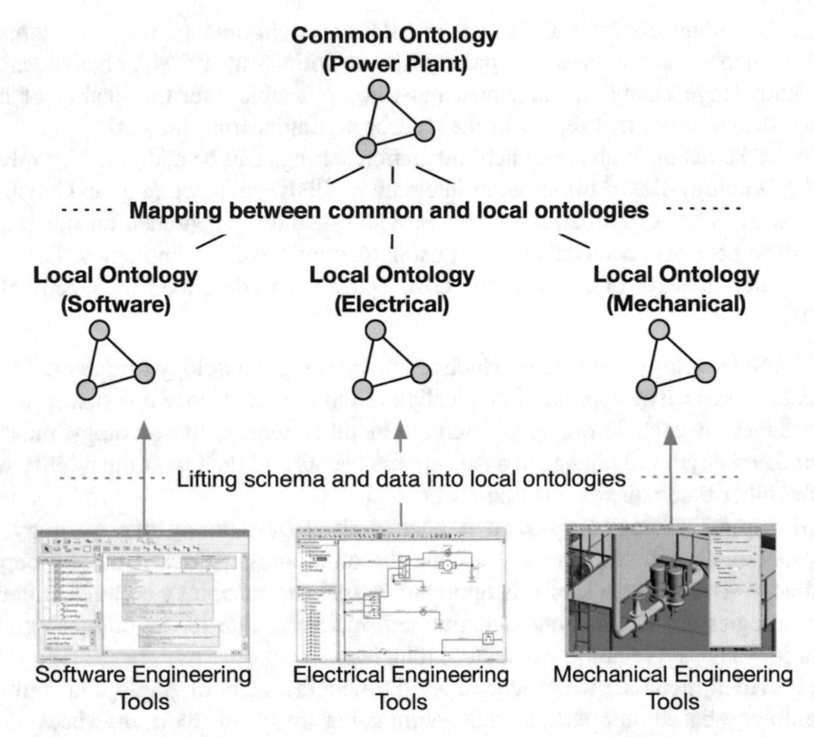

Fig. 7.3 Example of an OBII system for integrating engineering data from a power plant

KCMA requirements (e.g., change propagation to the overlapped data in other disciplines). In particular, an implementation of the KCMA process with ontology change management approaches would allow verifying their suitability to support KCMA in MDEng environments.

7.4.1 Ontology Change Management

The Semantic Web community has performed significant research in the area of ontology change management. In this section, we summarize works in this area based on our literature study (described in more detail in Ekaputra et al. 2015) and in terms of the requirements stated in Sect. 7.2. Table 7.2 provides an overview of the extent to which each requirement is addressed by each work we overview.

1. **Closely interlinked knowledge**. Closely interlinked knowledge is a condition where changes in one ontology within a system of interlinked ontologies may require propagation of the changes to the linked ontologies to maintain the global validity of the knowledge. This is not the typical setting for KCMA in Semantic Web community, which primarily focuses with open web data. This

Table 7.2 Semantic web generic solution alternatives for addressing knowledge change requirements within MDEng environment

	Type of interlinked knowledge			Amount of data	Changes in schema and instances		Change validation		Changes definition and detection		Support for data evolution and versioning	
	Single ontology	Loosely interlinked knowledge	Closely interlinked knowledge	Handle >1 M data points	Instance changes	Schema changes	Semi-automatic	Automatic	Low-level changes	High-level	Ontology evolution	Ontology versioning
MDEng KCMA			X	X	X	X	X		X	X	X	X
Klein (2004)		X				X					X	
Stojanovic (2004), Stojanovic et al. (2002)		X	(X)		X	X	X	X			X	X
Noy et al. (2006)	X					X			X		X	
Papavassiliou et al. (2009)	X			(X)	X	X			X	X	X	
Gröner et al. (2010)	X					X			X	X	X	
Zablith (2011)	X					X		X			X	
Vander Sande et al. (2013)		X			X				X			X
Horridge et al. (2013)		X		X	X	X	(X)		X		X	(X)
Graube et al. (2014)	X			(X)	X				X			X

[a] *X* fully supported; *(X)* partially supported

difference is reflected within most of traditional KCMA that focused on change management in a single ontology (Graube et al. 2014; Gröner et al. 2010; Noy et al. 2006; Papavassiliou et al. 2009; Zablith 2011) or multiple loosely interlinked ontologies (Horridge et al. 2013; Klein 2004; Redmond et al. 2008; Vander Sande et al. 2013), where changes in an ontology are independent and do not have to be propagated in order to maintain the validity of overall knowledge within a system. The work of Stojanovic is an exception to this trend, where she provided an attempt to propagate changes to relevant ontologies (Stojanovic 2004). However, her work is not further continued.

2. **Large amounts of data**. Horridge et al. provide an answer to the large-scale challenge of the changed data by introducing binary formats for storing ontology data and differences between ontology versions (Horridge et al. 2013). Their approach is claimed to handle more than one million triples. A different approach is adopted by Graube et al., where named graphs are used to store changes and ontology versions (Graube et al. 2014). Their approach did not scale well for change data analysis, since the query performance on the change data dropped significantly after several thousands of triples. Papavassiliou et al., on the other hand, successfully experimented their approach on almost 200 k triples (Papavassiliou et al. 2009). While the current approaches seem promising, given the closely coupled nature of the engineering data, these approaches need to be reevaluated in order to asses their feasibility.
3. **Changes in schema and instances**. Instance changes are required for addressing changes in the underlying data sources, whereas schema changes are crucial for addressing changes in the represented domains and changes in the intended use of the models and data, as previously mentioned in Sect. 7.2. Several ontology change management approaches are already able to deal with both schema and instance level changes (Horridge et al. 2013; Papavassiliou et al. 2009; Stojanovic 2004), where other approaches either focus on schema (Gröner et al. 2010; Noy et al. 2006; Zablith 2011) or focus on instances (Graube et al. 2014; Vander Sande et al. 2013).
4. **Change validation support**. This aspect of validation provides a mechanism to ensure the validity of data changes according to a predefined set of validation rules (automatic validation) or in combination with domain experts' involvement according to certain workflows (semi-automatic validation). Several approaches already support automatic change validations (Horridge et al. 2013; Stojanovic 2004). Furthermore, there are approaches from general Semantic Web concerning data validation and linked data quality, e.g., RDFUnit (Kontokostas et al. 2014) and Shape Expression (Boneva et al. 2014) that can be adapted to support ontology change validation. In the direction of semi-automatic validation, Stojanovic et al. proposed a mechanism to involve domain experts to check the semantic validity of ontology changes over multiple ontologies (Stojanovic et al. 2002). This involvement of stakeholders is indeed important in the MDEng environment due to the mission-critical characteristic of the domain, as we previously mentioned in Sect. 7.2.1.

5. **High-level change definition and detection**. One of the goals of KCMA is to provide stakeholders with a better decision support system. The high-level change definition and detection process helps to achieve this goal by providing a mean to detect and encapsulate atomic changes into more meaningful and higher level changes in terms of domain concepts, which are easier to understand, especially to non domain experts. In this regards, Papavassiliou et al. have developed an algorithm to support the detection of high-level changes from low-level changes, simplifying the effort to analyze changes in large datasets and without compromising performance (Papavassiliou et al. 2009). Alternatively, Gröner et al. used a subset of OWL-DL reasoning to recognize high-level change patterns (Gröner et al. 2010). One of the prerequisites for high-level change definition and detection is the formalization of low-level changes. This formalization can be achieved using triple patterns (Papavassiliou et al. 2009; Gröner et al. 2010; Vander Sande et al. 2013; Horridge et al. 2013; Graube et al. 2014) or specialized ontologies (Papavassiliou et al. 2009; Palma et al. 2009).
6. **Support for data evolution and versioning**. Ontology evolution (i.e., how to modify the data according to relevant changes in the surrounding) and versioning (i.e., management of different versions of data schema and instances caused by ontology evolution) are both important to the KCMA process and should be available and easily accessible by relevant stakeholders. Most of the ontology change management approaches focus either on ontology evolution (Gröner et al. 2010; Klein 2004; Noy et al. 2006; Zablith 2011) or on ontology versioning (Graube et al. 2014; Vander Sande et al. 2013). The rest of the approaches we surveyed try to address both ontology evolution and versioning (Stojanovic 2004; Horridge et al. 2013).

To conclude, parts of KCMA requirements in MDEng environment are already well explored in Semantic Web research. Schema and instance changes, for example, are addressed already by most approaches. Likewise, approaches for change detection, ontology evolution and ontology versioning are well researched and reported, providing ample options to choose from. However, due to the open nature of web data, approaches for ontology changes in closely interlinked knowledge settings are rarely investigated. Similarly, approaches for handling changes of large amounts of data and validating changes are currently limited, probably since these aspects are not the focus in current ontology change management research. There are options to use general ontology validation approaches for ontology change validation, i.e., by adapting RDFUnit (Kontokostas et al. 2014) or Shape Expression (Boneva et al. 2014) approach, but these adaptations are not yet seen as integral part of general ontology change management approaches. We therefore see the need to advance and combine existing approaches such that all KCMA requirements are sufficiently addressed. The KCMA reference process presented next provides a conceptual framework to guide this process.

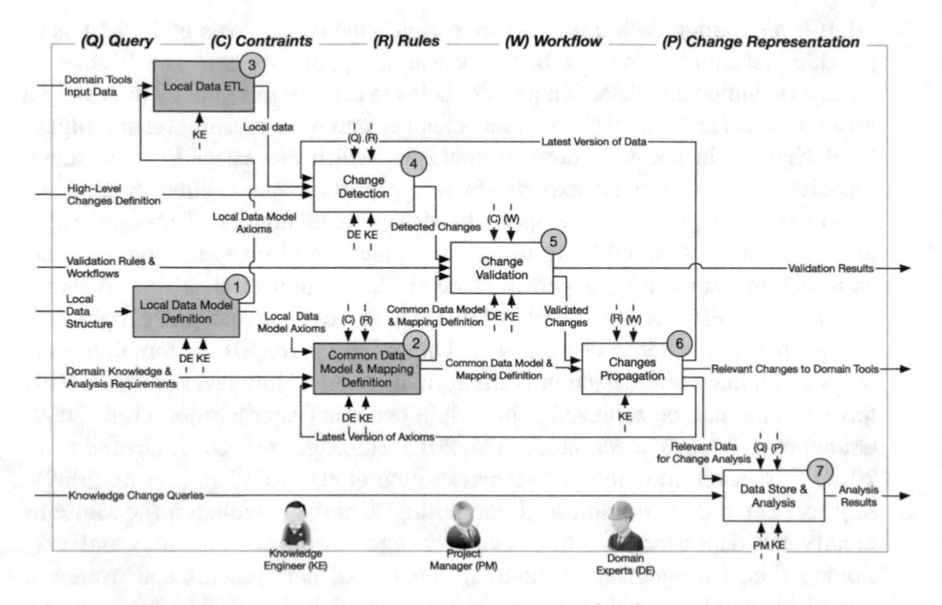

Fig. 7.4 Reference process model for KCMA in the MDEng environment. The *white boxes* extend a typical OBII process (*gray boxes*)

7.5 Reference Process for KCMA in MDEng Environment

In order to address the challenges of providing support for KCMA in MDEng environments, we propose a generic reference process shown in Fig. 7.4. This is a technology agnostic process that could be implemented with technologies drawn from any of the three areas we previously discussed. Implementations using different techniques but following this reference process will be easier to compare and will support a more objective comparison of the strengths and weaknesses of the available technologies. There are KCMA requirements for which we cannot cater at the process level but which should be considered during the implementation of the reference process (e.g., dealing with large amounts of data).

The reference process was derived by adapting and extending the Ontology-Based Information Integration approach (Wache et al. 2001; Calvanese et al. 2001) thus closely connecting process steps for data integration and change management. The reference process is technology agnostic: for this, we replaced all Semantic Web specific terms with general terms, e.g., “ontologies” with “data models.” The extension of OBII consisted in adding four more phases. These phases were derived from relevant related work and requirements and are shown as white boxes in the Fig. 7.4 while the original phases are shown as gray boxes. We utilize an IDEF-0[3] style diagram to structure the proposed approach, in which

[3]http://www.idef.com/idefo-function_modeling_method/.

processes are shown as boxes and resources are shown as directed arrows. The diagram clearly defines input (incoming arrows from the left-hand side of the box), output (outgoing arrows to the right-hand side of the box), consumable resources and stakeholders (input arrows from the bottom of the box) and standards (incoming arrows from the top of the box) used in the reference process.

There are three domain expert roles involved in the framework: Knowledge Engineer (KE), Project Manager (PM), and Domain Expert (DE). Input and output of the system is shown in the left- and right-side of the diagram, respectively. In the following, we explain the seven main phases of the KCMA reference process:

1. **Local Data Model Definition**. This phase requires the Knowledge Engineer and Domain Experts to translate the local tools data structure (e.g., MCAD model for mechanical engineer) to the local data model instance definition.
2. **Common Data Model and Mapping Definition**. KE and DE will define the common data model and its mappings to the local data models. To support this goal, vocabularies and standards are required to formalize the data model and mapping.
3. **Local Data Model Extraction, Transformation, and Load (ETL)**. With regards to the heterogeneous domain tools and their data formats within the MDEng environment, we need to provide the suitable extract, transform, and load (ETL) functions phase to produce the data in the required data model formats.
4. **Change detection**. This phase focuses on the detection of low-level (i.e., triples) and high-level (e.g., semantic and domain-specific) changes between two versions of engineering data. An important point to consider within this phase is to balance the expressiveness of high-level changes defined as input and the computational complexity of the detection algorithm, as mentioned in (Papavassiliou et al. 2009).
5. **Change validation**. The phase of change validation requires the definition of constraints for preserving the validity of data in the local (e.g., mechanical engineering) and global data models (e.g., power plant). Workflow definition is another important element, in order to configure involvement of validation components (e.g., constraint validation engine and domain experts) in the validation process.
6. **Change propagation**. Changes in the MDEng environment need to be propagated to the relevant components (i.e., common data model and other relevant local data models). This phase requires the common data model and mapping definitions, as well as validated changes. The knowledge engineer will configure the propagation based on the mapping definitions to make sure that no corrupted or irrelevant data is included in the propagation process.
7. **Data Store and Analysis**. The goal of this phase is to enable relevant stakeholders (e.g., project manager) to access and analyze the data and its changes within the projects. The changed data will be stored within a designated data store. Examples of queries that will be relevant to this data are: (1) Provenance

information of the changes (e.g., committer, date, reasons of change), (2) Change overview on specific objects, and (3) Analysis of completeness and inconsistencies over changes.

7.6 A Potential Semantic Web-Based Implementation of the KCMA Reference Process

We hereby provide an explanation of how the seven phases of the proposed reference process approach could be implemented with tools and techniques from the Semantic Web community. The framework draws on several standards and technologies, (e.g., SPARQL[4] for querying, PROV-O to represent provenance information) which can be used for structuring and implementing the approach.

1. **Local Data Model Definition**. Based on our experience in the domain, the W3C standard of RDF(S)[5] and OWL[6] languages are sufficient to define local data models. For a more detailed explanation of data model definition and some examples of engineering ontologies, we refer readers to Chap. 5.
2. **Common Data Model and Mapping Definition**. Similar to the case of local data models, we assumed that the RDF(S) and OWL languages are sufficient for common data model definitions. The mapping between common and local data models, however, is more complicated due to the unavailability of W3C standards for this task. There are several proposed approaches, e.g., SPIN[7] and EDOAL[8], as well as the generic SPARQL construct that could be utilized to define mappings between ontologies. However, they have to be investigated further to determine their suitability in the domain. Chapter 6 provides a detailed investigation of technologies for establishing mappings between ontologies.
3. **Local Data Model Extraction, Transformation, and Load (ETL)**. To extract data from heterogeneous data formats, ontology programming frameworks, e.g., Apache Jena[9] or OpenRDF Sesame[10] can be used. Furthermore, there are a number of applications for extracting data from various specific data formats, e.g., RDF123 (Han et al. 2008) and for spreadsheets and XMLTab Protégé plugin[11] and for XML-based documents. Chapter 5 contains information about data extraction from legacy data sources in engineering.

[4]http://www.w3.org/TR/sparql11-overview/.
[5]https://www.w3.org/TR/2004/REC-rdf-primer-20040210/; https://www.w3.org/TR/rdf-schema/.
[6]https://www.w3.org/TR/owl-ref/.
[7]https://www.w3.org/Submission/spin-overview/.
[8]http://alignapi.gforge.inria.fr/edoal.html.
[9]http://jena.apache.org/.
[10]http://rdf4j.org/.
[11]http://protegewiki.stanford.edu/wiki/XML_Tab.

4. **Change detection**. Generic programming frameworks, e.g., Apache Jena or OpenRDF Sesame, are typically able to detect low-level changes between two ontologies at the level of triples. Specialized algorithms are however required to detect high-level changes, either based on heuristics (Noy and Musen 2002), structural differences (Papavassiliou et al. 2009; Redmond and Noy 2011), or OWL reasoning (Gröner et al. 2010). Changes are represented either as triples (e.g., DBPedia change representation from Stadler et al. 2010) or specialized ontologies (e.g., change representation for OWL2 (Palma et al. 2009) and CHAO (Noy et al. 2006)).
5. **Change validation**. To formulate the constraints in a Semantic Web based solution, one possible option is to utilize the upcoming Shapes Constraint Language (SHACL[12]) by SHACL W3C working group, which aims to provide the standard constraint vocabulary for RDF graph data. SHACL strives to aggregate previous efforts of ontology constraint language definition and ontology quality assessment, e.g., RDFUnit (Kontokostas et al. 2014) and Shape Expression (Boneva et al. 2014). Chapter 13 contains examples of using SHACL.
6. **Change propagation**. The change propagation phase is closely related to the ontology mappings defined in the second phase. An example of possible Semantic Web-based solutions in this phase is to develop a change propagation solution based on SPIN mapping or SPARQL constructs.
7. **Data Store and Analysis**. From the Semantic Web perspective, one possible data storage solution is to utilize an RDF triplestore (e.g., OpenRDF Sesame). The PROV-O W3C standard can be used for representing change provenance information. Another important aspect in this phase is to provide different types of stakeholders with comfortable access for analyzing the data (Ekaputra et al. 2013). An example of end-user interface for ontology data analysis in the MDEng domain is presented in (Sabou et al. 2016).

To conclude, the current set of Semantic Web Technology tools and methodologies provides a relatively mature foundation for the implementation of the KCMA process. Semantic modeling, for instance, is required for *local data model definition* and *common data model and mapping definition phases*. This process can be achieved by relying on widely accepted W3C standards such as RDF(S) and OWL. The current set of ontology programming frameworks (e.g., Apache Jena and OpenRDF Sesame; required for *ETL* and *change detection phase*) could also be considered stable and well-maintained, while storage options for RDF graph data shows encouraging signs with more than 15 tools (commercial and open source) available at the moment[13] (required for *data store and analysis phase*).

However, there are critical issues to address for Semantic Web-based implementations of MDEng KCMA, namely the standardization of ontology constraint and mapping languages. Ontology constraint languages are required for the *change*

---

[12]https://w3c.github.io/data-shapes/shacl/.

[13]http://db-engines.com/en/ranking/rdf+store. Accessed on 16.02.2016.

validation phase. The upcoming SHACL standard is an important development in this direction and will be important for the ontology constraint aspect of the KCMA. Ontology mapping languages are required for the *common data model and mapping definition* and for the *change propagation* phases. Their development, however, is still yet to converge to W3C standards. EDOAL, SPIN, and SPARQL construct are currently the best options available. To overcome the lack of standards for mapping definitions, the adaptation of mapping methodologies from other research communities, (e.g., Triple Graph Grammars (Schürr and Klar 2008) from Model Driven Engineering) could be considered as additional options.

7.7 Summary and Future Work

In this book chapter, we have defined the context and challenges of KCMA in MDEng environments. To address the challenges, we have identified key requirements of KCMA in MDEng and provided an overview of techniques from three relevant research areas, namely Database Systems, Model-Based Engineering, and Semantic Web Technologies.

Our contribution in this book chapter is to generalize and extend the OBII approach, previously proposed for the purposes of data integration, to become a generic reference process of KCMA in MDEng environment. This generic and technology agnostic reference process is meant to lay the foundation towards a solution for providing a fully functional OBII-based KCMA solution for MDEng.

As our future work, we plan to provide a first concrete implementation of the reference process utilizing a selection of Semantic Web Technologies as discussed in Sect. 7.6. Additional concrete implementations using techniques from other research area (i.e., database and MBE) are also planned to enable comparison of the strengths and limitations of these different technologies for the KCMA tasks. We also plan to generalize the approach to address similar problem settings in other application domains, such as scholarly data management in empirical software engineering (Biffl et al. 2014).

Acknowledgments This work was supported by the Christian Doppler Forschungsgesellschaft, the Federal Ministry of Economy, Family and Youth, and the National Foundation for Research, Technology and Development in Austria.

References

Berardinelli, L., Biffl, S., Mätzler, E., Mayerhofer, T., Wimmer, M.: Model-based co-evolution of production systems and their libraries with automationML. In: Proceedings of the 20th IEEE International Conference on Emerging Technologies and Factory Automation (ETFA 2015), pp. 1–8 (2015)

Biffl, S., Kalinowski, M., Ekaputra, F.J., Serral, E., Winkler, D.: Building empirical software engineering bodies of knowledge with systematic knowledge engineering. In: Proceedings of the 26th International Conference on Software Engineering and Knowledge Engineering (SEKE 2014), pp. 552–559 (2014)

Boneva, I., Gayo, J.E.L., Hym, S., Prud'hommeau, E.G., Solbrig, H., Staworko, S.: Validating RDF with shape expressions. Technical Report (2014). arXiv:1404.1270

Calvanese, D., De Giacomo, G., Lenzerini, M.: Ontology of integration and integration of ontologies. Description Logics **49**, 10–19 (2001)

Ekaputra, F.J., Serral, E., Winkler, D., Biffl, S.: An analysis framework for ontology querying tools. In: Proceedings of the 9th International Conference on Semantic Systems (iSEMANTICS 2013), pp. 1–8. ACM (2013)

Ekaputra, F.J., Serral, E., Sabou, M., Biffl, S.: Knowledge change management and analysis for multi-disciplinary engineering environments. In: Proceedings of the Posters and Demos Track of 11th International Conference on Semantic Systems (SEMANTiCS 2015) (2015)

Goring, M., Fay, A.: Modeling change and structural dependencies of automation systems. In: Proceedings of the 17th Conference on Emerging Technologies and Factory Automation (ETFA 2012), pp. 1–8. IEEE (2012)

Graube, M., Hensel, S., Urbas, L.: R43ples: Revisions for triples. In: Proceedings of the 1st Workshop on Linked Data Quality co-located with 10th International Conference on Semantic Systems (SEMANTiCS 2014) (2014)

Gröner, G., Parreiras, F.S., Staab, S.: Semantic recognition of ontology refactoring. In: Proceedings of the 9th International Semantic Web Conference (ISWC 2010), pp. 273–288. Springer (2010)

Han, L., Finin, T., Parr, C., Sachs, J., Joshi, A.: RDF123: From Spreadsheets to RDF. In: Proceedings of the 7th International Conference on the Semantic Web (ISWC 2008), pp. 451–466. Springer (2008)

Horridge, M., Redmond, T., Tudorache, T., Musen, M.A.: Binary OWL. In: Proceedings of the 10th International Workshop on OWL: Experiences and Directions (OWLED 2013), co-located with 10th Extended Semantic Web Conference (ESWC 2013) (2013)

Kehrer, T., Kelter, U., Taentzer, G.: Propagation of software model changes in the context of industrial plant automation. Automatisierungstechnik **62**(11), 803–814 (2014)

Klein, M.: Change management for distributed ontologies. Ph.D. thesis, Vrije Universiteit Amsterdam (2004)

Kontokostas, D., Westphal, P., Auer, S., Hellmann, S., Lehmann, J., Cornelissen, R., Zaveri, A.: Test-driven evaluation of linked data quality. In: Proceedings of the 23rd International Conference on World Wide Web (2014), pp. 747–758 (2014)

McBrien, P., Poulovassilis, A.: Schema evolution in heterogeneous database architectures, a schema transformation approach. In: Proceedings of the 14th International Conference of Advanced Information Systems Engineering (CAiSE 2002), pp. 484–499. Springer (2002)

Meyers, B., Vangheluwe, H.: A framework for evolution of modelling languages. Sci. Comput. Program. **76**(12), 1223–1246 (2011)

Mordinyi, R., Serral, E., Winkler, D., Biffl, S.: Evaluating software architectures using ontologies for storing and versioning of engineering data in heterogeneous systems engineering environments. In: Proceedings of the 15th Conference on Emerging Technologies and Factory Automation (ETFA 2012), pp. 1–10. IEEE (2014)

Noy, N.F., Klein, M.: Ontology evolution: not the same as schema evolution. Knowl. Inf. Syst. **6** (4), 428–440 (2004)

Noy, N.F., Musen, M.A.: Promptdiff: a fixed-point algorithm for comparing ontology versions. In: Proceedings of the 18th National Conference on Artificial Intelligence (AAAI/IAAI 2002), pp. 744–750 (2002)

Noy, N.F., Chugh, A., Liu, W., Musen, M.A.: A framework for ontology evolution in collaborative environments. In: Proceedings of the 5th International Conference on the Semantic Web (ISWC 2006), pp. 544–558. Springer (2006)

Palma, R., Haase, P., Corcho, O., Gómez-Pérez, A.: Change representation for OWL 2 ontologies. In: Proceedings of the 6th International Workshop on OWL: Experiences and Directions (OWLED 2009) (2009)

Papavassiliou, V., Flouris, G., Fundulaki, I., Kotzinos, D., Christophides, V.: On detecting high-level changes in RDF/S KBs. In: Proceedings of the 8th International Conference on the Semantic Web (ISWC 2009), pp. 473–488 (2009)

Redmond, T., Noy, N.: Computing the changes between ontologies. In: Proceedings of the Joint Workshop on Knowledge Evolution and Ontology Dynamics (EVODYN 2011), pp. 1–14 (2011)

Redmond, T., Smith, M., Drummond, N., Tudorache, T.: Managing change: an ontology version control system. In: Proceedings of the 5th International Workshop on OWL: Experiences and Directions (OWLED 2008) (2008)

Roddick, J.F.: A survey of schema versioning issues for database systems. Inf. Softw. Technol. **37** (7), 383–393 (1995)

Sabou, M., Ekaputra, F.J., Kovalenko, O., Biffl, S.: Supporting the engineering of cyber-physical production systems with the AutomationML analyzer. In: Proceedings of the Cyber-Physical Production Systems Workshop (CPPS 2016) (2016)

Schürr, A., Klar, F.: 15 years of triple graph grammars. In: Proceedings of the 4th International Conference on Graph Transformations (ICGT 2008), pp. 411–425. Springer, Berlin (2008)

Serral, E., Mordinyi, R., Kovalenko, O., Winkler, D., Biffl, S.: Evaluation of semantic data storages for integrating heterogenous disciplines in automation systems engineering. In: Proceedings of the 39th Annual Conference of Industrial Electronics Society Conference (IECON 2013), pp. 6858–6865. IEEE (2013)

Stadler, C., Martin, M., Lehmann, J., Hellmann, S.: Update strategies for DBpedia live. In: Proceedings of the Sixth Workshop on Scripting and Development for the Semantic Web (SFSW 2010) (2010)

Stojanovic, L.: Methods and tools for ontology evolution. Ph.D. thesis, Karlsruhe Institute of Technology (2004)

Stojanovic, L., Maedche, A., Motik, B., Stojanovic, N.: User-driven ontology evolution management. In: Proceedings of the 13th International Conference on Knowledge Engineering and Knowledge Management (EKAW 2002), pp. 285–300. Springer (2002)

Taentzer, G., Ermel, C., Langer, P., Wimmer, M.: A fundamental approach to model versioning based on graph modifications: from theory to implementation. Softw. Syst. Model. **13**(1), 239–272 (2014)

Vander Sande, M., Colpaert, P., Verborgh, R., Coppens, S., Mannens, E., Van de Walle, R.: R&Wbase: git for triples. In: Proceedings of the Linked Data on the Web Workshop (LDOW 2013) (2013)

Vogel-Heuser, B., Legat, C., Folmer, J., Rösch, S.: Challenges of parallel evolution in production automation focusing on requirements specification and fault handling. Automatisierungstechnik **62**(11), 758–770 (2014)

Wache, H., Voegele, T., Visser, U., Stuckenschmidt, H., Schuster, G., Neumann, H., Hübner, S.: Ontology-based integration of information: a survey of existing approaches. In: IJCAI-01 Workshop: Ontologies and Information Sharing, pp. 108–117 (2001)

Waltersdorfer, F., Moser, T., Zoitl, A., Biffl, S.: Version management and conflict detection across heterogeneous engineering data models. In: Proceedings of the 8th International Conference on Industrial Informatics (INDIN 2010), IEEE, pp. 928–935 (2010)

Zablith, F.: Harvesting online ontologies for ontology evolution. Ph.D. thesis, The Open University, UK (2011)

Part III
Intelligent Applications for Multi-disciplinary Engineering

Chapter 8
Semantic Data Integration: Tools and Architectures

Richard Mordinyi, Estefania Serral and Fajar Juang Ekaputra

Abstract This chapter is focused on the technical aspects of semantic data integration that provides solutions for bridging semantic gaps between common project-level concepts and the local tool concepts as identified in the Engineering Knowledge Base (EKB). Based on the elicitation of use case requirements from automation systems engineering, the chapter identifies required capabilities an EKB software architecture has to consider. The chapter describes four EKB software architecture variants and their components, and discusses identified drawbacks and advantages regarding the utilization of ontologies. A benchmark is defined to evaluate the efficiency of the EKB software architecture variants in the context of selected quality attributes, like performance and scalability. Main results suggest that architectures relying on a relational database still outperform traditional ontology storages while NoSQL databases outperforms for query execution.

Keywords Ontology · Semantic data integration · Versioning · Performance · Multidisciplinary projects

---

R. Mordinyi (✉) · F.J. Ekaputra
Institute of Software Technology and Interactive Systems, CDL-Flex,
Vienna University of Technology, Vienna, Austria
e-mail: richard.mordinyi@tuwien.ac.at

F.J. Ekaputra
e-mail: fajar.ekaputra@tuwien.ac.at

E. Serral
Leuven Institute for Research on Information Systems (LIRIS), KU Leuven,
Naamsestraat 69, 3555, 3000 Louvain, Belgium
e-mail: estefania.serralasensio@kuleuven.be

S. Biffl and M. Sabou (eds.), *Semantic Web Technologies for Intelligent Engineering Applications*, DOI 10.1007/978-3-319-41490-4_8

8.1 Introduction

In large-scale systems engineering projects, like power plants, steel mills, or car manufactures, the seamless cooperation and data exchange of expert knowledge from various engineering domains and organizations is a crucial success factor (Biffl et al. 2009a). This environment consists of a wide range of engineering systems and tools that differ in the underlying technical platforms and the used data models. Each domain or organization usually prefers using their own well-known models, from now on referred as local tool models. In order to successfully develop projects, it is essential to integrate important knowledge of different domain experts. However, these experts usually prefer using their well-known local tool models. In addition, they want to access data from other tools within their local data representation approach (Moser and Biffl 2012). The standardization of data interchange is one of the most promising approaches (Wiesner et al. 2011) to enable efficient data integration that allows experts to continue using their familiar data models and formats. This approach is based on agreeing on a minimal common model for data exchange that represents the common concepts shared among different disciplines on project level.

Chapter 2 presented main use cases with typical process steps during the engineering phase within the life cycle of production systems. Selected scenarios focused on the capability to interact appropriately within a multidisciplinary engineering network while pointing out the need for a common vocabulary over all engineering disciplines involved in an engineering organization. The described challenges in the context of engineering data integration referred to a consistent production system plant model in order to support quality-assured parallel engineering, and the ability to access and analyze integrated data, e.g., for project progress and project quality reports. Versioning of exchanged information helps to improve change management and team collaboration over the course of the engineering project. As part of an efficient data management it is essential for process observations, project monitoring, and control across engineering disciplines (Moser et al. 2011b).

As a common baseline it can be concluded that it is necessary to clearly distinguish between local concepts of engineering tools and common concepts (Moser and Biffl 2010) (i.e., data sets representing heterogeneous but semantically corresponding local data elements) at project level. Consequently, interoperability between heterogeneous engineering environments is only supported if the semantic gap between local tool concepts and common project-level concepts can be properly bridged. The Engineering Knowledge Base (EKB) (Moser and Biffl 2010) (see Chap. 4) provides the means for semantic integration of the heterogeneous models of each discipline using ontologies, and thus facilitates seamless communication, interaction, and data exchange. Semantic technologies are capable of linking corresponding local concepts of engineering tools with each other via common project-level concepts representing the data integration needs of engineering disciplines at their interfaces.

With respect to the defined scenarios, conducted interviews with experts from industry (Biffl et al. 2012), and the application of two types of concepts in an engineering network, we have identified three fundamental requirements relevant for taking decisions regarding the software architecture of an integration platform

- Data insertion in the local tool ontologies should minimize effort and time for validating changes and assuring data consistency (Mordinyi et al. 2011, 2012), and versioning of exchanged data for sophisticated query support.
- Data transformations between local engineering concepts through common concepts should reduce the effort for the quality controlled propagation of changes across the different disciplines within the project (Winkler et al. 2011).
- Queries across different local data models using the syntax of common concepts should support collaboration and management at a project level by relying on capabilities for analysis of process data and facilitating project monitoring and control (Moser et al. 2011b).

Research progress in semantic integration for efficient data management provides the foundation and the mechanisms to define mappings between local and common engineering concepts for integrating engineering knowledge from different domain in the project (Wiesner et al. 2011), as well as the necessary transformations following these mappings for sharing data (Moser and Biffl 2012). Using ontologies and OWL[1] (a standardized ontology language) is a common approach to explicitly specify semantics of the underlying data models of engineering tools and the relations among them (Moser et al. 2011a; Wiesner et al. 2011; Moser and Biffl 2012). However, main scenarios for ontology applications focus on the classification of data and knowledge gathering. Typical scenarios refer to minimal number of individual adaptations along with the support for high number of queries and knowledge reasoning. In fact, for specific application scenarios, OWL 2 profiles[2] have been additionally introduced to gain implementation benefits for querying and reasoning. Consequently, this comes at the expense of a restricted OWL DL specification.

For an Engineering Knowledge Base (EKB) that efficiently addresses the requirements listed before, it is an important decision factor which semantic storage should be used to store data of the local and common ontologies. Studies like (Bizer and Schultz 2001; Mironov et al. 2010; Haslhofer et al. 2011) have analyzed several implementations of semantic storages regarding, e.g., performance or memory consumption, but do not take the requirements of integration scenarios in multi-disciplinary engineering projects into account. Ontology storages utilized for data integration in such engineering projects need to support frequent changes of engineering data for facilitating efficient change management and synchronization mechanisms (Winkler et al. 2011) as well as versioning (Waltersdorfer et al. 2010) of the exchanged data to enable analysis and monitoring of project progress, and

[1]https://www.w3.org/2001/sw/wiki/OWL.

[2]OWL 2: http://www.w3.org/TR/owl2-overview/#Profiles.

detection of risks (Moser et al. 2011b). In (Serral et al. 2013, 2012) three different types of semantic storages have been evaluated regarding (i) data insertion, (ii) data transformation, and (iii) data querying in the context of integration use cases, but it misses evaluation of data transformation over common project-level concepts as well as data management and querying in the context of versioning.

In this chapter, four software architectures are analyzed and evaluated regarding the mentioned integration requirements: (i) insertion and versioning of engineering data, (ii) transformation between local and common concepts, and (iii) querying of common concepts. Each of the architectures makes use of ontologies in different ways

(a) **Semantic Data Store**: since semantic data integration is facilitated using ontologies, the first choice is an architecture with an ontology component (SWAD-Europe 2002) that stores and manages both concepts and versions of individuals.
(b) **RDB2RDF Mapper**: even though ontology stores have been heavily investigated in the recent past, they are fairly new in comparison to databases which have a long research history and robust implementations. The second architecture variant consists of an ontology store that manages concepts only. Individuals are stored and versioned in a database while SPARQL[3] queries are automatically transformed into SQL queries using relational database to RDF mappers (Sequeda et al. 2012).
(c) **NoSQL Graph Databases**: such database type as a component in a software architecture for integration seems reasonable for two aspects: (1) NoSQL DBs provide better horizontal scalability (Cattell 2011) and are more flexible for schemaless operations when compared to traditional relational databases, and (2) they offer hierarchical representations of structures similar to those used in ontologies. Consequently, dependencies and relations between models stored in an ontology may be represented in the same way for model instances in the database while avoiding potential schema mismatches.
(d) **Versioning System**: in accordance with the requirements, the fourth software architecture variant utilizes a proper versioning management system for administrating individuals. Similar to the three previous variants, an ontology component copes with the concepts while an indexing framework with the individuals supporting the execution of SPARQL queries.

Based on the aforementioned limitations of current ontology storage implementations and the requirements of multidisciplinary engineering projects, we derive the following research issues:

RI-1—Data Models: Investigate how a metamodel for versioning engineering data models, common concepts, and relations need to be modeled in order to support efficient insertion, transformation, and querying

[3]https://www.w3.org/TR/rdf-sparql-query/.

RI-2—Historical Data Analysis: Investigate to what extent the storage concepts are capable of executing SPARQL queries for analyzing versioning information for, e.g., project monitoring purposes and evaluate their performance
RI-3—Prototype Evaluation: Evaluate the performance, memory consumption, and storage size of the storage concept in the context of instance versioning and concept transformations. How do they behave in case of high amount of exchanged data and type of operation (i.e., insertion, deletion, or update)? Investigate the performance relation between graph databases, relational databases, and ontology-based stores. Which factors for complexity and characteristic variables influence the performance of the solution?

This chapter focuses on analyzing and evaluating the different types of software architecture candidates and their strategies to organize knowledge and data with respect to the previously identified requirements. On the basis of a real-world industrial scenario and driven by use cases as presented in Chap. 2, we describe the advantage and limitations of each of the proposed EKB software architecture variants, compare gained evaluation results regarding performance, memory, and disk usage, and discuss their complexity in the context of usability, maintenance, and scalability.

The remainder of this chapter is structured as follows: Sect. 8.2 summarizes related work on semantic integration and storage capabilities while Sect. 8.3 presents a use case from an industrial partner to emphasize requirements Sect. 8.4 illustrates the different EKB software architecture variants and their usage of ontologies. Section 8.5 investigates the evaluation design and the results of the comparison. Section 8.6 discusses them, while Sect. 8.7 concludes the chapter and identifies further work.

8.2 Related Work

In this section, we summarize the most important related work on (a) Semantic Web technologies, (b) semantic data integration for general problem space illustration, (c) the Engineering Knowledge Base (EKB) as a specific framework used in the evaluation scenarios to facilitate semantic integration, (d) semantic data stores for managing ontology data, (e) NoSQL graph databases, and (f) versioning capabilities.

8.2.1 Semantic Web Technologies

The Semantic Web is the vision of Tim Berners-Lee to have World Wide Web contents understandable for both humans and machines (Berners-Lee et al. 2001). To achieve the Semantic Web vision, a set of supporting tools and technologies (hence called Semantic Web Technologies) is developed. We will briefly explain two main concepts of Semantic Web technologies, namely Ontology and SPARQL query language. For a general introduction of Semantic Web Technologies used in the domain, we refer the reader to the Chap. 3 of this book.

Ontology is an explicit specification of a conceptualization to enable knowledge sharing and reuse (Gruber 1995). Ontology holds an important role as the main pillar of Semantic Web Technologies (Berners-Lee et al. 2001). To represent the ontology for Semantic Web, W3C develop several recommendations Semantic Web languages (also called vocabularies), such as Resource Description Framework (RDF[4]) and RDF Schema (RDFS[5]). Web Ontology Language (OWL[6]) is later introduced to provide definition for a more complex knowledge and relations between concepts.

SPARQL[7] is a W3C recommendation for querying and manipulating RDF graph data and is widely used in the Semantic Web community. SPARQL can be used for querying and manipulating RDF graph data across different data sources as well as a single source. Besides to be used directly to query RDF data, SPARQL engines are used to build several other technologies, such as constraint checking (e.g., SPIN, SHACL) and transformation (e.g., SPIN,[8] using SPARQL Construct).

8.2.2 Semantic Data Integration

Software systems developers use a wide range of tools from software vendors, open-source communities, and in-house developers to develop their systems. To make these tools working together to support developers' development process in an engineering environment remains challenging, as these tools follow a wide variety of standards (IEEE 2007). Any integration approach has to address the levels of technical heterogeneity, i.e., how to connect systems that use different platforms or protocols to enable messages exchanges (Hohpe and Woolf 2003; Chappell 2004); and semantic heterogeneity, i.e., how to translate the content of the messages between heterogeneous systems that utilize different local terminologies for common concepts in their domain of discourse, so these systems can understand each other and conduct a meaningful conversation (Aldred et al. 2006; Hohpe 2006; Moser et al. 2009a).

Semantic integration is defined as the solving of problems originating from the intent to share data across disparate and semantically heterogeneous data (Halevy 2005). These problems include the heterogeneous data schemas alignment, the

[4]RDF: http://www.w3.org/TR/2004/REC-rdf-primer-20040210/.

[5]RDFS: http://www.w3.org/TR/rdf-schema/.

[6]OWL2: http://www.w3.org/TR/owl2-overview/.

[7]SPARQL 1.1: http://www.w3.org/TR/sparql11-query/.

[8]SPIN: http://www.w3.org/Submission/spin-overview/.

duplicate entries detection, the inconsistencies reconciliation, and the complex relations modeling in different data sources (Noy 2004; Noy et al. 2005). One of the most important and most actively studied problems in semantic integration is establishing semantic correspondences (also called mappings or alignment) between vocabularies of different data sources (Doan et al. 2004).

In (Noy 2004), three major aspects of the ontologies applications were identified for supporting semantic integration: (a) the (semi-) automatic mappings finding tasks, (b) the declarative formal representation of these mappings, and (c) reasoning using these mappings. Approaches comprising heuristics-based or machine learning techniques that use various characteristics of ontologies (e.g., structure, concepts, and instances) may be used to find mappings, or alternatively developers of different applications may agree upon a general upper ontology. These approaches are similar to approaches for mapping XML schemas or other structured data (Bergamaschi et al. 1999). Naturally, mappings definition between ontologies, whether it is conducted automatically, semi-automatically, or interactively, is not a goal in itself. The goal is to use the resulting mappings for various integration tasks: e.g., data transformation and query answering. Since ontologies are often used for reasoning, it is only natural that many of these integration tasks involve reasoning over source ontologies and mappings (Noy et al. 2005).

8.2.3 Engineering Knowledge Base

Data integration defined as "the problem of combining data residing at different sources, and providing the user with a unified view of these data" (Lenzerini 2002). Heterogeneous data sources with incompatible data models or even incompatible technologies and their data has to be offered to users through a uniform interface. Engineering Knowledge Base (EKB) framework has been introduced which acts as an approach for addressing challenges originating from data heterogeneity that can be applied for a range of domains (Moser et al. 2009a), e.g., in the production automation domain (Moser and Biffl 2010), automation systems engineering, and Software Engineering. More details on the framework can be found in Chap. 4.

The EKB, as shown in Fig. 8.1, could be seen as a Semantic Web-based data modeling approach (Moser 2009), which supports explicit modeling of existing knowledge in machine-understandable ontology syntax (e.g., using RDF/S or OWL as knowledge representation and SPARQL, SPIN, or EDOAL[9] as mapping representation between data models). The EKB focuses on providing links between local data structures and support of information exchange between these local data structures through a common data structure to make systems engineering more efficient and flexible. The EKB framework (Moser et al. 2009a) stores the local

[9]EDOAL: http://alignapi.gforge.inria.fr/edoal.html.

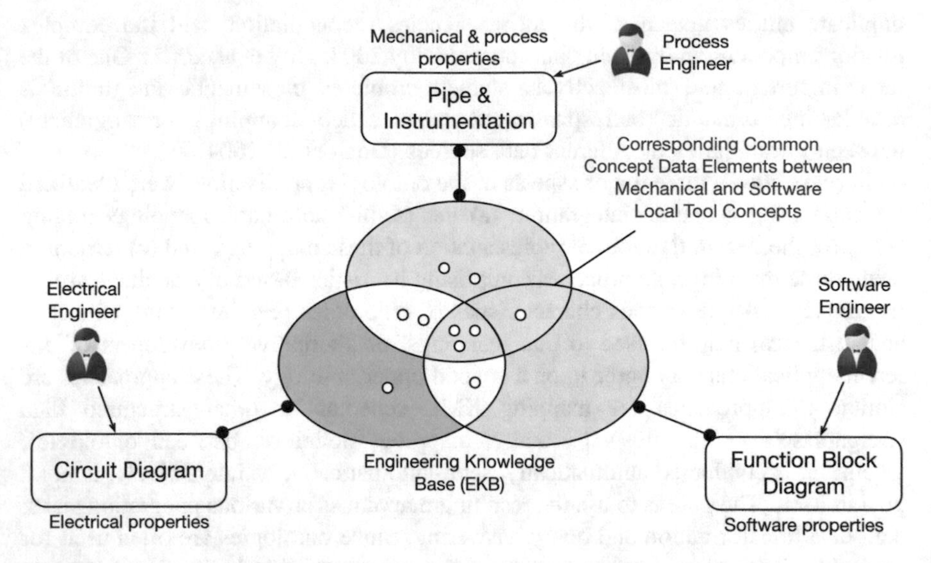

Fig. 8.1 EKB concept of mapping data elements (based on Mordinyi et al. 2012)

knowledge in ontologies and provides semantic mapping services to access design-time and run-time concepts and data.

The general mechanism of the EKB framework uses common concepts identified prior to the data integration process as basis for mappings between proprietary local knowledge and more generic domain specific knowledge. Due to the mappings definition between local and domain ontologies, data structures that are semantically equal can be identified (Moser et al. 2009b). Furthermore, the EKB also holds the mappings between the stored data models, allowing the definition of mapping between one or more attributes of a data model entity from source model to another attribute of a data model entity of a target data model. Based on these mappings, transformations (Biffl et al. 2009b; Moser et al. 2010) between two data models can be defined by the user.

8.2.4 Semantic Data Stores

Numerous semantic data storages have been developed to store and support the management of semantic data (Haslhofer et al. 2011). These storages can be classified as (a) File-based storage, (b) RDF triplestores (e.g., Jena TDB), and (c) Relational databases storages (e.g., Jena SDB).

8.2.4.1 Ontology in File Stores

Ontology File Storage is one or more files where the ontology and its data instances (i.e., so-called individuals) are stored together either in a single file or distributed among segments across several files.

For data management and querying, this approach loads the complete ontology model data (i.e., the engineering knowledge base definition and the data instances) into the memory. This allows the data to be accessed and managed using Ontology API (e.g., Apache Jena[10] or OpenRDF Sesame[11]), but it has the drawback of high memory consumption. In this regards, OWLIM (Bishop et al. 2011) also provides SwiftOWLIM, which is an in-memory RDF database. It uses optimized indexes and data structures to be able to process tens of millions of RDF statements on standard desktop hardware. Jena, Sesame, and SwiftOWLIM source code are provided free of charge for any purpose.

Some approaches, such as (De Leon Battista et al. 2007; Novák and Šindelář 2011) have successfully used this architecture; however, in both approaches the authors plan to use other more sophisticated storage solutions to make their approaches scalable to large models. More details about these approaches can be found in (Serral et al. 2012). The Ontology File Store (in-memory) store is considered very useful for tests or small examples, but in general it is not recommended for working with large models.

8.2.4.2 Ontology in Triple Stores

The engineering data instances can be also stored and managed using triple stores in the similar way that they stored in files and managed in memory. In a triple store, ontologies and the instances are specified using ontology languages—however, the instances (i.e., the individuals) are internally managed using special databases built specifically for storing triples (i.e., subject-predicate-object expressions, a specific form of representing data using an ontology). These databases are also called semantic stores or triple stores. In this way, the database management is transparent for users and the data can be accessed and managed.

The generic schema of these special databases corresponds to one table that contains three columns named Subject, Predicate, and Object. Thus, it reflects the triple nature of RDF statements. In practice, most of the RDF triplestores add the fourth column that stores the named graph of these RDF statements. The triple store can be used in its pure form (Oldakowski et al. 2005), but most existing systems add several modifications to improve performance or maintainability. A typical approach, the so-called normalized triple store, is adding two further tables to store resource URIs and literals separately, which requires significantly less storage space

[10]http://jena.apache.org.

[11]http://www.openrdf.org.

(Harris and Gibbins 2003). Furthermore, a hybrid of the normalized triple store can be used, allowing the values to be stored themselves either in the triple table or in the resources table.[12] Therefore, users can manage the data and use SPARQL queries having a better performance than ontology file stores thanks to the use of the databases using this approach.

Some relevant examples of these stores are Apache Jena TDB[13] and BigOWLIM (Lu et al. 2007), OWLIM-SE,[14] and Bigdata.[15] Apache Jena TDB supports the full range of Jena APIs and TDB performs and scales well. BigOWLIM is designed for large data volumes and uses file-based indices that allow it to scale, positioning it as an enterprise-grade database management system that can handle tens of billions of statements. OWLIM-SE is native RDF rule entailment and storage engine and the commercial edition of OWLIM. OWLIM-SE is implemented in Java and packaged as a storage and inference layer for the Sesame RDF framework. Loading, reasoning, and query evaluation, even against huge ontologies and knowledge bases, can be executed very quickly. Bigdata is an open-source general-purpose scale-out storage. This storage is fully persistent, is compliant to Sesame 2, and includes a high-performance RDF database supporting RDFS and limited OWL inference. Some examples of applications of this architecture are (Klieber et al. 2009; Miles et al. 2010). More details about these approaches can be found in (Serral et al. 2012).

8.2.4.3 Ontology in Relational Databases

Relational Database Storages typically store the ontology (terminological component or T-Box) separately using various methods (e.g., as mapping configuration, or RDF file), while the data instances (assertion components or A-Box) are stored within relational databases. In this case, only the T-Box is extracted into a memory while instances have to be accessed by ordinary database queries.

Furthermore, SPARQL queries formulated for the ontology have to be translated into the query language associated with the relational database. An approach applying this architecture is for instance the Apache Jena SDB.[16] It allows the data of the model to be stored in a relational database, like Oracle, PostgreSQL,[17] MySQL,[18] or HSQLDB.[19] A Jena SDB storage can be accessed and managed with the Jena API and can be queried using SPARQL. Sesame also facilitates relational

[12]http://jena.sourceforge.net/DB/layout.html.

[13]Jena TDB: http://jena.apache.org/documentation/tdb/index.html.

[14]OWLIM-SE: http://www.ontotext.com/owlim.

[15]Bigdata: http://www.systap.com/bigdata.htm.

[16]Apache Jena SDB: https://jena.apache.org/documentation/sdb/.

[17]PostgreSQL: http://www.postgresql.org.

[18]MySQL: https://www.mysql.com/.

[19]HSQLDB: http://hsqldb.org/.

database support, like on PostgreSQL (see footnote 20) or MySQL but enables only direct execution of SPARQL queries. This type of storage adopts binary tables for the database, mapping the triples of the RDF graph to these binary tables (Lu et al. 2007).

A slightly different approach is D2RQ (Bizer and Seaborne 2004), which is an RDF-based platform that is used to access the content of relational databases without having to replicate it into RDF storage. The D2RQ is open-source software published under the Apache license. The implementation belongs to the family of relational database to RDF mappers, a platform to query databases as Virtual RDF Graphs. Other implementations are Quest (Rodriguez-Muro et al. 2012), Virtuoso (Dupont et al. 2011), or Ultrawrap (Sequeda and Miranker 2013).

Minerva (Zhou et al. 2006) is a component of the IBM Integrated Ontology Development Toolkit (IODT) that supports SPARQL querying. Using Minerva, one can store multiple large-scale ontologies in different ontology stores, launch SPARQL queries, and obtain results listed in tables or visualized as RDF graphs. Currently, Minerva can cope with IBM DB2, Apache Derby[20] and HSQLDB (see footnote 22) as the back-end database. Other examples are Oracle 10 g RDBMS, Sesame on PostgreSQL, and DLDBOWL.[21] This type of storage adopts binary tables for the database, mapping the triples of the RDF graph to these binary tables. The most common schema representation of an ontology is composed by a table for each class (resp. each property); each class table stores all instances belonging to the same class and each property table stores all triples which have the same property (Lu et al. 2007). This architecture has been successfully applied in several projects such as (Tinelli et al. 2009; Calvanese et al. 2011; Wiesner et al. 2011). More details about these approaches can be found in (Serral et al. 2012).

8.2.5 NoSQL Graph Databases

We identified two major standards for interfacing with graph databases: (a) the Blueprints API by TinkerPop (TinkerPop) and (b) the W3C Linked Data Specification.

The Blueprints API is part of TinkerPop, an open-source graph computing framework. It is not a graph database per se, but provides useful tools for implementing graph databases, like the Gremlin query language. The advantages for graph database vendors in implementing the API is the possibility to combine it with tools provided by TinkerPop without further effort and its interoperability with other databases from other vendors. Property Graph model is used by TinkerPop as a base structure for its data, which is stored by vertices and edges (represented as

[20]http://incubator.apache.org/derby.

[21]http://swat.cse.lehigh.edu/downloads/dldb-owl.html.

directed arcs). A vertex can have a set of key/value pairs called properties. Edges connect exactly two vertices and might feature a property map as well (TinkerPop).

Major implementations of the W3C Specification on Linked Data are based on the Resource Description Framework (RDF). A set of triples builds the underlying structure of RDF itself. Each triple consists of a subject, predicate as well as an object and might be represented as a directed graph. Apache Jena, OpenRDF Sesame, or AllegroGraph are example APIs that are able to manage the set of triples, and provide parsers for loading and exporting RDF graphs in different formats, as well as providing storage solutions for the triples as well as support for reasoning and querying RDF data via the SPARQL query languages. The triple-based approach is simpler and more flexible compared to the Property Graph model used by TinkerPop. Nevertheless, the number of triples (e.g., nodes and edges) stored in the triple-based storage approach is much higher than that of the property graph model, because a new triple is needed for each additional attribute.

TinkerPop provide support for interoperability with RDF by providing two components for interfacing with RDF data in the OpenRDF Sesame framework. First, TinkerPop SailGraph can be used as mapper from the TinkerPop stack to a Sail implementation provided by OpenRDF. Sail is an interface that abstracts the storage layer used in OpenRDF. Thus, SailGraph allows persisting a Blueprints graph via OpenRDF. Second, TinkerPop GraphSail another implementation of the Sail interface enables TinkerPop to store RDF data from OpenRDF into a TinkerPop database.

8.2.6 Versioning

In software engineering, version control is an important challenge in application development and maintenance, i.e., the task of keeping a software system consisting of many components with different versions and configurations well organized (Tichy 1985). The objective of a version control system is to maintain and control transformations made to a software system during its development and maintenance. Controlling access to source files, storing, identifying, retrieving different versions of source and target files, and rebuilding target files when source files change (Korel et al. 1991) are some of the challenges of a version control system. Prominent implementations of a versioning system are for instance SVN (Pilato 2004) and Git (Chacon 2009). The fundamental difference between SVN and Git is where revision information is stored and how the operations are executed. In case of SVN, revisions are stored on a central server, requiring an established network connection every time an operation (e.g., commit and update) is executed. On the other hand, Git provides the entire history within the repository on the local hard drive, thus Git operations mainly run locally.

Implementations with focus on ontologies in the context of versioning systems are for instance SemVersion (Völkel and Groza 2006), Bigdata, or owl2vcs (Zaikin and Tuzovsky 2013). SemVersion is an RDF-centric versioning approach, which provides structural and semantic versioning for RDF models and RDF-based ontology languages. It functions as a layer for a TripleStore that enable exchanges of concrete TripleStores. Owl2vcs is a family of tools that supports the collaborative development of web ontologies by means of versioning systems. Rather than relying on the textual representation of information used in various versioning system implementations, it supports version control on an ontology structure. Bigdata is an implementation for high concurrency, and very high aggregate IO rates of ontology storage. It provides a so-called retention history that can be considered also a temporal storage. By default no historical data is retained, but developers can specify a period to enable access to historical data. Once the period finished, the data is lost. Bigdata also offers a Read/Wrote store approach that "[…] may be used as a time-bounded version of an immortal database […]". For a generic reference process of versioning and knowledge changes in multidisciplinary engineering, we refer readers to Chap. 7.

In context of graph databases, a widely used naive approach is to model any changes of vertices as a linked list. The current revision is copied and only afterwards updated every time a vertex should be updated. Furthermore, timestamps or other metainformation can be stored at the "prev" edges to indicate when or which changes occurred. In general, since most graph databases support traversal in both directions, it is not required to have additional "next" edges. This approach does usually not introduce additional complexity for querying the current revision. This is particularly true, if versioning for edges is not required. However, when query complexity should not be compromised, versioning of vertices as well as edges seems much more difficult. The time-based versioning[22] approach makes use of timestamps focuses on the separation of structure and state. It aims for a good balance between write costs and read performance for both querying the current version as well as reading of historical data. A new record can be added through the creation of two vertexes and a single edge. Additionally, only a few operations are required for updates and deletions, which is very simple. However, query performance and simplicity are limited, because all queries, independent of the involvement of historical data, depend on checking the timestamps of the relationship on each traversal. In an approach introduced by (Castelltort and Laurent 2013), operational and versioned data are kept in two separate graphs. Every element of the operational data graph is represented by a "TraceElement" in the version graph, which are connected to a revision model that keeps track of both property and relational changes for the referenced node.

[22]Robinson, Ian. Time-Based Versioned Graphs. http://iansrobinson.com/category/neo4j/, 2014. Accessed: 2015-11-08.

8.3 Use Case: A Steel Mill Plant Engineering

Chapter 2 illustrates usage scenarios with key challenges of system engineers and managers in the transition from traditional to Cyber-Physical Production System (CPPS) engineering environments. The chapter provides a deeper view on needs for semantic integration during typical engineering steps within the life cycle of production systems in the context of a lab-size production system. This section presents a more specific use case from an industrial partner building a steel mill engineering plant. The intention of the use case is to point out methods, tools, and practices for systems engineering. In particular, the concepts of tools (e.g., versioning of source code) and methods (e.g., continuous integration), which have been successfully used for software development, should be adopted for systems engineering domains, like automation systems development projects.

In this use case, the engineering development environment deploys several engineering tools like EPlan[23] (electrical engineering), logi.CAD[24] (software engineering), EtherCAT ET9000,[25] or SYCON.Net[26] (mechanical engineering). In most cases of a software engineering project, the main "data model" is globally defined using selected programming language; in systems engineering every tool applies its own data model reflecting its own specific view on the system. For instance, Fig. 8.2 shows a simplified UML class diagram for describing the hardware configuration of an engineering plant as used by the ET9000. On the other hand, logi.CAD implements the IEC61131-3[27] standard for programmable logic controllers represented by the TC6[28] schema.

Similar to software development, where changes of source code elements are checked in into a revision system and then distributed to other team members through update operations, changes made in one engineering tool mentioned above should also be propagated to other engineering tools to provide a consistent view on the system in all involved deployed tools.

The number of exchanged data records strongly depends on the size of the system. In the context of our industry partner and from a hardware configuration point of view, a typical engineering plant consist of about 6 million data records in a single engineering project.

However, as described in Sect. 8.2.3 only changes relevant to common concepts are propagated, leaving about 20 % of its data (about 1.2 million data records in a typical case) to be exchanged with other engineering tools. While this number of data records is shared with other tools (e.g., with logi.CAD tool for software engineering design) and therefore it does not increase the number of data elements,

[23]EPLAN: http://www.eplanusa.com.

[24]Logicals: http://www.logicals.com.

[25]Ethercat: http://www.ethercat.org.

[26]Sycon.NET: http://www.hilscher.com.

[27]PLC Open: http://www.plcopen.org/pages/tc1_standards/iec_61131_3.

[28]TC6: http://www.plcopen.org/pages/tc6_xml/.

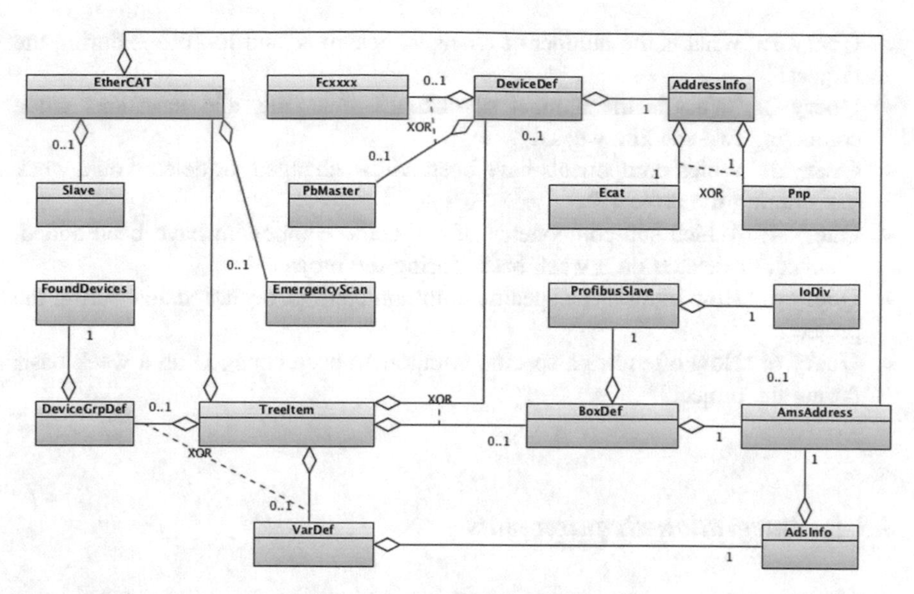

Fig. 8.2 UML class diagram (without properties) for describing hardware configuration used by engineering tools like ET9000 or SYCON.Net (based on Mordinyi et al. 2014)

the other tools will provide additional data records (about 30–40 % of the exchanged data, which roughly about 400 k data records in a typical case) that are shared with other specific engineering discipline and also needs to be managed.

Considering that about 20 different types of engineering tools from several engineering disciplines will be required for the development of engineering plants, the overall number of data records describing the engineering plant and shared among the engineering tools is considerably higher.

Typically, processes synchronization of engineering discipline views occur at fixed points in time in the traditional automation systems engineering (Winkler et al. 2011; Moser et al. 2011b). It is recommended to make changes visible to the entire project (i.e., to propagate changes) once a specific and cohesive task/requirement has been realized, since the intention is to adopt software engineering concepts and practices. Assuming that such an engineering project lasts about 18 months in average and that project engineers make their changes available once a day, the overall number of data records exchanged, transformed, processed, and versioned during the project would be once again significantly higher.

There is the need for near-time analysis of the project progress based on real change data (Moser et al. 2011b) due to the distributed parallel activities requirement of engineering processes. Based on current and systematically integrated data, project managers need to see between milestones the overview on project progress. Relevant queries formulate intentions to: (a) operation types (insert, update, delete) to identify the volatility of common concepts, (b) specific components of the engineering plant, or (c) both in the context of a time frame. Based on (Moser et al. 2011b) examples of relevant queries for, e.g., project managers are

- **Query 1**: "What is the number of changes, deletions, and insertions during the project?"
- **Query 2**: "What is the number of changes, deletions, and insertions when comparing two specific weeks?"
- **Query 3**: "Which components have been added, changed, or deleted on a week basis during the project?"
- **Query 4**: "Which sub-components of a specific component have been added, changed, or deleted on a week basis during the project?"
- **Query 5**: "How often has a specific common concept been updated during the project?"
- **Query 6**: "How often has a specific component been changed on a week basis during the project?"

8.3.1 Integration Requirements

In this sub-section, we describe the explicit integration requirements of data storages for automation systems development projects based on the use case definition that we explained previously in this section and general use case of engineering in Chap. 2. The requirement focuses on three aspects, namely data insertion, data transformation, and data querying. We will use the previously presented case study as a typical example for explaining these requirements.

8.3.1.1 Data Insertion

Consistent project data in all the discipline-specific tools is an important aspect in the context of integrating engineering tools from heterogeneous engineering disciplines. Traditional automation systems engineering processes follow a basic sequential process structure with distributed parallel activities in specific project phases. However, these processes are lacking systematic feedbacks to earlier steps and suffer from inefficient change management and synchronization mechanisms.

Local tool changes must be committed to make changes available to all project participants and their tools in order to improve these processes. To minimize inconsistencies between the tools, the approach encourages engineers to commit their changes as often as possible. A commit can also be interpreted as a bulk operation referencing to a set of data management operations. The bulk operations should be faster than the sequential execution of commands a bulk operation refers to. Nevertheless, the performance limits are set by the time response that users consider acceptable, since the user triggers any of the operations. Furthermore, the consumption of memory has to be limited to a reasonable amount that does not harm the entire system when loading the ontology or when large bulk operations are executed. Additionally, it is important that the stored data can be easily accessed and are readable in order to support data migration and to ensure data ownership. Finally, in order to facilitate maintenance of the system at semantic level, semantic facilities have to be provided.

8.3.1.2 Data Transformation

As previously stated, the first step towards a consistent view on an automation system project is to commit local changes. However, every engineering tool uses its own concepts, which is not always compatible to other tools' concepts. Therefore, during change propagation the data provided by the tools has to be transformed to the common concepts (i.e., assuming that there are defined mapping between common concepts and local tools concepts) and then propagated to the other tools. By transforming information (provided by a specific) tool to every other tool, information is shared among project participants and a consistent view is facilitated.

Basically, a transformation consists on transforming data from one to another ontology. Data transformation may vary from simple string operations, such as contact or substring, to complex operations, where external services are needed or where transformation executions depend on specific attributes' values. For a complete reference of semantic mapping types used in engineering, we refer the readers to Chap. 6 of this book. It is essential that the transformation executions do not compromise the system performance since several transformations have to be performed for propagating changes. In addition, provided techniques for transformation should make sure that memory consumption is kept to a minimum. The data storage is required to support mapping and transformation mechanisms like SPIN and EDOAL; and it is favored to make use of them as built-in tools rather than external tools in order to minimize unnecessary additional maintenance efforts. Also, the mappings and transformed data needs to be easily accessed and readable.

8.3.1.3 Data Query

Automation systems development projects typically manage a huge amount of data spread over large number of heterogeneous data sources originated from different engineering disciplines. This characteristic hampers data analysis and querying across the project, which could lead to emergence of inconsistencies in the project data. If such inconsistencies are not identified early, they may cause costly corrections during the commission phase or even failures during the operation phase.

In the interchange standardization approach, the usage of the common model as a mediator (Wiederhold 1992) is a common technique for querying purposes. In this case, a single access point and common conceptualization for querying local data sources is provided by the common concepts. Thus, the query formulation becomes independent of the mediated local data sources. Individual users do not need to have a detailed knowledge on all the concepts used in all project disciplines, and could focus instead on the common concepts that are relevant for the whole project.

Two important aspects for querying are performance and memory consumption. Acceptable response time for a query will strongly depend on a specific application. If a big volume of data must be processed and the accuracy of results is crucial in the applications, a longer execution time might be acceptable, e.g., for consistency checking, an important aspect in the use case. However, other applications, like navigation to specific parts in the project or analysis of subsets of data (e.g., analyzing the created signals in a specific time period) require fast response times (e.g., 0.1–1 s) to keep the user's attention focused (Nielsen 1993). In regards with memory consumption, it is essential to not load all affected data into memory for not causing memory exceptions, since it is needed to manage a large volume of data. Some techniques for avoiding this are, e.g., exploiting indexing techniques or query optimization (Gottlob et al. 2011).

The support for SPARQL (Pérez et al. 2006), which is currently the W3C standard for querying ontologies, is another important requirement for providing proper storage usability. Thus, SPARQL queries can be used for querying project data through the common concepts. SPARQL makes the queries more compact and easier to describe, while reducing the debugging time, since its syntax makes virtually all join operations implicit. In the case of data storages that exploit relational databases to manage local tools data, support for automated query transformation from SPARQL into SQL must be provided. It is also noted that a build-in query transformation is favored rather than external tools, in order to minimize additional maintenance efforts. In addition, inference capabilities have to be supported by the semantic storages since they can considerably simplify required queries.

8.4 Engineering Knowledge Base Software Architecture Variants

This section illustrates the four introduced Engineering Knowledge Base (EKB) software architecture variants in detail.

8.4.1 Software Architecture Variant A—Ontology Store

The first EKB software architecture variant (SWA-A) uses a single ontology component that stores and manages both concepts and individuals (see Fig. 8.3) Engineering tools use the component to insert, update, or delete data, e.g., using the Sesame API, on their local tool data models. Mappings, implemented in SPARQL, describe the relation between two concepts, perform transformations on the provided data, and update the instance set of the targeted model. Queries formulated

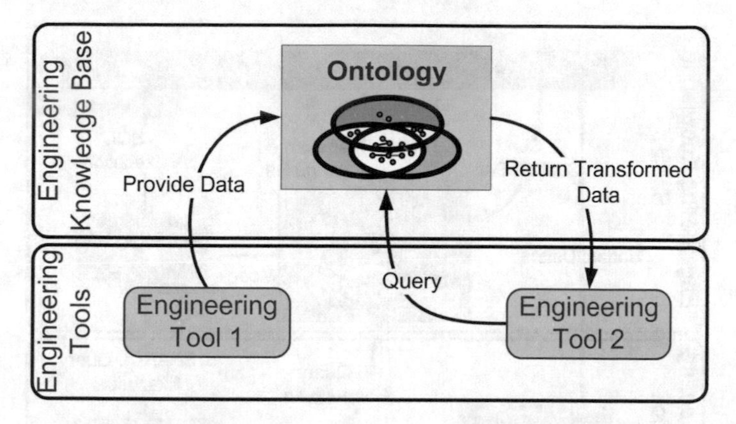

Fig. 8.3 Concepts and instances are stored in a single ontology store (based on Mordinyi et al. 2014)

and executed by other engineering tools will retrieve the transformed data. Details on the process can be found in (Moser 2009).

Versioning of instances is executed according to the publicly available *changeset vocabulary*.[29] The used vocabulary defines a set of terms for describing changes to resource descriptions.

8.4.2 *Software Architecture Variant B—Relational Database with RDF2RDB Mapper*

The second EKB software architecture variant (SWA-B) manages and stores concepts and instances in two different storage components. As shown in Fig. 8.4, the ontology store manages concepts only while individuals are stored in a relational database. Engineering tools insert, update, or delete data (Step 1) using SQL within their designated tables that reflect the same model as described in the ontology. Adaptations on that table trigger a process (Step 2) that requests the ontology to transform the change set to the models it is mapped to. The transformed change set (Step 3) is used to update the corresponding tables of the identified models.

In case an engineering tool wants to request information from the system, it formulates a SPARQL query based on the models described in the ontology (Step 4), hands it over to the RDF2RDB Mapper (Step 5). After transforming the SPARQL query to SQL, the mapper executes it on the relational database, and returns the result to the application. Versioning of instances is performed according to the schema as in SWA-A.

[29]Changeset vocabulary: http://vocab.org/changeset/schema.html.

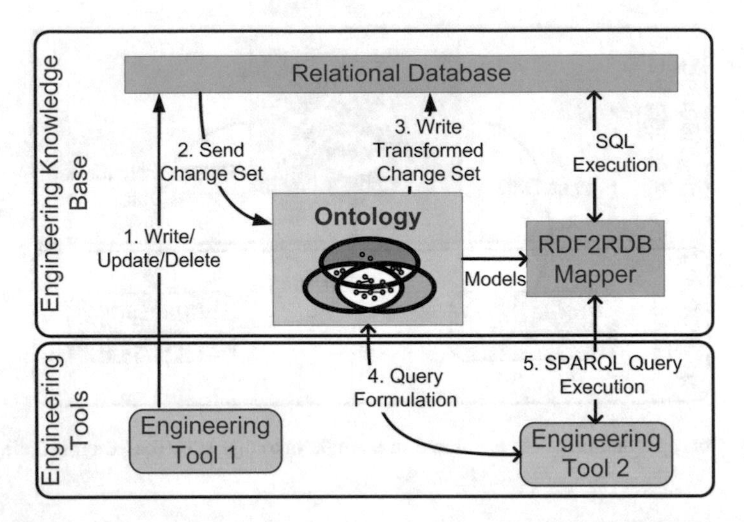

Fig. 8.4 Application of a mapping configuration and relational database (based on Mordinyi et al. 2014)

8.4.3 *Software Architecture Variant C—Graph Database Store*

The third EKB software architecture variant (SWA-C) relies on graph databases. The authors of (Mordinyi et al. 2015) provide details of how to represent complex models with the database concept as well as how to map ontology concepts on graph database schema for efficient execution of operations.

Since graph database implementations do not provide versioning per se, a versioning model had to be developed (Mordinyi et al. 2015). Figure 8.5 illustrates the schema used to handle multiple versions of an individual *Sample*. For each individual the approach creates a current, revision, and history node in the database. Additionally, a *Commit* node is created that stores metainformation of the operation, such as a timestamp or the committer.

Current nodes (e.g., *Sample*) represent the latest state of an individual including its attributes. They are created when an entity is inserted into the database, updated when an individual gets changed, and removed when deletion of the entity is requested. *History* node (e.g., *SampleHistory*) is created only when an entity is inserted. It never is never removed and can therefore be used to query already deleted individuals. The history node enables access to all versions of a particular entity and can be accessed by following the revision link(s). Those links can also be used to track changes. *Revision* nodes (e.g., *SampleRevision*) store the state of a particular individual at a specific point in time. Whenever an individual is changed, a new revision node is created and the current node is updated.

Figure 8.6 shows the EKB software architecture variant that uses an ontology component to store and manage concepts and a NoSQL graph database to store and

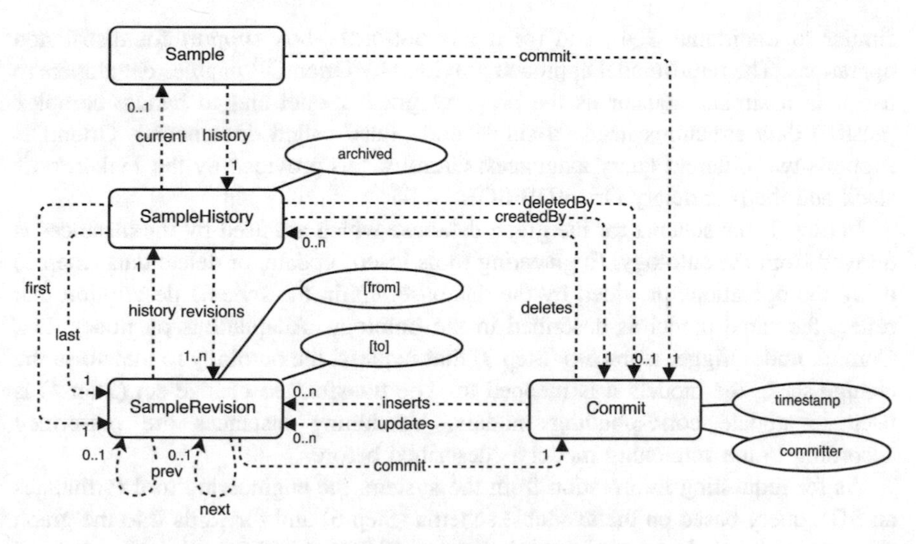

Fig. 8.5 Data Model for Versioning (based on Mordinyi et al. 2015)

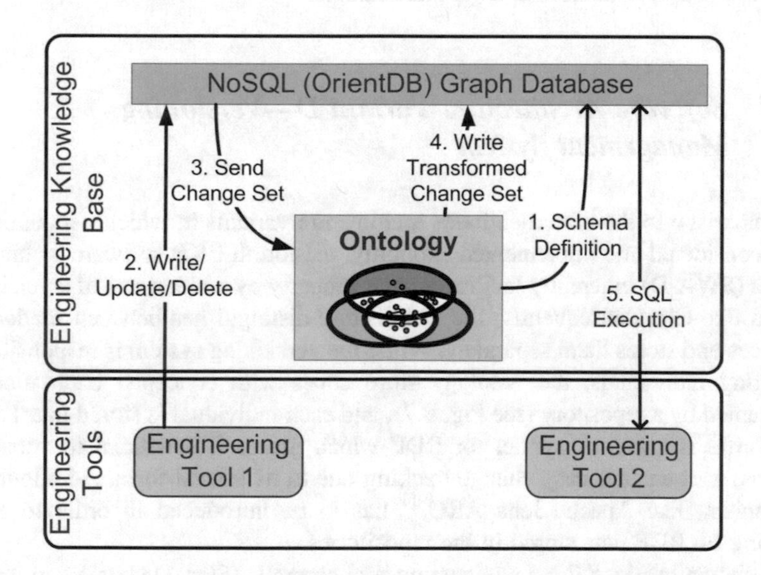

Fig. 8.6 Concepts are stored in Ontology, while instances are managed in a NoSQL Graph Database (based on Mordinyi et al. 2015)

version individuals. The back-end relies on the OrientDB[30] implementation and was selected because (a) it supports a multi-model approach (i.e., it combines graph and document databases), (b) it provides an easy to use API and query language

[30]OrientDB: http://www.orientdb.org.

similar to traditional SQL, and (c) it has out-of-the-box support for distributed operations. The multi-model approach provided by OrientDB enables developers to use it in a similar manner as the property graph model and to handle complex (nested) data structures inside a single node (also called documents). OrientDB supports two different query languages: Gremlin[31] as provided by the TinkerPop[32] stack and the proprietary OrientDB SQL.

In Step 1, the schema for the graph database which are used by the instances is derived from the ontology. Engineering tools insert, update, or delete data (Step 2) using the operations provided by the database within the schema description that reflect the same model as described in the ontology. Adaptations on nodes (i.e., Commit node) trigger a process (Step 3) that requests the ontology to transform the change set to the models it is mapped to. The transformed change set (Step 4) is used to update corresponding models. Versioning instances are performed according to the versioning model as described before.

As for requesting information from the system, the engineering tool formulates an SQL query based on the available schema (Step 5) and forwards it to the graph database. Orientdb is compatible with a subset of SQL ANSI-92, while for complex queries it provides additional SQL-like features.

8.4.4 *Software Architecture Variant D—Versioning Management System*

In comparison to the aforementioned architecture variants in which versioning has to be considered and implemented explicitly, the fourth EKB software architecture variant (SWA-D) inherently facilitates such property by making use of a versioning system like Git. Consequently, the architecture distinguishes between models and instances and stores them separately. While the versioning system is responsible for managing individuals, the ontology store copes with concepts. Each model is represented by a repository (see Fig. 8.7), and each individual is stored in a Turtle[33] file. Turtle is a textual syntax for RDF which makes it beneficial to versioning systems, as it enables easy change tracking due to its textual form. Additionally, a mechanism, like Apache Jena ARQ,[34] has to be introduced in order to enable querying on RDF data stored in the repositories.

As shown in Fig. 8.7, an engineering tool commits (Step 1) changes to its local repository, which is automatically pushed (Step 2) by a so-called hook to the master repository of the model. Triggered by a hook on the master repository of the model, the transaction manager identifies correlating models (Step 3, 4) and requests the

---

[31]Gremlin: https://github.com/tinkerpop/gremlin/wiki.

[32]Apache TinkerPop: http://tinkerpop.incubator.apache.org/.

[33]Turtle: http://www.w3.org/TR/turtle/.

[34]Apache Jena ARQ: http://jena.apache.org/documentation/query/index.html.

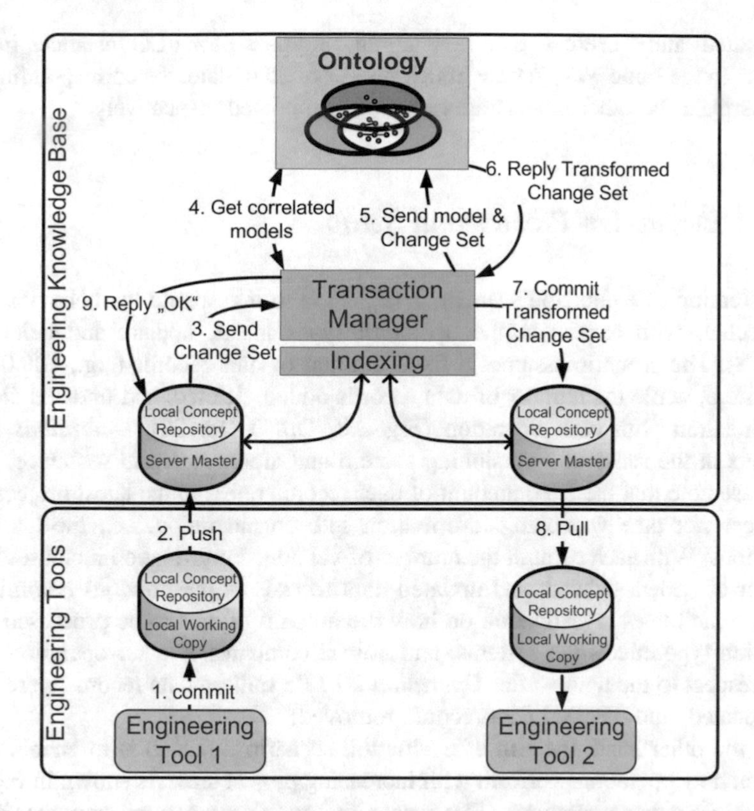

Fig. 8.7 Versioning system and an ontology-based store (based on Mordinyi et al. 2014)

ontology to transform (Step 5, 6) the change set as defined in the mappings between the concepts. Finally, the result is committed (Step 7) in the master repository of affected models, from which the engineering tool pulls (Step 8) the latest changes.

8.5 Evaluation

In order to demonstrate the characteristics of the four different EKB software architecture variants, we evaluated performance and used memory (disk space and memory consumption) in the context of two evaluation scenarios (see Sect. 8.5.1) and with respect to querying versioning information. The scenarios used for evaluation are a simplified version of the constellation of engineering tools as illustrated in Sect. 8.3 and in (Moser and Biffl 2012). The integration setup consists of three interconnected models (Serral et al. 2013): (a) model representing electrical plans (EPL), (b) model representing programmable logic control (PLC), and (c) the common concept signal which defines a relation between the two aforementioned models. This means, that in case of a new EPL instance, the data is transformed, propagated, and "inserted" as a new signal, and as a new PLC instances into the storage. In the same way, in case of deletion or of an update, the corresponding data records from the other models are removed or updated respectively.

8.5.1 Evaluation Process and Setup

The intention of evaluation scenario 1 (ES-1) is to investigate the behavior of the architecture with respect to the operation types insert, update, and delete (see Fig. 8.8). The scenario assumes a fixed amount of data records (e.g., 100.000) in the system, while the number of data records added, deleted, and updated changes in correlation with each operation (Fig. 8.8, Op. 0–Op. 20). *Operations* reflect commits in the sense of a versioning system and are executed in sequence.

Please note that the fixed amount of data records refers to the latest project state, and does not take versioning information into consideration, i.e., the history of operations. With each commit the number of versioned information increases by the number of added, deleted, and updated data records of the previous commit. The scenario facilitates investigation on how the amount of data to be processed by an operation type effects the systems, and how it compares to other operation types. With respect to the figure, after *Operation 20* 1.05 million data records were added and updated, and 950.000 data records removed.

On the other hand, the aim of evaluation scenario 2 (ES-2) is to enable evaluation on how operations perform with increasing project size. As shown in Fig. 8.9, there are no delete operations. This means that with each commit new data added, the size of the project and the amount of versioning information increases. With respect to the figure, after *Operation 0* there are 10.000 data records in the storage, while after *Operation 13* 101.000 data records were processed (considering versioning information).

The evaluations were performed on an consumer laptop, with an Intel® Core™ i7-3537U Processor 2 GHz, 10 GB RAM, and 256 GB SSD hard disk running Ubuntu 12.04 64 bit, OpenJDK 64 bit, and JRE 7.0_25 with a java heap size of 8 GB RAM. For evaluating SWA-A, originally Bigdata version 1.2.3 was picked,

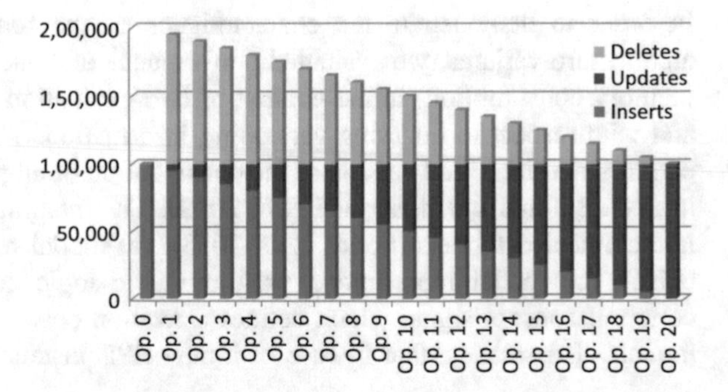

Fig. 8.8 Amount and type of operations in evaluation scenario 1

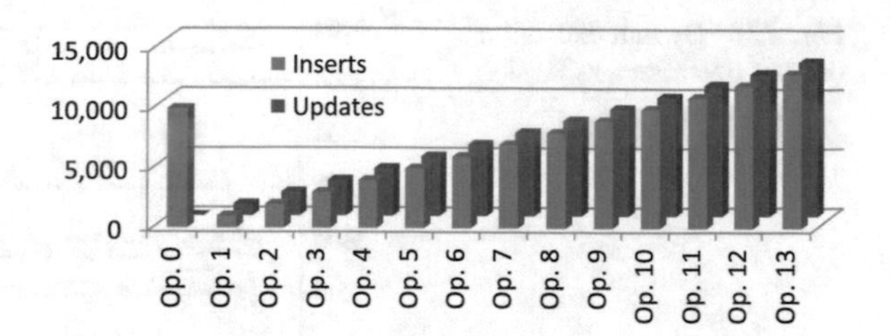

Fig. 8.9 Amount and type of operations in evaluation scenario 2

but due to high memory consumption resulting in OutOfMemoryExceptions both scenarios could only perform four operations. Instead, Sesame Native store v2.6.3 was chosen. In case of a SWA-B D2RQ (Bizer and Seaborne 2004) version 0.8.1 along with mysql 5.5 was deployed, for SWA-C OrientDB version 2.0.5 was selected, while in case of a SWA-D the version control system Git and Apache Jena ARQ 2.11.0 were selected. In the following SWA-A, B, C, and D will refer to the concrete implementations of the described EKB software architectures.

8.5.2 *Evaluation of Data Management Capabilities*

This section provides the evaluation results of the two evaluation scenarios applied to the software architecture variants.

8.5.2.1 Performance Results of Evaluation Scenario 1

Figure 8.10 illustrates the performance in terms of the time required for inserting, updating, and deleting a set of instance as defined by ES-1 for all four software architecture variants. The figure shows that SWA-B is the fastest, while SWA-C the slowest. The reason is the different approach on how versioning is done. As described in Sect. 8.4.3 SWA-C stores information several times in order to keep revisions transparent to query formulation, while SWA-B does it only once. Additional updates on graph structures lead to additional performance costs. From the figure we can also observe the fluctuations in time in case of SWA-A, while SWA-D provides an almost continuously constant execution time. This results due to the fact that at each *Operation* there is always the full amount of files to be processed by the system.

Comparing the various architecture variants, using SWA-A over SWA-B increases overall execution time by 2–3 times, by around 3–6 times in case of SWA-C. While SWA-D is still slower than SWA-A or SWA-B, it is still faster than SWA-C. Nevertheless, the main drawback of SWA-D is the huge number of files the file system has to cope with. Although at *Operation 20,* Git had to process 2.2 million turtle files, the system needed to manage an additional 24.1 million files for

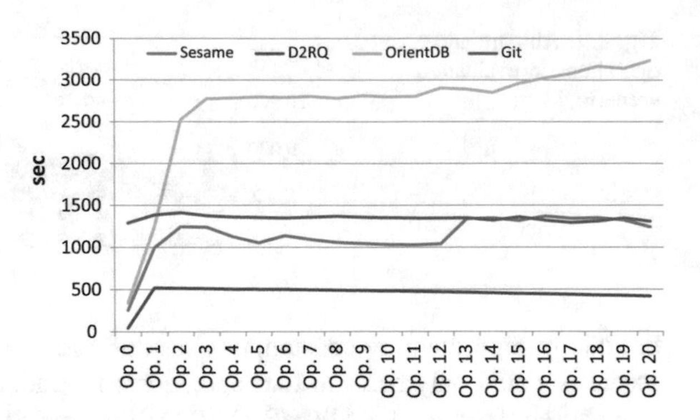

Fig. 8.10 Overall execution time of operations in ES-1

version control. As a matter of fact, the inode table (Tanenbaum 2007) of the file system had to be increased to 54 million in order to cope with such amount of files.

8.5.2.2 Performance Results of Evaluation Scenario 2

Figure 8.11 shows the results of evaluation scenario 2 for all four software architecture variants. Comparing the approaches at absolute time, it is evident that SWA-B is faster than the others, while SWA-D the slowest. SWA-D is almost two to three times slower than SWA-A and almost 4–6 times slower than SWA-B.

The figure also illustrates the minimal difference between the solution variant SWA-B and SWA-C. Since ES-2 does not contain any *delete operations*, it seems that by taking the results of the previous scenario into account, such operations costs the most time for SWA-C. The reason is the high number of steps needed to reorganize the versioning tree in case of a delete operation. Update operations on the other hand are not of a major influence.

Memory Consumption: Comparing the memory consumption of the three approaches (Table 8.1), it has been monitored that in ES-1 SWA-B used constantly 150 MB RAM, SWA-D consumed a maximum of 290 MB RAM while SWA-C had occupied 2.1–3.4 GB RAM and in case of SWA-A over 5.7 GB RAM. In case of ES-2, SWA-B used constantly 150 MB RAM, SWA-D needed a maximum of

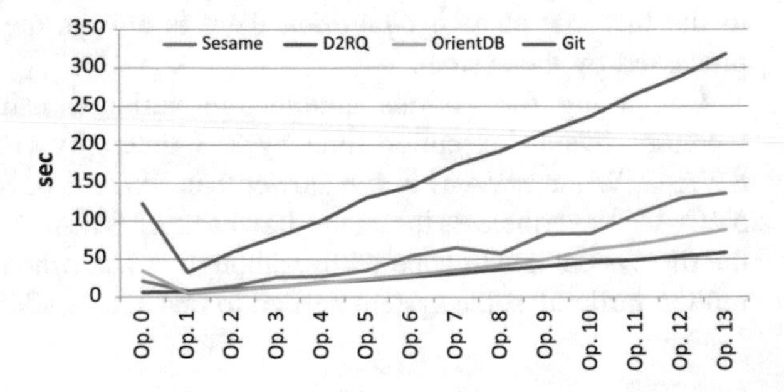

Fig. 8.11 Overall execution time of operations in ES-2

148 MB RAM, while SWA-C needed about 1.8 GB RAM and SWA-A required 4.6 GB RAM.

Disk Usage: Comparing the disk usage of the three approaches, it has been measured that in case of ES-1 SWA-A had to manage about 300 million triples occupying about 20 GB of disk space, SWA-B used about 1 GB of space, SWA-C about 3 GB, while SWA-D reserved about 4 GB of disk space. In case of ES-2, SWA-A had to manage 103 million triples using about 6 GB of disk space, SWA-B occupied 352 MB disk space, SWA-C about 500 MB, and SWA-D about 400 MB (Table 8.1).

8.5.3 *Evaluation of Historical Data Analysis Capabilities*

In order to evaluate the performance time of query execution, six industry relevant queries (see Sect. 8.3) were implemented, and executed after each *Operation*, i.e., after executing all read, write, and delete operations.

The results of the first query are shown in Fig. 8.12. The query intends to identify the amount of changes, deletions, and insertions during the project. The figure shows that for all software architecture variants the execution time of query 1 increases, the further the project progresses. However, in case of SWA-C *Operation*

Table 8.1 Summary of memory consumption and disk usage in the context of evaluation scenario 1 and 2

SWA variant	Evaluation scenario 1		Evaluation scenario 2	
	Memory	Disk (GB)	Memory	Disk
SWA-A	5.7 GB	20	4.6 GB	6 GB
SWA-B	150 MB	1	150 MB	352 MB
SWA-C	2.1–3.4 GB	3	1.8 GB	500 MB
SWA-D	290 MB	4	148 MB	400 MB

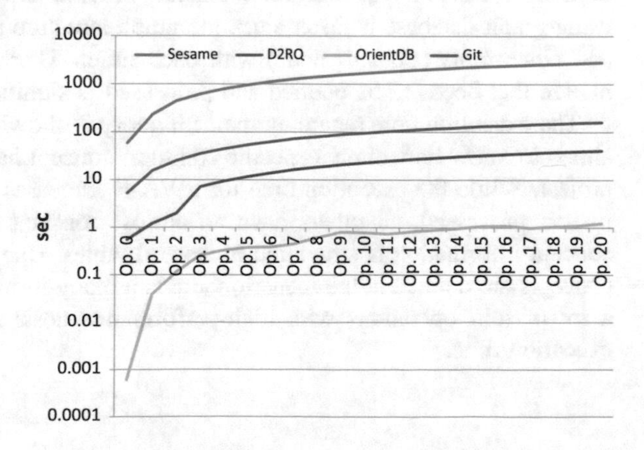

Fig. 8.12 Query execution time in the context of evaluation scenario 1 for query 1

Fig. 8.13 Query execution time in the context of evaluation scenario 1 for query 2

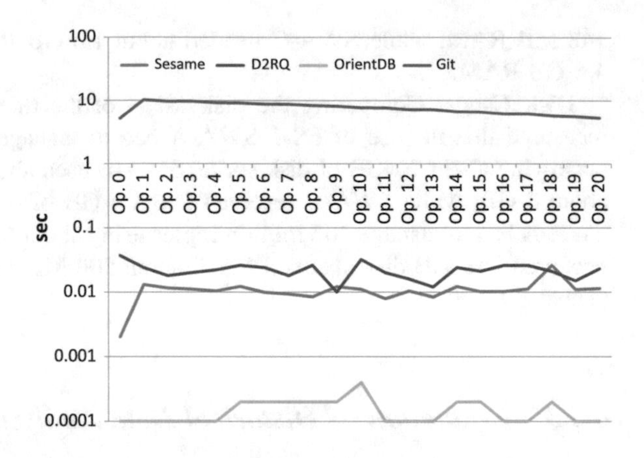

information already contained in *Commit* nodes are linked with each other, therefore the amount of information to be processed is minimal in comparison to the other variants.

Figure 8.13 illustrates the query execution time for query 2 which intends to compare two *Operations* (i.e., 2 weeks) based on the number of changes, deletions, and insertions. For SWA-A, SWA-B, or SWA-C it takes almost the same amount of time—please note the logarithmic scale of the y-axis. SWA-D lacks performance due to the high number of individual files putting pressure on the operating system and consequently of the change sets which need to be derived. The difference in time may be ascribed to inaccuracies at measurement.

The third query aims to identify those components which have been added, changed, or deleted on a week basis during the project. The intention is to investigate how group operations effect execution time. As shown in Fig. 8.14 the query execution time is very slow for database-related architecture variants, and 10–100 times higher for the other alternatives.

Figure 8.15 illustrates the results of query 4 which gathers all sub-components of a specific component that has been updated on a week basis during the project. Solution variants SWA-A and SWA-B have similar execution results. The variant using graph database is faster since the graph structure already links change types and *Operations* (*Commit* node) with each other. Therefore the amount of information that needs to be queried and processed is significantly lower.

The execution time regarding the fifth query is shown in Fig. 8.16. The query's aim is to know how often a specific common concept has been updated during the project. While the execution time for SWA-B increases with the amount of information processed, all others have an almost constant value. The reason is that concept information is structured in several tables. Due to the way how common concepts are defined in the scenario there is a many-to-many relation. This requires a lot of join operations with high-performance costs resulting in a high query execution time.

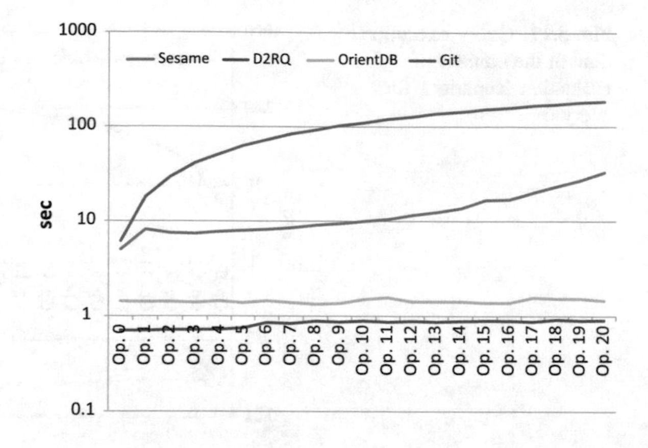

Fig. 8.14 Query execution time in the context of evaluation scenario 1 for query 3

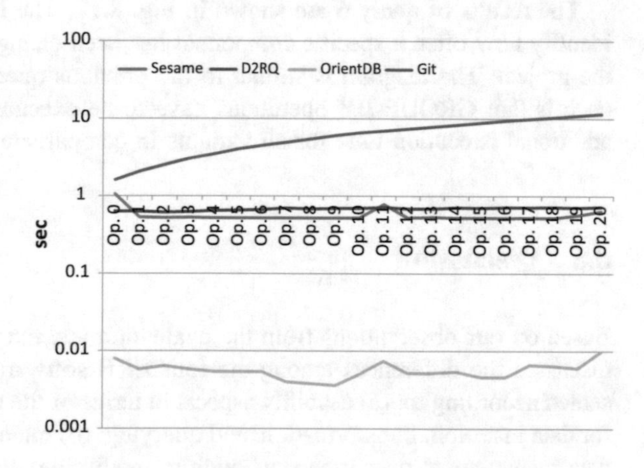

Fig. 8.15 Query execution time in the context of evaluation scenario 1 for query 4

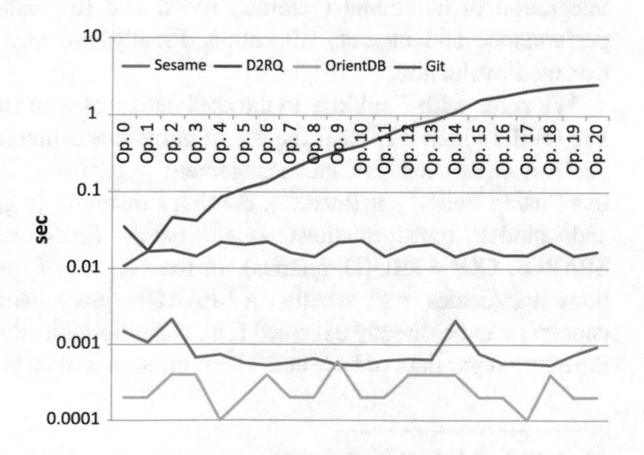

Fig. 8.16 Query execution time in the context of evaluation scenario 1 for query 5

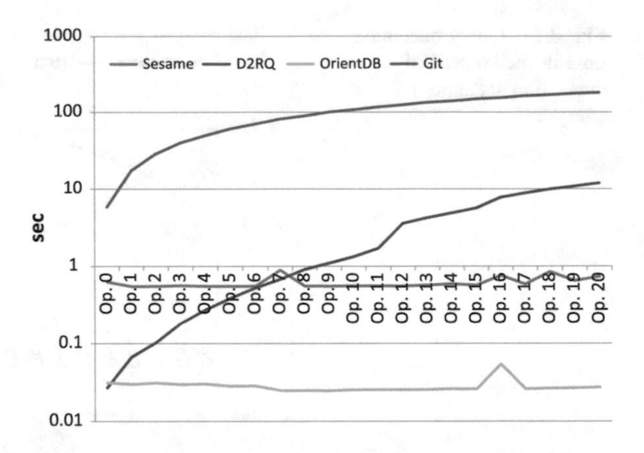

Fig. 8.17 Query execution time in the context of evaluation scenario 1 for query 6

The results of query 6 are shown in Fig. 8.17. The intention of the query is to identify how often a specific component has been changed on a week basis during the project. The diagram is similar to the previous query. The determining difference is that GROUP BY operations have to be executed, which requires a lot of additional execution time for all variants in comparison to the previous query.

8.6 Discussion

Based on our observations from the evaluation and existing literature, this section discusses the differences among the four EKB software architecture variants presented according to: (a) usability aspects in terms of the used semantic technologies for data insertion, transformation, and querying; (b) maintenance aspects that enable data management over time (i.e., adding, modifying, and removing data), and the integration of additional (external) tools; and (c) scalability aspects in terms of performance and memory allocation. Finally, we analyze the limitations of the executed evaluation.

We refer with Usability to the application of web standard semantic languages like OWL, which facilitates the use of numerous semantic technologies available to perform tasks such as data management (e.g., using Protégé or Jena), reasoning (e.g., using Pellet[35] or Racer[36]), ontology mapping (e.g., using OWL constraints), and model transformations (e.g., using SPIN rules, SWRL rules, and SPARQL CONSTRUCT queries). In the context of query and result transformations it indicates, e.g., whether a SPARQL query launched against the common concepts can be directly executed (i.e., can automatically process the project data of the query regardless of how and where the source data is represented) or it has to be

[35]Pellet: http://clarkparsia.com/pellet/.

[36]Racer: http://franz.com/agraph/racer/racer_features.lhtml.

transformed to other query/queries to be able to process the engineering tools data (i.e., the local data). If the query has to be transformed, then other transformation is also needed to return the results as asked in the SPARQL query, i.e., in terms of the common concepts.

SWA-A and SWA-D represent the data using standard semantic languages and therefore facilitate the application of existing semantic technologies. In addition, it allows SPARQL queries to be directly executed across ontology models preventing the need of transforming the queries (from common concepts to local tool concepts). However, applications in a SWA-B or SWA-C require: (a) the SPARQL queries to be transformed to the corresponding query language (e.g., SPARQL to SQL); and (b) the obtained results from the databases to be transformed in accordance to the data requested by the SPARQL query. In case there is no query transformation available, the query has to be manually formulated in the language of the target store, which increases the complexity of the application and its coupling to the back-end.

Maintenance is about facilities provided in order to perfect the system, to adapt the system and to correct the system (Lientz and Swanson 1980). In SWA-A both knowledge base definitions and data instances can be maintained using semantic tools (e.g., Protégé) and different middleware (e.g., Jena or Sesame). In the other software architecture variants, the knowledge base definition and the data instances are managed differently. Thus, while the ontologies' definition can be managed using semantic tools, the data instances have to be managed using relational or graph database tools at a lower level of abstraction. In case of SWA-D, even additional tool support is needed to manage a large set of files. In addition, for SWA-B the schema of the databases has to be modified (e.g., deleting or creating tables) when ontologies change. However, some relational database storages, such as SDB,[37] support this synchronization in an automatic way.

Scalability refers to the efficiency of data storage access (i.e., a longer response time indicates less efficiency) and how it scales up with large data applications in terms of performance and used memory. It is known that ontology files are very efficient for small models by reducing load and update time significantly (Shen and Huang 2010; Vysniauskas et al. 2011); however, for big volumes of data, this storage type becomes unsuitable. In the context of the presented software architectures, the performance of SWA-C considerably varies depending on the applied database (Shen and Huang 2010). According to the Berlin SPARQL Benchmark (Bizer and Schultz 2001), the comparison of the fastest triple storage with the fastest relational database storage shows that the latter one has a better overall performance with increasing dataset size, which can also be confirmed by the evaluations carried out in the context of this chapter.

Our practical evaluation shows that pure ontology-based stores do not provide sufficient performance and have very slow query execution times in comparison to the other variants. The results show that SWA-B (Sequeda et al. 2012) or SWA-C

[37]SDB: https://jena.apache.org/documentation/sdb/.

deliver better performance, however, they often require manual adaptations in case of complex models (Bornea et al. 2013). Furthermore, the resulting data model is not sufficient for efficiently executing queries concerning relationships between ontologies, like the one necessary to relate local tool and common concepts. These queries are either slow in terms of their execution time or difficult to express using SQL.

In general, the main advantage of SWA-A is its simplicity. On the one hand, since a single ontology component stores and manages both concepts and individuals the complexity for higher level applications is minimal. On the other hand, the evaluated solution provides very slow query execution times in comparison to the other solutions. Evaluations of commercial implementations will be considered as future work in order to get a more general and deeper overview on ontology storages.

The advantage of SWA-B and SWA-X is its acceptance by both research community (in sense of available research) and industry (in sense of robustness, performance and scalability). They offer also a higher confidence than for the first solution since they are well-known among system administrators. However, there is an additional added complexity because an ontology storage is used for managing concepts while a relational database is used for storing individuals. Deploying two components requires more logic in higher level applications to be able to determine when and how to access which storage component. During the performed evaluations we also encountered that the derived mapping configuration in SWA-B may be insufficient and often requires manual adaptations to improve performance in case of complex models. Additionally, D2RQ is limited to OWL-Light which restricts query expressiveness. Finally, consistency between ontology models and database models must be enforced in case of model adaptations.

The advantage of SWA-C is that it supports better expression of relations between concepts and faster execution of queries upon them. If a system uses many-to-many relationships, in a relational database, a JOIN table with foreign keys of both participating tables must be created, which further increases operation costs.

SWA-D provides more advantages for versioning and managing the progress of individuals since it uses Git, a well-established versioning system. Complexity of the solution increases as beside the separation between concepts and individuals an additional indexing framework has to deployed to enable query execution. In the solution each individual is kept in a single file to be able to use Git-supported functionality and thus to keep track of changes more efficiently. However, the performance of the architecture strongly depends on the used file system as it needs to be able to handle millions of files in a single folder.

Threat to validity. A lack of control variables in the evaluation is a threat to validity in our work. One example of this is the different representation of common concepts and change sets, which are slightly different for each variant's implementation. These differences may provide advantages or disadvantages towards specific implementation. Additionally, not all of the variants can be queried using the same querying language and engine (i.e., SWA-C can only use OrientDB SQL language for querying instead of SPARQL). Providing the common querying interface (e.g., via SPARQL) could improve this condition. Another option to improve the control variables aspect is to refer the variants' implementation toward a generic reference process (e.g., as proposed in Chap. 7). This could be done in order to provide more focus on the variants' comparison in certain aspects of the process.

8.7 Conclusion

In this chapter, we have described and investigated four selected software architectures (ontology store, relational database, graph database store, and versioning management system; see Sect. 8.4) embedding ontologies that satisfy the integration requirements of in typical multidisciplinary engineering projects: querying of common concepts, transformation between local and common concepts, and versioning of engineering data. Based on an industrial use case we have evaluated and compared four different possible software architectures to achieve semantic integration: (A) a single ontology component, (B) an ontology component and a relational database using RDF2RDB mapper, (C) an ontology component and a graph database, and (D) an ontology component and a proper versioning system for managing individuals. The results of the evaluation has shown that the software architecture using an ontology storage (variant A) or with a versioning management system (variant D) are better in terms of usability and maintenance; however, the software architecture using a relational database (variant B) outperforms the rest of variants in terms of scalability, specifically for insert performance, memory and disk usage, while the software architecture using a graph database (variant C) outperforms the rest of variants for query execution.

As future work we will extend our evaluations to commercially available solutions in a large-scale environment. Furthermore, we intend to use complex engineering tool data models like AutomationML (Drath et al. 2008; Lüder et al. 2010), to increase the number of mappings and therefore to have the need to support frequent data transformations.

Acknowledgments This work was supported by the Christian Doppler Forschungsgesellschaft, the Federal Ministry of Economy, Family and Youth, and the National Foundation for Research, Technology and Development in Austria.

References

Aldred, L., van der Aalst, W., Dumas, M., ter Hofstede, A.: Understanding the challenges in getting together: the semantics of decoupling in middleware. In: BPM Center Report BPM-06-19. http://www.BPMcenter.org (2006)

Bergamaschi, S., Castano, S., Vincini, M.: Semantic integration of semistructured and structured data sources. SIGMOD Rec. **28**, 54–59 (1999). doi:10.1145/309844.309897

Berners-Lee, T., Hendler, J., Lassila, O.: The semantic web. Sci. Am. **284**, 34–43 (2001). doi:10.1038/scientificamerican0501-34

Biffl, S., Mordinyi, R., Moser, T.: Anforderungsanalyse für das integrierte Engineering - Mechanismen und Bedarfe aus der Praxis. atp edition –Automatisierungstechnische. Praxis **54**, 28–35 (2012)

Biffl, S., Schatten, A., Zoitl, A.: Integration of heterogeneous engineering environments for the automation systems lifecycle. In: IEEE International Conference on Industrial Informatics (INDIN), pp. 576–581 (2009a)

Biffl, S., Sunindyo, W.D., Moser, T.: Bridging semantic gaps between stakeholders in the production automation domain with ontology areas. In: Proceedings of the 21st International Conference on Software Engineering & Knowledge Engineering (SEKE 2009), USA, pp. 233–239 (2009b)

Bishop, B., Kiryakov, A., Ognyanoff, D., et al.: OWLIM: a family of scalable semantic repositories. Semant. Web **2**, 33–42 (2011). doi:10.3233/SW-2011-0026

Bizer, C., Schultz, A.: The Berlin SPARQL benchmark. Int. J. Semant. Web Inf. Syst. **5**, 1–24 (2001). doi:10.4018/jswis.2009040101

Bizer, C., Seaborne, A.: D2RQ—treating non-RDF databases as virtual RDF graphs. In: Proceedings of the 3rd International Semantic Web Conference (ISWC2004) (2004)

Bornea, M.A., Dolby, J., Kementsietsidis, A., et al.: Building an efficient RDF store over a relational database. In: Proceedings of the 2013 International Conference on Management of Data—SIGMOD'13, p. 121 (2013)

Calvanese, D., De Giacomo, G., Lembo, D., et al.: The MASTRO system for ontology-based data access. Semant. Web **2**, 43–53 (2011). doi:10.3233/SW-2011-0029

Castelltort, A., Laurent, A.: Representing history in graph-oriented NoSQL databases: a versioning system. In: 8th International Conference on Digital Information Management, ICDIM 2013, pp. 228–234 (2013)

Cattell, R.: Scalable SQL and NoSQL data stores. ACM SIGMOD Rec. **39**, 12 (2011). doi:10.1145/1978915.1978919

Chacon, S.: Pro Git. Apress (2009)

Chappell, D.A.: Enterprise Service Bus. O'Reilly Media Inc. (2004)

De Leon Battista, A., Villanueva-Rosales, N., Palenychka, M., Dumontier, M.: SMART: a web-based, ontology-driven, semantic web query answering application. In: CEUR Workshop Proceedings (2007)

Doan, A.H., Noy, N.F., Halevy, A.Y.: Introduction to the special issue on semantic integration. ACM SIGMOD Rec. **33**, 11–13 (2004). doi:http://doi.acm.org/10.1145/1041410.1041412

Drath, R., Lüder, A., Peschke, J., Hundt, L.: AutomationML—the glue for seamless automation engineering. In: IEEE International Conference on Emerging Technologies and Factory Automation, ETFA, pp. 616–623 (2008)

Dupont, G.M., de Chalendar, G., Khelif, K., et al.: Evaluation with the VIRTUOSO platform: an open source platform for information extraction and retrieval evaluation. In: Proceedings of the 2011 Workshop on Data InfrastructurEs for Supporting Information Retrieval Evaluation, pp. 13–18 (2011)

Gottlob, G., Orsi, G., Pieris, A.: Ontological queries: rewriting and optimization. In: Proceedings—International Conference on Data Engineering, pp. 2–13 (2011)

Gruber, T.R.: Toward principles for the design of ontologies used for knowledge sharing. Int. J. Human-Comput. Stud. **43**, 907–928. (1995) http://dx.doi.org/10.1006/ijhc.1995.1081

Halevy, A.: Why your data won't mix? Queue 3:50–58 (2005). doi:http://doi.acm.org/10.1145/1103822.1103836

Harris, S., Gibbins, N.: 3store: Efficient bulk RDF storage. In: Proceedings of the 1st International Workshop on Practical and Scalable Semantic Systems (PSSS'03), pp. 1–20 (2003)

Haslhofer, B., Momeni, E., Schandl, B., Zander, S.: Europeana RDF store report. Library Europeana, pp. 447–458 (2011)

Hohpe, G.: 06291 Workshop report: conversation patterns. In: Leymann, F., Reisig, W., Thatte, S. R., van der Aalst, W. (eds.) The Role of Business Processes in Service Oriented Architectures.

Internationales Begegnungs- und Forschungszentrum für Informatik (IBFI), Schloss Dagstuhl, Germany (2006)

Hohpe, G., Woolf, B.: Enterprise Integration Patterns: Designing, Building, and Deploying Messaging Solutions (2003)

IEEE: IEEE Recommended Practice for CASE Tool Interconnection—Characterization of Interconnections (2007)

Klieber, W., Sabol, V., Kern, R., et al.: Using Ontologies for Software Documentation (2009)

Korel, B., Wedde, H., Magaraj, S., et al.: Version management in distributed network environment. In: Proceedings of the 3rd International Workshop on Software Configuration Management, pp. 161–166. ACM Press, New York (1991)

Lenzerini, M.: Data integration: a theoretical perspective. In: Proceedings of the Twenty-First ACM SIGMOD-SIGACT-SIGART Symposium on Principles of Database Systems, pp. 233–246. ACM, Madison (2002)

Lientz, B.P., Swanson, E.B.: Software Maintenance Management. Addison-Wesley, Boston (1980)

Lu, J., Ma, L., Zhang, L., et al.: SOR: a practical system for ontology storage, reasoning and search. In: Proceedings of the 33rd International Conference on Very Large Data Bases, pp. 1402–1405 (2007)

Lüder, A., Hundt, L., Keibel, A.: Description of manufacturing processes using AutomationML. In: Proceedings of the 15th IEEE International Conference on Emerging Technologies and Factory Automation, ETFA 2010 (2010)

Miles, A., Zhao, J., Klyne, G., et al.: OpenFlyData: an exemplar data web integrating gene expression data on the fruit fly *Drosophila melanogaster*. J. Biomed. Inf. **43**, 752–761 (2010). doi:10.1016/j.jbi.2010.04.004

Mironov, V., Seethappan, N., Blondé, W., et al.: Benchmarking triple stores with biological data. In: CEUR Workshop Proceedings (2010)

Mordinyi, R., Moser, T., Winkler, D., Biffl, S.: Navigating between tools in heterogeneous automation systems engineering landscapes. In: IECON Proceedings (Industrial Electronics Conference), pp. 6178–6184 (2012)

Mordinyi, R., Pacha, A., Biffl, S.: Quality assurance for data from low-tech participants in distributed automation engineering environments. In: IEEE International Conference on Emerging Technologies and Factory Automation, ETFA (2011)

Mordinyi, R., Schindler, P., Biffl, S.: Evaluation of NoSQL graph databases for querying and versioning of engineering data in multi-disciplinary engineering environments. In: 2015 IEEE 20th Conference on Emerging Technologies Factory Automation (ETFA), pp. 1–8 (2015)

Mordinyi, R., Serral, E., Winkler, D., Biffl, S.: Evaluating software architectures using ontologies for storing and versioning of engineering data in heterogeneous systems engineering environments. In: Proceedings of the 2014 IEEE Emerging Technology and Factory Automation, ETFA 2014, Barcelona, Spain, 16–19 Sept 2014, pp. 1–10. Vienna, Austria (2014)

Moser, T.: Semantic integration of engineering environments using an engineering knowledge base. Ph.D. thesis, Vienna University of Technology (2009)

Moser, T., Biffl, S.: Semantic tool interoperability for engineering manufacturing systems. In: Proceedings of the 15th IEEE International Conference on Emerging Technologies and Factory Automation, ETFA 2010 (2010)

Moser, T., Biffl, S., Sunindyo, W.D., Winkler, D.: Integrating production automation expert knowledge across engineering domains. Int. J. Distrib. Syst. Technol. **2**, 88–103 (2011). doi:10.4018/jdst.2011070106

Moser, T., Biffl, S., Sunindyo, W.D., Winkler, D.: Integrating production automation expert knowledge across engineering stakeholder domains. In: Barolli, L., Xhafa, F., Vitabile, S., Hsu, H.-H. (eds.) Proceedings of the 4th International Conference on Complex, Intelligent and Software Intensive Systems (CISIS 2010). IEEE Computer Society (2010)

Moser, T., Mordinyi, R., Mikula, A., Biffl, S.: Making expert knowledge explicit to facilitate tool support for integrating complex information systems in the ATM domain. In: International

Conference on Complex, Intelligent and Software Intensive Systems (CISIS'09), pp. 90–97. IEEE Computer Society, Fukuoka, Japan (2009a)
Moser, T., Mordinyi, R., Winkler, D., Biffl, S.: Engineering project management using the Engineering Cockpit: a collaboration platform for project managers and engineers. In: IEEE International Conference on Industrial Informatics (INDIN), pp. 579–584 (2011b)
Moser, T., Schimper, K., Mordinyi, R., Anjomshoaa, A.: SAMOA—a semi-automated ontology alignment method for systems integration in safety-critical environments. In: Proceedings of the 2nd IEEE International Workshop on Ontology Alignment and Visualization (OnAV'09), International Conference on Complex, Intelligent and Software Intensive Systems (CISIS'09), pp. 724–729. Fukuoka, Japan (2009b)
Moser, T., Biffl, S.: Semantic integration of software and systems engineering environments. IEEE Trans. Syst. Man Cybern. Part C Appl. Rev. **42**, 38–50 (2012). doi:10.1109/TSMCC.2011.2136377
Nielsen, J.: Usability Engineering, vol. 44, p. 362. Morgan Kaufmann, Pietquin O and Beaufort R (1993). doi:10.1145/1508044.1508050
Novák, P., Šindelář, R.: Applications of ontologies for assembling simulation models of industrial systems. In: Lecture Notes in Computer Science (including subseries Lecture Notes in Artificial Intelligence and Lecture Notes in Bioinformatics), pp. 148–157 (2011)
Noy, N.F.: Semantic integration: a survey of ontology-based approaches. SIGMOD Rec. 33:65–70. ST—Semantic integration: a survey of onto. doi:http://doi.acm.org/10.1145/1041410.1041421 (2004)
Noy, N.F., Doan, A.H., Halevy, A.Y.: Semantic integration. AI Mag. **26**, 7–10 ST—Semantic Integration (2005)
Oldakowski, R., Bizer, C., Westphal, D.: RAP: RDF API for PHP. In: Workshop on Scripting for the Semantic Web at 2nd European Semantic Web Conference (ESWC) (2005)
Pérez, J., Arenas, M., Gutierrez, C.: Semantics and complexity of SPARQL. Semant. Web—ISWC **4273**, 30–43 (2006). doi:10.1007/11926078
Pilato, M.: Version Control with Subversion. O'Reilly and Associates, Inc. (2004)
Rodriguez-Muro, M., Hardi, J., Calvanese, D.: Quest: efficient SPARQL-to-SQL for RDF and OWL. In: Demos of the 12th International Semantic Web Conference (ISWC 2012) (2012)
Sequeda, J.F., Arenas, M., Miranker, D.P.: On directly mapping relational databases to RDF and OWL. In: Proceedings of the 21st International Conference on World Wide Web—WWW'12, p. 649. ACM Press, New York (2012)
Sequeda, J.F., Miranker, D.P.: Ultrawrap: SPARQL execution on relational data. J. Web Semant. **22**, 19–39 (2013). doi:10.1016/j.websem.2013.08.002
Serral, E., Kovalenko, O., Moser, T., Biffl, S.: Semantic integration data storage architectures: A systematic comparison for automation systems engineering (2012)
Serral, E., Mordinyi, R., Kovalenko, O., et al.: Evaluation of semantic data storages for integrating heterogenous disciplines in automation systems engineering. In: IECON Proceedings (Industrial Electronics Conference), pp. 6858–6865 (2013)
Shen, X., Huang, V.: A framework for performance study of semantic databases. In: Proceedings of the International Workshop on Evaluation of Semantic Technologies (IWEST 2010). http://www.ceur-ws.org (2010)
SWAD-Europe: SWAD-Europe deliverable 10.1—scalability and storage: survey of free software/open source RDF storage systems (2002)
Tanenbaum, A.S.: Modern Operating Systems, 3rd edn. Prentice Hall Press, Upper Saddle River (2007)
Tichy, W.F.: RCS—a system for version control. Softw.-Pract. Exp. **15**, 637–654 (1985). doi:10.1002/spe.4380150703
Tinelli, E., Cascone, A., Ruta, M., et al.: I.M.P.A.K.T.: an innovative semantic-based skill management system exploiting standard SQL. In: ICEIS 2009—11th International Conference on Enterprise Information Systems, Proceedings, pp. 224–229 (2009)
TinkerPop: TinkerPop—An Open Source Graph Computing Framework. http://www.tinkerpop.com. Accessed 8 Apr 2015

Völkel, M., Groza, T.: SemVersion: an RDF-based ontology versioning system. In: Proceedings of IADIS International Conference on WWW/Internet (IADIS 2006), pp. 195–202 (2006)

Vysniauskas, E., Nemuraite, L., Paradauskas, B.: Hybrid method for storing and querying ontologies in databases (2011)

Waltersdorfer, F., Moser, T., Zoitl, A., Biffl, S.: Version management and conflict detection across heterogeneous engineering data models. In: IEEE International Conference on Industrial Informatics (INDIN), pp. 928–935 (2010)

Wiederhold, G.: Mediators in the architecture of future information systems. Computer **25**, 38–49 (1992). doi:10.1109/2.121508

Wiesner, A., Morbach, J., Marquardt, W.: Information integration in chemical process engineering based on semantic technologies. Comput. Chem. Eng. **35**, 692–708 (2011). doi:10.1016/j.compchemeng.2010.12.003

Winkler, D., Moser, T., Mordinyi, R., et al.: Engineering object change management process observation in distributed automation systems projects. In: Proceedings of 18th European System and Software Process Improvement and Innovation (EuroSPI 2011), pp. 1–12 (2011)

Zaikin, I., Tuzovsky, A.: Owl2vcs: Tools for distributed ontology development. In: Proceedings of 10th OWL: Experiences and Directions Workshop. http://www.CEUR-WS.org (2013)

Zhou, J., Ma, L., Liu, Q., et al.: Minerva: a scalable OWL ontology storage and inference system. Seman. Web 429–443 (2006). doi:10.1007/11836025_42

Chapter 9
Product Ramp-up for Semiconductor Manufacturing Automated Recommendation of Control System Setup

Roland Willmann and Wolfgang Kastner

Abstract Predictable and fast production launch of new products (product ramp-up) is a crucial success factor in the production industry in general, and for the production of integrated circuits (ICs) in particular. During the ramp-up phase of the product there is, inter alia, the need for product-specific configuration of a wide range of software systems that control the production process in a fully automated manner. This collection of software systems is sourced from several vendors and has therefore to be configured in different ways. Moreover, configuration has to be orchestrated along the whole production process, in accordance with the needs of the new product. This is a complicated, error-prone, and time-consuming task for product engineers, process engineers, and application engineers. The approach described in this chapter avoids such efforts and risks through a semiautomated generation of configurations of software systems. It uses a knowledge base which provides a unified configuration schema across all involved software systems. The approach applies automated reasoning of new configuration content based on the knowledge about new products' characteristics and knowledge about the existing production environment. The knowledge base is described by ontology models and based on Semantic Web technologies. The described approach is the basis for IT professionals of IC factories and of factories with comparable IT infrastructure to standardize the configuration of software systems being in charge for production control and process control. It contributes to accelerate the launch of new products and to make their ramp-up phase more deterministic.

Keywords Knowledge management • Product ramp-up • Ontology • Ontology mapping and matchmaking • Multidisciplinary engineering

---

R. Willmann (✉)
Institute of Software Technology and Interactive Systems, CDL-Flex, Vienna University of Technology, Vienna, Austria
e-mail: r.willmann@speed.at

W. Kastner (✉)
Technische Universität Wien, Treitlstraße 1-3, 1040 Vienna, Austria
e-mail: k@auto.tuwien.ac.at

S. Biffl and M. Sabou (eds.), *Semantic Web Technologies for Intelligent Engineering Applications*, DOI 10.1007/978-3-319-41490-4_9

9.1 Introduction

Launching and ramping up a new product in a production line is a highly interdisciplinary engineering task which involves experts from several domains (ramp-up team). Knowledge about currently produced products and the underlying production processes of a production line has to be combined with the specific needs of the new product. During the product ramp-up phase, a collection of product-specific instructions has to be created which describes the overall production flow (process plan), the details of each single step of the production flow (process step), and interactions between process steps. The validity of each instruction has to be evaluated. Depending on the complexity of the product and the production process, the ramp-up team has to master a challenging but also error-prone task which is sometimes also based on trial and error. The complexity of the product is driven by the number of components which are used to compose the new product or by high quality demands, while the complexity of the production process depends on the number of single process steps which have to be performed with continuously high precision.

However, there are limits concerning the available budget and duration which must not be exceeded by the ramp-up team. After completion of the ramp-up phase, the resulting instructions must enable the production resources of the production system (operators, equipment, suppliers, or control software) to produce instances of the new product with repeatable quality on a certain level. This quality level (yield) limits the count of instances of the new product which are allowed to be scrapped or reworked because of missed quality criteria. The majority of companies still struggles to perform the ramp-up of new products within the planned costs or budget, or to achieve the planned yield after ramp-up (Slamanig and Winkler 2012, p. 488). In the past, almost two-thirds of the companies were unable to meet their time-related targets, nearly 60 % of the companies failed to achieve their cost-related goals, and 47 % of the companies stated that they could not attain their objectives in process quality. Altogether, the results revealed that the companies within the industries being investigated by Slamanig and Winkler's study lack considerable knowledge and expertise in managing their product change projects in their supply chain networks.

A significant proportion of the problem is caused by poor planning and information exchange. With increasing complexity of the product and the production process the problem is also valid within a production system and not only across the supply chain. Such complex products are, for instance, integrated circuits (ICs) and their production processes which are said to be the most complex ones one can imagine in today's manufacturing industry.

For this reason, the ramp-up of a new product is still an individual project (ramp-up project) instead of a routine process, although some companies have developed technical concepts and business models in order to lower the risk of product ramp-up. Such measures include aspired quality gates during the ramp-up project and a modularization of product design in order to maximize the reuse of existing knowledge (knowledge reuse).

The trend toward individualized products and shorter product life cycles, which is expected to be accelerated by the subject *Industrie 4.0* enforces companies to improve the performance of product ramp-up projects. Similarly to many other industries, IC manufacturers have to deal with shortening product life cycles and an increasing number of new products in their production lines (Kurttila et al. 2010; Fransoo and de Kok 2007). As a consequence, product ramp-up projects have to be better predictable with respect to their costs and durations, but also concerning the achieved yield. Moreover, costs and duration of ramp-up projects have to be minimized while the achieved yield has to be maximized.

In Sect. 9.2, the challenges and performed measures of companies are discussed in more depth. A knowledge-based ramp-up process (K-RAMP) is described which addresses those challenges. Section 9.3 explains the challenge of product ramp-up in an IC production in more depth. The IC production was chosen, because of the high complexity of IC products and the underlying production process, but also because of the quality of available production data and other measures which have been already taken in these companies. The section is of particular interest because the architecture of the production-IT of IC manufacturers represents an almost ideal status quo with respect to production control and process control. Moreover, there are already concepts in place which maximize opportunities for reuse of elements of currently produced products (forerunner products).

Section 9.4 introduces K-RAMP from its process perspective. It is explained how the approach determines reusable elements of forerunner products or of existing process plans for the creation of forerunner products. Based on this process view on K-RAMP, the requirements to be satisfied are introduced in Sect. 9.5. In Sect. 9.6, the underlying ontology models and the software architecture of the knowledge base are described. Section 9.7 is completing the description of the approach by addressing the reuse and adaptation of setups of the process control software (further described as production-IT). This part is particularly essential as process control is crucial in order to meet the quality targets at the end of the ramp-up phase.

9.2 Definition of Product Ramp-up

9.2.1 In-Depth Insight into the Product Ramp-up

The time span within which a new product is ramped up after the end of product development to stable volume production is called the ramp-up phase (Fig. 9.1). In accordance to Terwiesch et al., time-to-volume attracted reasonable more attention than time-to-market already at the beginning of the century (Terwiesch et al. 2001, p. 435). “The fundamental difference between time-to-market and time-to-volume is

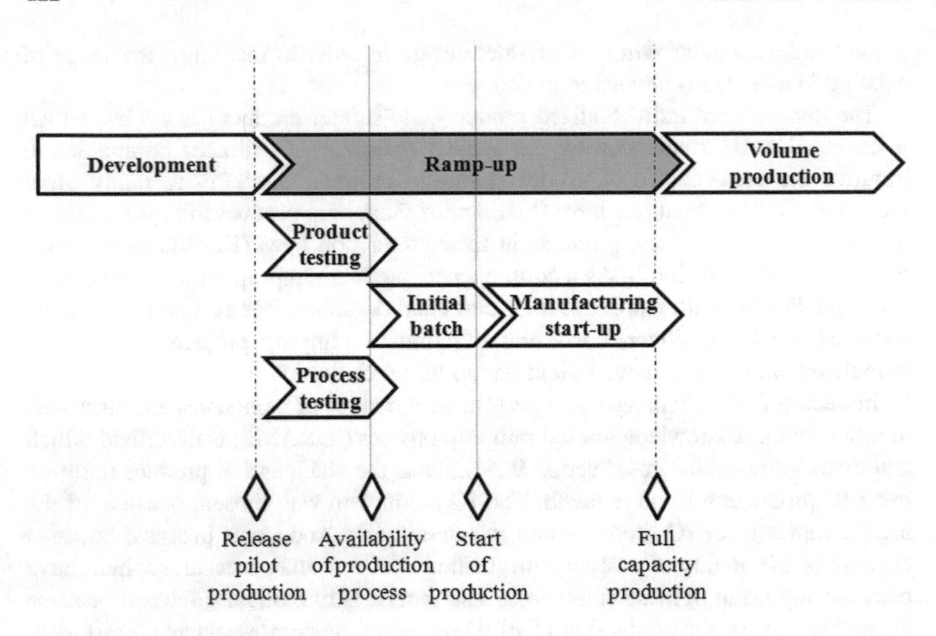

Fig. 9.1 Location of the ramp-up phase of a new product based on Slamanig and Winkler (2012, p. 484)

that the former ends with the beginning of commercial production whereas the latter explicitly includes the period of production ramp-up. Production ramp-up is the period during which a manufacturing process makes the transition from zero to full-scale production at target level of cost and quality."

The ramp-up phase involves the adjustment of all elements of the production system, so that instances of the new product (workpieces) can be produced with the required quality, in the planned quantities and within the proposed production costs per piece. This is a complex and still error-prone multidisciplinary task which involves engineers of several domains of expertise. As shown in the sequel, the major issue is related to knowledge management within the production system and along the supply chain of a production process.

The ramp-up phase usually begins as soon as the development phase ends. From the development phase, a so-called *bill of materials* (BOM) of the new product is obtained. This is the hierarchical structure of all subparts or subproducts which compose the final product, including the detailed characteristics of each ingredient of the subproducts (e.g., size, weight, uniformity of surface).

Consider a chocolate cake (and later its production) as a running example. The chocolate cake is composed of the chocolate gloss and the layered structure of dough and jam. The chocolate gloss consists of liquid chocolate and the gloss, while the layered structure of dough and jam is assembled by an upper layer of dough, a lower layer of dough, and a layer of jam in between. Liquid dark chocolate is a mixture of hot water and solid chocolate. This decomposition of the chocolate cake is continued until further decomposition is out of scope of the bakery.

For instance, the bakery has no interest in the production of solid dark chocolate because it is acquired from a supplier (the solid chocolate is therefore called consumable material). Each consumable material and each subproduct has to follow specific characteristics because it influences the function of the product—like the portion of cacao fat, sugar, and milk of chocolate influences the expected taste of the chocolate gloss and the reflectivity of its surface.

After the new variety of chocolate cake has been successfully created, the baker has to ensure that other occurrences of this chocolate cake can be produced with repeatable quality. There should be as less as possible variation with respect to quality from one instance to the other. In case of a chocolate cake, the quality might be specified by functions like the taste, the uniformity of the surface of the gloss, or the softness of the dough. In order to ensure this low variation, an accurate plan is required. This plan is known as the recipe of the cake. In more general terms of manufacturing, this is entitled a device-independent process plan.

Why is it a device-independent process plan? Usually the recipe of the cake does not consider exactly the handling instructions of the used kitchen tools, like the mixer, the oven, or the hot plate. This set of handling instructions would be too specific as it focuses on a particular tool set. After developing the recipe of the cake (i.e., the generic process plan) it must be possible that the cake can be baked with low variation of quality, with large volumes in different dislocated baker's shops, and using different tool sets. Therefore, device-specific handling instructions have to be specified for each tool set, using (1) the knowledge about the BOM, (2) the device-independent process plan, and (3) the experience about the capabilities and usage of the local tool set.

These device-specific handling instructions comprise of (1) the device-specific processing instructions for locally available devices (such as oven-specific programs and time periods of heating phases which need to be considered during the baking process), as well as (2) the device-specific control instructions, in order to decide when a process step is completed or needs to be readjusted (such as a decision when the baking process has been completed based on the color of the dough in the oven). The first kind of instructions includes the setup of local devices in order to treat occurrences of a particular product (e.g., a chocolate cake) properly, while the second kind of instructions enables device-specific decision-making during one particular execution of a process step (process job) while treating the occurrences of a product.

Considering the chocolate cake, the strength of the bakery might be more than creating a new variety of a cake sometimes and to produce hundreds of instances of it. Let us assume that a bakery has put focus on the creation of customized cakes. Instead of hundreds of occurrences only a few "copies" are created upon customer requests by specifying the taste, the structure, the color, or the size of the cake. In such a business model, the bakery must be able to develop appropriate handling instructions as fast and as accurate as possible. Speed is essential since the cake should be delivered to the customer as fast as possible. The tools in the baker's shop need to be utilized to be able to pay for their investment. And the personnel resources have to be utilized in the best way. Accuracy is essential to avoid failures during baking and therefore losses due to time, wasted material or energy.

This business model is exactly the one which is also faced by most manufacturing industries today. Concepts which are describing this situation are *shortening of product lifecycles* or *lot size 1*. The evolution of *Industrie 4.0* will further accelerate this trend. The faster and the more accurate the setup of the devices of a production system's can be adopted in order to produce occurrences of new varieties of products, the lower are the costs to be faced during the ramp-up phase. Cost drivers during the ramp-up phase are personnel costs, scrapped material, energy consumption, the utilization of devices which cannot be used for treating forerunner products during these times, or setup times which force devices even to idle.

The development phase provides a device-independent process plan. The plan may not address the specific setup of each production machine or measurement system (both in common are named equipment) in the production system. This need is caused by the differences between the equipment of the volume production line (the targeted production system) where occurrences of the product shall be produced and the equipment of the pilot line (original production system) where only the first occurrences of the new product were created for development and evaluation purposes. The differences are caused by equipment, which was acquired from different suppliers during different times. Equipment variations are therefore caused by the variation of equipment structure across vendors, as well as by the age and thus the different stages of technical progress of equipment.

In order to face this challenge, some companies are therefore following the strategy of *copy-exactly*, where every production system is an exact copy of a production system template which is following a corporate design rule. Using this approach, the complexity of a product ramp-up project is reduced significantly, because the information of the original production system simply needs to be transferred without any modifications to the targeted production system. No additional assumptions need to be performed because equipment and control software in both production systems use exactly the same configuration.

However, although the *copy-exactly* approach sounds attractive it is coming with a price. Copy-exactly requires identical production equipment for every production system of a company in order to reduce the complexity of change in case of transfer of products between production systems. As a consequence, for complex production processes, either significant investment is needed for leading-edge production equipment in all production systems simultaneously or increasingly outdated equipment has consequently to be used. The latter would be then used for production of new leading-edge products as well. Probably not all companies want or can deal with such restriction. Also admitted by Terwiesch et al. (Terwiesch and Xu 2003, p. 4), most semiconductor manufacturers still favor a much more aggressive process change during the ramp and do not follow the copy-exactly approach. Therefore, the discussed approach is beneficial for a broad ecosystem of industrial manufacturing.

Summarizing the previous sections, it can be concluded that the outcome of a product ramp-up project is still difficult to predict. The essential task during a ramp-up project is the specification and evaluation of handling instructions, which can be used to produce occurrences of the new product by utilizing production resources of the targeted production system. The complexity of a ramp-up project increases with the complexity and quality demands of the product and therefore the complexity and required precision of the underlying production process. Cost, duration, and the achieved initial yield are the metrics for measuring the success of a ramp-up project. Most manufacturers have to deal with heterogeneous equipments across the original production system and the targeted production system. For this reason, they are not able to copy information with regard to a new product exactly between both production systems.

9.2.2 A Knowledge System Based Product Ramp-up (K-RAMP)

Essentially K-RAMP uses a hierarchical composition structure for the product and another hierarchical composition structure for the process which creates instances of the corresponding product (Fig. 9.2).

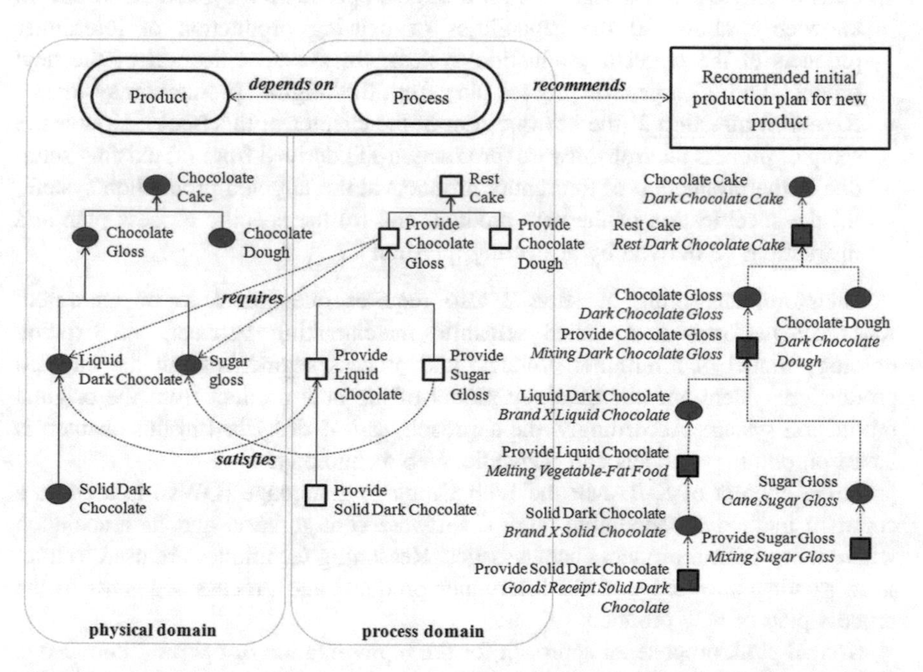

Fig. 9.2 Relation between the product and the production process and how it results in a recommended process plan

The composition structures of products and processes need not to be identical. However, each node of the production process' composition structure depends on nodes of the product's composition structure for two reasons.

- Each node of the production process has to satisfy the characteristics of a node within the product's composition structure on the same composition level.
- Each node of the process plan requires one or more nodes of the product's composition structure on the next lower level as input in order to fulfill its objectives.

The result of K-RAMP represents a recommended initial, device-specific process plan for the creation of a product from the original production system using production resources of the targeted production system. The recommended process plan comprises the sequence of nodes from the process tree, by considering their dependencies on nodes of the product tree. Within the recommended process plan each node is enhanced by detailed device-specific handling instructions which fit to the production resources of the targeted production system. These handling instructions for production resources are derived from existing information about the production of forerunner products at the targeted production system. Through the sequel of this chapter the nodes of the process plan will be called process segments. The questions to be answered by K-RAMP are therefore as follows:

- **Research question 1**: How is a new device-specific process plan and instructions of single process segments for a targeted production system derived from knowledge about (a) the capabilities of existing production of forerunner products at the targeted production system, (b) the specification of the new product, and (c) a generic process plan from the original production system?
- **Research question 2** (the key question of the chapter of this book): How is the setup of process control software (production-IT) derived from (a) existing setup due to the production of forerunner products at the targeted production system, (b) the specification of the new product, and (c) the specific process plan and instructions as derived by answering question 1?

Therefore, answering question 2 also requires question 1 to be answered. K-RAMP performs a so-called semantic matchmaking between the existing ontology model of forerunner products and process segments from the targeted production system and the ontology model of the new product from the original production system. Accordingly, the approach, as it is described in this chapter, is based on ontology models and Semantic Web technology.

In the context of K-RAMP the Web Ontology Language (OWL) is used as a common method for modeling control software configuration and its association with quality-critical product characteristics. Reasoning techniques are used to infer from existing knowledge about forerunner products and process segments to the process plan of new products.

Konrad et al. propose an approach for the representation of operator knowledge (Konrad et al. 2012). Their approach is based on semantic technologies aiming for ramping up of a new production process. It allows interactive knowledge input with respect to information gathered from dynamic data (machine events) and static data (specification of products, processes, and production systems as well as relations between them). Semantic Web technology contributes through specialization and reasoning. Their approach assumes that static knowledge is widely entered manually, while reasoning allows deriving new assumptions from an existing base of static and dynamic knowledge.

However, the approach of Konrad et al. does not address the creation of new classes of process segments based on the capability of existing ones in order to grow a taxonomy model of reusable process segments. However, this activity has to be performed as part of a product ramp-up project, as will be shown in the sequel. Beside others, the proposed K-RAMP approach shall reason new concepts of process segments and proposes them to the ramp-up team for evaluation and qualification.

K-RAMP contributes twice to the ramp-up of a new product.

- K-RAMP immediately derives a device-specific process plan and handling instructions, in order to produce the new product by utilizing the capabilities of the targeted production system's resources. Moreover, it also highlights the gaps which cannot be solved through reuse of existing knowledge. Being aware of such gaps, the ramp-up team members may focus their work on such missing links and encountered additional actions, which have to be performed. Such activities usually comprise of the development and qualification of missing process segments. These activities particularly may have an impact on the duration and costs of the product ramp-up project. Due to the knowledge about such gaps at the very beginning of the ramp-up project, the ramp-up team is aware about such additional efforts immediately.
- K-RAMP provides a permanently updated status of the ongoing ramp-up project and the possibility to query the current status of completion of the device-specific process plan, its details (handling instructions) and still existing gaps at any time throughout the ramp-up. Therefore, K-RAMP also drives the ramp-up team actively through the project.

In the previous sections, K-RAMP was outlined as a function which uses knowledge about the new product as well as knowledge about the production of forerunner products of the targeted production system as input for recommending an initial process plan as an output. For this purpose, K-RAMP attempts to determine appropriate subproducts and subprocesses, which can be reused for the creation of the new product. In the next section, some measures of manufacturing companies are discussed which still allow them to reuse existing production knowledge of the targeted production system in order to reduce complexity of ramp-up projects. It is essential to be aware about such measures as they also support the success of the discussed K-RAMP approach.

In the next section, also some concepts of IC production are introduced. It is beneficial for several reasons to take a product ramp-up use case from IC production for the discussion of K-RAMP. One reason is the complexity of IC products and the IC production process which also makes a product ramp-up project a very complex task. If K-RAMP is successful in IC production then it will be successful for any other discrete production as well. Another reason is the availability of rather good quality of collected production data, which can be utilized for the purpose of machine learning and for determining reuse of information. Other industries still struggle with the minor quality and completeness of their collected production data. Finally, there is a comprehensive production-IT established for the purpose of automated monitoring and control of the production processes, which enforces the motivation for automatically generated product-specific setup during the ramp-up phase. All in all, IC production therefore provides a perfect case study of *smart production*, as it is requested by *Industrie 4.0* for future production systems.

9.3 Challenge of IC Production—Prerequisites for Efficient Product Ramp-up

In the following sections, some challenges of IC production are discussed and how IC manufacturers master the complexity of production ramp-up and large product mix in their production systems. It is also shown that other industries are following similar concepts in order to enforce reusability of existing production knowledge during ramp-up of new products. These measures are discussed to some detail as they are a good foundation for a successful application of K-RAMP.

The production of ICs on silicon which are used for microprocessors or memory chips in our computers, cell phones and an increasing list of other electronic devices is one of the most demanding discrete production processes. There are several reasons why this production process is so demanding. Since the first commercial silicon-based ICs in the 1960th feature sizes permanently shrank. Feature sizes of currently available IC products reach 22 nm and IC manufacturers are researching for 14 nm capable production processes. Moreover, it is the large number of significantly more than 100 consecutive process steps which need to be exactly synchronized to lead to products of high quality. Aside thousands of product characteristics with quality tolerances on the level of micrometers and nanometers, the vertical matching of the individual layers which build up millions of electrical components on areas of square millimeters have to be ensured. It is therefore valid to talk about very complex products and production processes.

Many ICs are produced simultaneously on circular very thin monocrystalline silicon disks, also called wafers, which are sawn in the end. The common sizes of silicon wafers in today's factories are either 200 or 300 mm. A typical 300 mm wafer factory produces around 10,000 wafers per month. Many IC products are manufactured simultaneously in a production line (production system). Handling instructions for production resources have to be changed automatically for wafers which are assigned to a particular IC product in order to ensure proper treatment. At the same time the yield (i.e., the ratio of good ICs compared to all produced ICs) has to be around 98 % for the production system to be profitable. To achieve this goal, every single process step has to be executed in a flawless manner (Xiao 2012).

Therefore, in modern IC factories, almost every single process step is followed by a metrology step which measures samples and production-IT which adjusts the process setup in real time and triggers corrective actions immediately and automatically at the event of excessive process variations or error-related process changes. Throughout the remaining chapter, an arbitrary sequence of process steps followed by one or more metrology steps shall be named process segment. Referring to the previous sections, we remember that a process segment represents a single node within the hierarchical composition structure of the production process.

Over the years, the production-IT was enhanced by a variety of powerful control components, which run generally fully automated. These control components usually require a product-specific control setup as it was introduced with the simple example of checking the color of the cake while baking it in the oven in the introduction Sect. 9.2.1. The most common control components currently in use are shown in Fig. 9.3 and provide the following functionality:

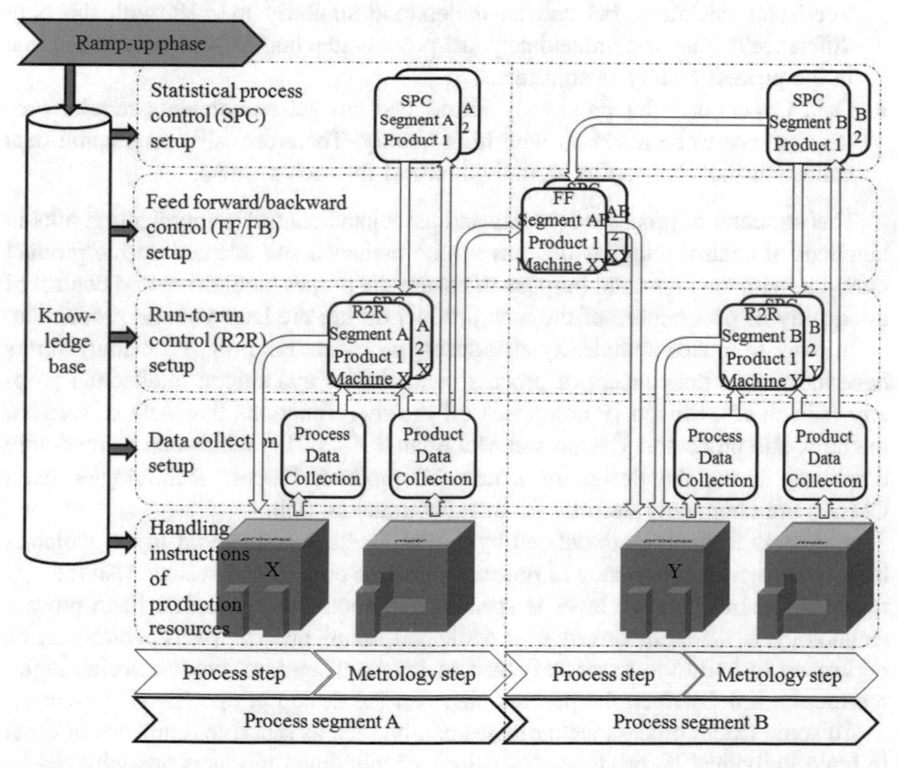

Fig. 9.3 Assignment of data collection and control components to process segments and production resources (e.g., equipment)

- Statistical process control (SPC) aligns data of measured samples with historical trends and control limits which are based on observed statistical data. Notifications to process engineers or immediate corrective actions are triggered automatically if an instable behavior of the controlled process is recognized (Dietrich and Schulze 2003).
- Run to run control (R2R) complements SPC. While SPC monitors metrics and triggers corrective actions in case of deviations, R2R acts through readjustment of processes before a deviation occurs. R2R uses pre- and post-process measurements of samples (individual wafers in case of IC production) which are taken out of the controlled process, as well as the expected quality target (e.g., the thickness of a metallization layer which was deposited to the wafer) and knowledge about the correlation between this quality target and a few setup parameters of the controlled process (e.g., the duration of the deposition process and the temperature in the process chamber during metal deposition).
- Feed forward control/feed backward control (FF/FB) considers dependencies along the process sequence. For example, the layer thickness of the photo resistor after the lithography process can affect the subsequent reactive ion etch process (Ruegsegger et al. 1998). An FF model holds the information, how the reactive ion etch process recipe has to be adjusted for compensation of photoresistant thickness. FB can be understood similarly to R2R with the only difference that not the immediately last process step but the setup of another one in the process history is adjusted.
- Data Collection—the previously introduced control components require production data to be recorded with high quality. Therefore, also production data collection models need to be individualized for each product.

This structure of process segments and associated control components results in hundreds of control setups which have to be reviewed and adapted during product ramp-up, with the important purpose to ensure the proper monitoring and control of the quality of occurrences of the new product during the later volume production.

In order to reduce complexity also during a product ramp-up, IC manufacturers have introduced the concept of process technologies and silicon intellectual property (silicon IP). Silicon IP comprises off-the-shelf functions like A/D converters, memory, and processors (Nenni and McLellan 2013, p. 19) which can be randomly combined during the design of a new IC product. Process technologies (e.g., CMOS) are commonly used by IC manufacturers as well.

A process technology is defined by a specific stack of material layers which is built by a particular sequence of process segments on a silicon wafer. Also the size range of features on each layer is specific for process technologies. Each process technology is therefore linked to a dedicated set of process plans which can be performed to build the requested stack of layers. Therefore, process technologies represent a link between the process plan and the design of specific IC products.

To some extent process technologies can be seen as reusable templates in order to build individual IC products. Variations of individual products are achieved by modification of the layout of single layers, the thicknesses of layers, controlled impurities of semiconductor material, or other means of parameterization.

Because of this branch-specific method of process technologies and modularization of products by the use of silicon IP, the semiconductor industry provides a good pattern for customization of products using templates and unified underlying production processes. Some other industries, like the production of printed circuit boards (PCBs) apply this method as well (Macleod 2002).

Also the automotive manufacturing industry uses the concept of product templates. In this domain, product templates are called platforms. An outstanding example of a platform approach is Toyota's policy concerning its car models. Toyota is currently launching new generations of its successful car models that utilize more than 70 % of their forerunners' components, and the platforms of these cars have remained largely constant through successive car generations (Hüttenrauch and Baum 2008; Slamanig and Winkler 2012, p. 486).

The concepts of process technologies and of product platforms can be considered similar. It has a general validity in the manufacturing industry for reduction of the complexity of management of product variants and therefore the complexity of product ramp-up projects. Facing the emerge of *Industrie 4.0*, this method will be rather likely adopted by other manufacturing industries as well.

Driven by the complexity and quality demands of IC production a comprehensive set of control components was developed as part of the production-IT. Setting up those control components in order to meet product-specific needs is an essential task during a product ramp-up project. In order to master the complexity of a production process and therefore of product ramp-up projects, concepts of process technologies or product platforms were introduced. Such concepts support the reuse of existing production knowledge for new products.

From the perspective of K-RAMP, the approach of templates or platforms helps to develop an algorithm which determines reusable subproducts or process segments of forerunner products while planning or performing ramp-up projects. In the next sections, K-RAMP is introduced as a process for gathering such existing knowledge and matching it with the needs of a new product.

9.4 The Process Perspective of K-RAMP

In the sequel of this section, an overview of K-RAMP is provided. This section is followed by a discussion of the involved ontology models and the necessary architecture of software components which envelopes a Semantic Web-based knowledge base. An essential part is the mapping of the product design—in particular quality-relevant characteristics of the product design—to the configuration settings of control software components. However, in order to determine reusable configuration settings, appropriate and thus reusable process segments and subproducts have to be found in advance. The corresponding process is described in the next sections.

Successful searching of reusable subproducts of forerunner can be realized by matching them with the quality characteristics of the new product. Searching for reusable process segments is supported by the mapping between the composition structure of products and the composition structure of the production process as introduced in Sect. 9.2.2.

The principle idea of this mapping was introduced as axiomatic design by Suh (1990, 2001). It describes the technique of mapping (Fig. 9.4) between concepts of the customer domain, the functional domain, the physical domain, and the process domain. In particular, customer attributes (CA) of the customer domain are mapped to functional requirements (FR) of the functional domain, design parameters (DP) of the physical domain are mapped to FRs, and process variables (PV) of the process domain are mapped to DPs. This mapping is performed on several levels of a tree-like composition structure. The mappings between CAs and FRs, as well as between FRs and DPs are closely related to quality function deployment (QFD) (Breyfogle III 2003; Said El-Haik 2005). However, the mapping between CAs and FRs is not covered within K-RAMP as it is assumed to be part of the preceding product design phase.

The essential parts of the axiomatic design which have been considered with K-RAMP are the mapping between the functional domain and the physical domain in order to determine reusable subproducts, and the mapping between the physical domain and the process domain in order to determine reusable process segments.

In the sequel of this chapter, particularly the mapping of the physical domain and the process domain is in the focus. According to Suh, it is the association between

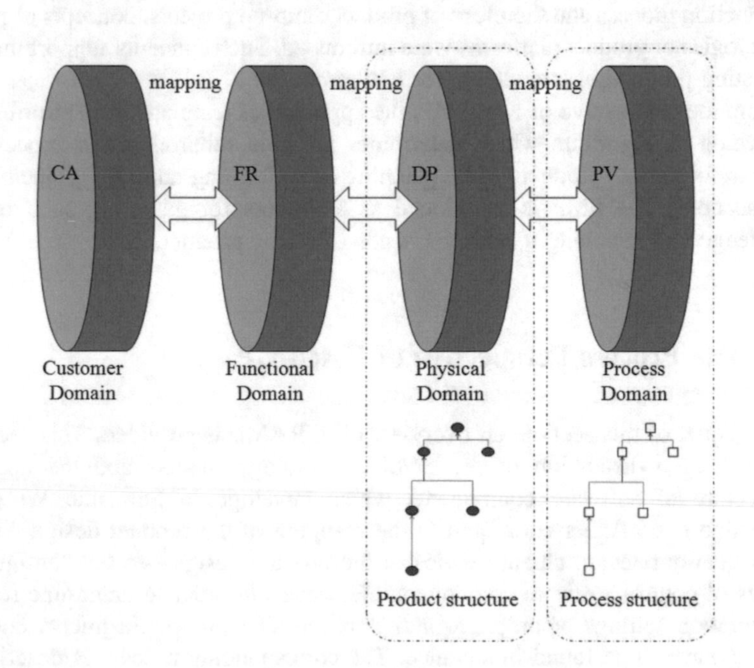

Fig. 9.4 Mapping between domains of axiomatic design based on Suh (1990)

DPs and PVs. The concept of DPs of axiomatic design shall be equivalent with the concept of products or subproducts, and the concept of PVs shall be equivalent with the concept of process segments during the following discussion.

The focus of this chapter is the recommendation of control system setup as outcome for the ramp-up of a new product. Automated recommendation of control system setup for new products requires knowledge about the currently applied setup. From Fig. 9.3, it can be seen that control system setup is associated with the process segments which shall be controlled. K-RAMP assumes that every process segment is capable to produce a particular category of subproducts and that the applied control models are equivalent for all subproducts of this category, at least on a certain level of generalization. Deducing associations between new subproducts and already existing process segments is therefore essential for an automated recommendation of the setup of control software components (Question 2). Ahead of this particular topic, the approach for gathering reusable process segments from a new product's design has to be discussed (Question 1). How K-RAMP solves Question 1 is illustrated in this section.

Generally, in all three relevant knowledge domains of K-RAMP (i.e., the functional domain, the physical domain and the process domain) each node of the composition structure is specified by a set of characteristics. Moreover, nodes of the composition trees of mapped knowledge domains correspond to each other. For instance, a particular node of the process structure (a process segment) requires an arbitrary number of nodes of the product structure in order to satisfy a particular other node of the product structure on the next higher composition level (see discussion in context of Fig. 9.2).

The V-model in Fig. 9.5 shows the overall domain of production knowledge as it is contained within the K-RAMP knowledge base. It covers a design perspective where elements of the functional domain, the physical domain, and the process domain are mapped to each other, and it covers a control side where the connection to the physical representatives of the production-IT is modeled. On the design side, the nodes of the composition structures of each domain are specified by sets of

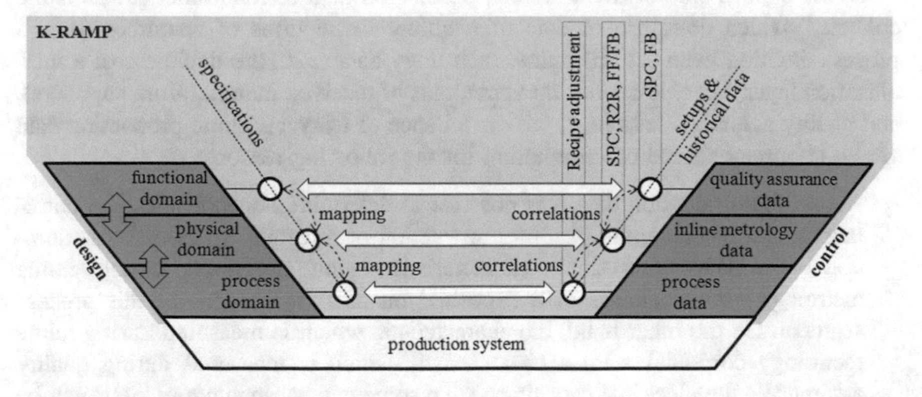

Fig. 9.5 V-Model of design and control in manufacturing

characteristics (specifications). By the use of these specifications, it is possible to determine weights of the associations *satisfies* and *requires* as used in Fig. 9.2. It is for instance possible to determine the coverage (satisfaction) of a subproduct's specification through a particular specification of a process segment.

Each domain of the design has a counterpart on the control side of the V-model. The control side is implemented in the production system through the use of control software components. These components ensure that the specifications of the design side are fulfilled. The specifications of the design side are therefore related to the setups of the control software components on the control side of the V-model.

The elements of the K-RAMP knowledge base do not provide any mappings to the layout of the physical production system. These equipment engineering topics are covered by other research papers, like Moser et al. (2010), Winkler and Biffl (2010), Moser et al. (2011).

The specifications of the process domain interact with the setups of each process segment. For instance, there might be a specific processing duration specified for a heating process. The recipe adjustment as it is implemented in the production system uses the specified processing duration in order to combine it with a couple of other settings of an equipment recipe. All settings together are controlling an automated procedure of the equipment or a machine.

The specifications of the physical domain are associated with the setups of control software components in the inline metrology area. For instance, the specification of a subproduct may require a layer thickness within a particular range of tolerance. The inline metrology of the production system measures this thickness. The setup of the SPC software component or the R2R software component ensures that appropriate actions are performed if the thickness is running out of range.

FRs of the functional domain are associated with the quality assurance of the production system. For instance, the electrical capacitance of a layer is specified by an FR in the functional domain. The measurement of this electrical capacitance is performed during the quality assurance of the production system. Through product-specific setup of the SPC software component and appropriate actions, the electrical capacitance is kept within the limits during volume production.

At the control side of the V-model, production data of forerunner products are collected, which comprise of data of handling instructions of resources (e.g., a process duration been set up), inline metrology data (e.g., the thickness of a metallization layer, but also quality measurements of received material from suppliers), and quality assurance data (e.g., the capacitance of a layer). These production data are an important source of information for the following reasons:

- Across the production data it is possible to determine dependencies and correlations. For instance, the handling instruction of a resource—the process duration—correlates with some characteristic being measured during inline metrology—the thickness of a deposited metallization layer—of this process segment. On the other hand, the characteristic which is measured during inline metrology correlates with a characteristic which is measured during quality assurance—the electrical capacitance. To some extent, such correlations can be used to control the result of process segments or the overall production process and thus the function of the final product.

- Between the control side of the V-model and the design side of the V-model it is possible to determine whether the production resources, process segments, or subproducts of the targeted production system are capable to produce instances of the new product in accordance to its specification.

These dependencies are used by K-RAMP to answer Questions 1 and 2. In the sequel of this section, the process perspective of K-RAMP is explained. This process perspective is equivalent to the one which has to be followed by the ramp-up teams manually today. The result of this discussion is a list of requirements which has to be satisfied by a knowledge-based approach.

K-RAMP is performed recursively on every level of composition of the structure of FRs, the (sub-)products, and the process segments. The first and most obvious step is to determine existing (sub-)products which already match the characteristics of a new (sub-)product, or which satisfy the FRs of the new (sub-)product.

Matching between existing (sub-)products and new (sub-)products is based on the left side of the V-model only. However, Semantic Web's open world assumption (OWA) only uncovers not-matching (sub-)products, but it is not possible to determine a set of matching (sub-)products. Therefore, an external matchmaking algorithm has to support the reasoning.

The coverage of new FRs by existing (sub-)products can be rather likely derived by overlapping quality assurance data of existing (sub-)products (right side of the V-model) and the specifications of new FRs (activity A in Fig. 9.6).

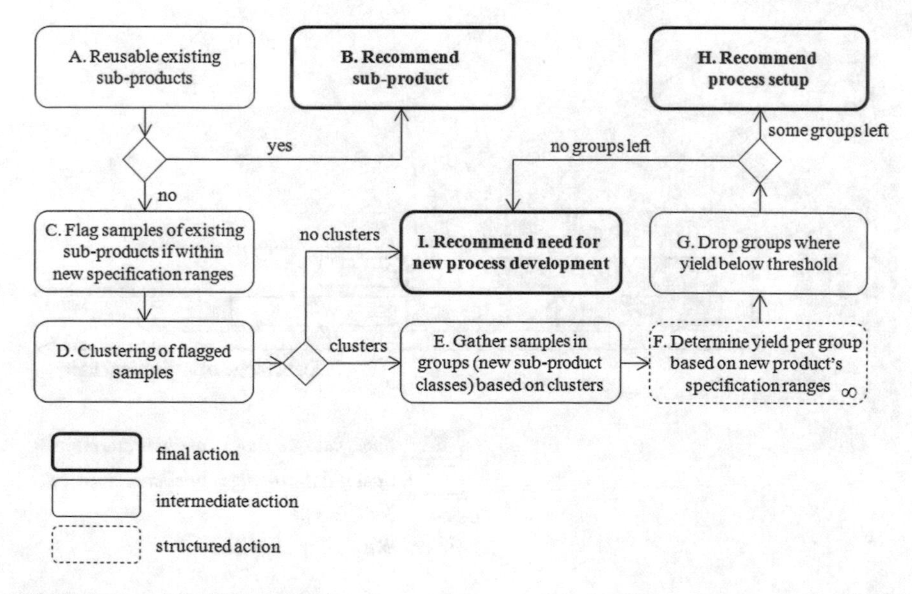

Fig. 9.6 Overview of K-RAMP

Before detailed studies on the relationship between process segments and characteristics of (sub-)products, K-RAMP tries to find existing (sub-)products whose characteristics overlap with new (sub-)products. If the specification bands of all characteristics of an existing (sub-)product are enclosed by the specification bands of the corresponding characteristics of a new (sub-)product it can be implied that every occurrence of the existing (sub-)product is also a member of the enclosing new (sub-)product. Consequently, the process segment which creates the existing (sub-)product should be also capable to produce the new (sub-)product.

In Fig. 9.7, there are three (sub-)products P_X, P_Y, and P_Z, all being members of the same product class P. Product classes, for instance, are chocolate or copper metallization layer, while the (sub-)products represent specific brands of chocolate vendors or specific interlayer dielectric layers (ILDs) of an IC's process technology (Xiao 2012, pp. 371–374). Each produced occurrence (or sample) is a member of a (sub-)product. Each (sub-)product in Fig. 9.7 has characteristics C_1 and C_2 which are in common with other (sub-)products—particularly if they are of the same product class. However, for each (sub-)product these characteristics have different specification ranges, meaning the value ranges which are allowed without harming the quality of the respective (sub-)product. In Fig. 9.7, specification ranges are written in a form where the subscripted letter refers to the (sub-)product and the superscripted number refers to the characteristic.

R_Y^1 is the specification range of characteristic C_1 for (sub-)product P_Y.

In Fig. 9.7, there is also a subclass of (sub-)product P_Z named $P_{Z'}$ which shall be discussed later as well.

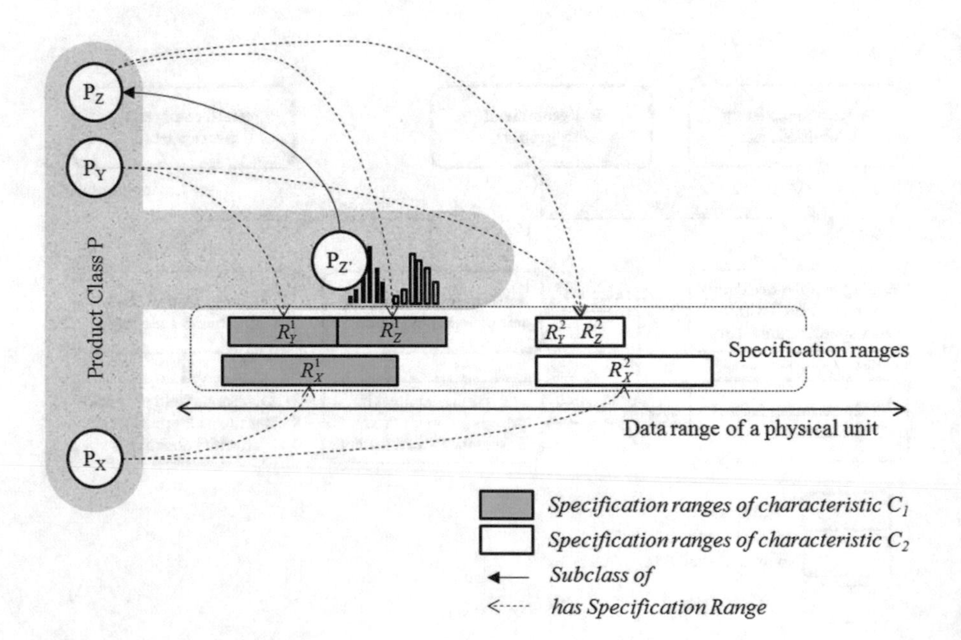

Fig. 9.7 Examples of mutually enclosing specification ranges

How is reusability in the product domain determined? It is not useful to try matching between all existing and all new (sub-)products. Each new (sub-)product is used to determine related existing (sub-)products instead. For this purpose, the ontology model of K-RAMP provides a generalization structure of (sub-)product classes. Based on the set of related (sub-)product classes their members have to be found which are enclosed by the new (sub-)product. The list of characteristics and the limits of their specification ranges are the defining elements of memberships in (sub-)product classes. A (sub-)product class is more specific, the more characteristics are necessary to specify its members. A (sub-)product class C is a subclass of another (sub-)product class D if the list of characteristics of C is a superset of the list of characteristics of D and the specification range of each common characteristic in D encloses the respective characteristic's specification range in C.

According to Fig. 9.7, P_Y is enclosed by P_X because R_X^1 fully encloses R_Y^1, and R_X^2 fully encloses R_Y^2. Another already existing subproduct P_Z is not enclosed. The specification ranges R_X^1 and R_Z^1 are only overlapping each other. This is the most trivial decision-making to determine reusable (sub-)products and, consequently, associated process segments being capable to produce the (sub-)products.

Having, for instance, a look to P_Y and P_X, the decision is obvious. Occurrences of the forerunner product P_Y are already produced with large volume and sufficient high yield. For this reason, the majority of measured characteristics C_1 and C_2 which are taken from samples in a process segment while producing the subproduct P_Y are within the specification ranges R_Y^1 and R_Y^2. So it is rather likely, that if P_X is produced by the same process segment the majority of measured C_1 and C_2 of samples of P_X will be within the even wider specification ranges R_X^1 and R_X^2.

However, even (sub-)products whose specification ranges are not fully enclosed by the equivalent specification ranges of a new (sub-)product can be potential candidates for reuse. This situation is demonstrated by (sub-)product P_Z in Fig. 9.7. Occurrences of the (sub-)product P_Z—which are represented by measured samples of the related process segment—might be within the specification range R_Z^1 of (sub-) product P_Z and within the specification range R_X^1 of the new subproduct P_X as well. It is now the challenge of K-RAMP (activities C and D in Fig. 9.6) to find a systematic cause why parts of the samples also fit to the new specification range of P_X while others do not. If no such cause is found the ramp-up team has to be aware about this knowledge gap and has to plan the development of a new process segment (activity I in Fig. 9.6) which covers the needs of P_X independently of existing process segments.

However, if such a systematic cause is determined (e.g., a specific range of the process duration or the applied heating temperature) it can be used as supplemental constraint of a new handling instruction which is derived from the handling instruction of the existing process segment. This new handling instruction can be the baseline for the new process segment. The occurrences (samples) of subproduct P_Z which are produced due to the systematic cause become members of (sub-) product $P_{Z'}$ which is a subclass of P_Z. Both are still members of the same product class P.

The coverage of a FR by (sub-)products shall be above an appropriate threshold in order to determine them as appropriate reusable (sub-)products with some certainty. With respect to this coverage, Suh recommends to calculate the entropy as introduced by Shannon. In K-RAMP, the theoretical possible first pass yield (FPY) with a range between 0 and 100 % is used to determine how much the members of $P_{Z'}$ and therefore the derived handling instruction of the new process segment would cover the specification range R_X^1. If the theoretical FPY is above a certain threshold the existing subproduct can be considered for reuse.

The FPY is calculated as the portion of defect-free parts of a specific (sub-)product which are passing a particular process segment of the overall production process (Wappis and Jung 2010, pp. 179–180) at the first pass (without rework). Defect-free parts are within the respective specification ranges of all given characteristics of a (sub-)product. During production of the forerunner products, this condition is evaluated against the specification ranges of the forerunner's (sub-) products. However, K-RAMP uses the historical measurements of forerunner products and evaluates them against the specification ranges of the new (sub-) products.

It may happen that there is no (sub-)product available above the threshold and the search result is empty. However, if there is enough information available concerning the correlation of data from inline metrology of individual characteristics and the characteristics of functional requirements (from quality assurance) then it may be possible to determine a regression function. If this regression function is part of the knowledge base as well, it can be used to calculate the ideal adjustment of the existing (sub-)product's characteristics in order to maximize the coverage of the FR of a new (sub-)product. A similar approach is also possible by considering the correlation between existing (sub-)product's characteristics and setup parameters from the handling instruction of the related process segment.

Reuse of an existing (sub-)product or a process segment by adjustment is considered in activity F of Fig. 9.6. If, for instance, $P_{Z'}$ does not perfectly cover the specification ranges of P_X it can be tried to evaluate potential shifting of setup parameters by the use of correlation models in order to improve the coverage.

For instance, as demonstrated in Fig. 9.8, by keeping a dedicated process temperature and particular deposition material—both are parts of the handling instruction of a process segment—it is possible to alter the *Deposition time* for the purpose of layer thickness (axis *d*) adjustment. As a part of the continuously ongoing quality monitoring, the layer thickness measurements of samples of forerunner (sub-)products are used to determine the shape of the distribution function of this process segment and to calculate the average and the standard deviation. Based on these indicators and the specification range of the layer thickness (upper specification limit USL_1 and lower specification limit LSL_1) it is possible to calculate the FPY of the forerunner (sub-)product with respect to the layer thickness.

The regression function of the deposition time and the layer thickness, as well as the shape of the distribution function and its standard deviation is now used by K-RAMP. An adjusted deposition time for the layer thickness of the new (sub-)

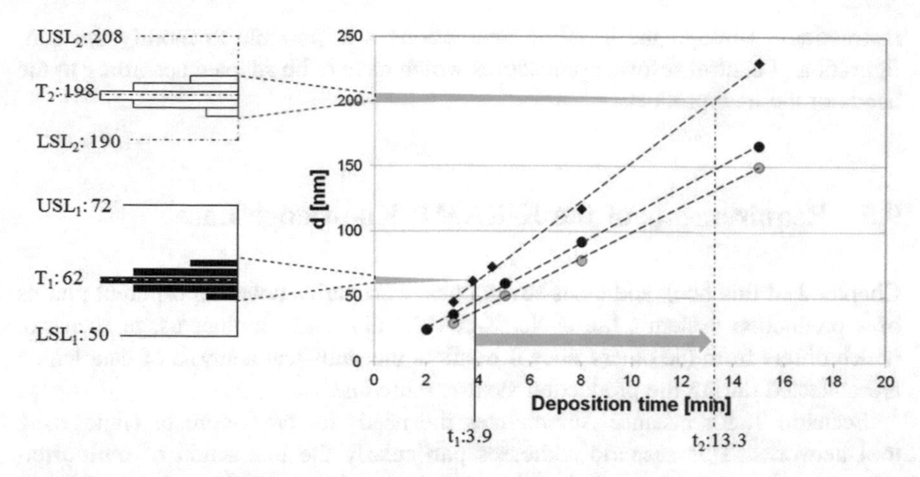

Fig. 9.8 Impact of shifting process segment settings on shifting of product design based on (Smietana et al. [2014])

product (specification range LSL_2 to USL_2) is determined. Knowing the specification limits of the new subproduct's layer thickness it is possible to calculate (1) the adjusted deposition time by applying the regression function on the new layer thickness and (2) the attempted FPY by shifting the distribution function in a way that the average is equal to the target T_2 of the new subproduct's layer thickness. It is important to mention that there might be more than one characteristic, like the layer thickness in our example, and the regression function could be multivariate accordingly.

By performing activity F of Fig. 9.6, it is therefore even possible to determine appropriate process segments and their handling instructions if there is no immediate coverage visible. If adjustment is no option because the resulting quality (determined FPY for new subproduct) is below a certain threshold then a gap was considered and the ramp-up team must be notified (activity I in Fig. 9.6).

Based on available information about existing subproducts and process segments, it is thus possible to perform matchmaking with the new subproduct. A similar process of decision-making needs to be performed by ramp-up teams during the ramp-up of every new product in every branch of the manufacturing industry. In addition, all new or adjusted handling instructions of process segments also require appropriately modified setup of aforementioned control software procedures. An increasing number of process segments and (sub-)products increase the efforts with respect to information which needs to be collected, and it increases the risk of failures which are caused by the lack of information and knowledge exchange. K-RAMP comprises the activities of ramp-up teams in a semiautomated process as it was described in the previous sections.

Keeping the recursive approach of K-RAMP in mind, the final situation is a structured sequence of device-specific process segments including handling instructions. Through the handling instructions it is possible to modify the configuration of control software procedures which have to be adapted according to the needs of the new product.

9.5 Requirements of the K-RAMP Knowledge Base

Chapter 2 of this book addresses several usage scenarios covering different phases of a production system's life cycle. K-RAMP illustrates another usage scenario, which differs from the others since it requires the statistical analysis of data which are collected during the production system's lifetime.

Scenario 1, for instance, summarizes the needs for cross-domain engineering tool networks. This scenario addresses particularly the interaction of multidisciplinary engineering teams during the planning and construction of a production system. Typically, this usage scenario applies once during the construction of a production system and later in case of modifications. As already mentioned, the ramp-up of a new product is also a challenge including the interaction of different engineering tools. As stated for usage scenario 1, the needs for semantic support are assumed to be the same as in the case of a product ramp-up project.

The usage scenario 2 of Chap. 2 is of particular interest from the perspective of a new product's ramp-up. Exemplarily for plant engineering, it discusses the needs of a BOM of a conveyer belt and the fact that single types of components (e.g., screws, bolts) can be reused independently from the individual size, width, length, or transport speed of a conveyor belt. This idea concerning reuse of subparts is one important aspect which is also considered as a first step during the ramp-up of a new product which shall be produced in a targeted production system. For a new product, it has to be determined which types of subparts are already used in forerunner products. Therefore, neither the particular type of subpart nor the product segment which produces these subparts need to be described specifically for the new product. The needs for semantic support are also valid for the product ramp-up use case. For mastering product ramp-up, appropriate needs and the common view on best practices of architectures for production system components have to be supported through adequate techniques of generalization or metamodeling.

The usage scenario 3 addresses a flexible organization of the production system. It must be ensured that the production system can produce a maximum variety of products. The authors explain that the production systems need to provide information about their capabilities, the access paths, and the control-related features. However, throughout the discussion of K-RAMP, some assumptions are made which are caused by achievements of the production infrastructure of IC factories. It is assumed that those achievements will be also the target of the development of classical production systems toward *Industrie 4.0*. Due to these achievements, the immediate knowledge about the production system's capabilities, its access paths or even the control-related features is not so relevant during the ramp-up of a new product.

Capabilities of components of a production system are a combination of the equipment hardware and well-formed instructions (equipment recipes, handling instructions). Equipment recipes, for instance, are loaded fully automated by the equipment for the purpose of configuration which is performed by function blocks of control components. Alternatively, by means of handling instructions human operators are instructed to turn knobs and adjust dials as provided by the equipment interface. Those actions again lead to the configuration of the control logic inside the equipment.

Through the changes in the instructions, the same equipment can take significantly different capabilities. For example, different valves can be opened or closed, and thereby the distribution of process fluids can be controlled. These process fluids may have significant impact on the equipment's capability and the product being produced. For instance, a production system might be generally capable to coat silicon wafers, but the specific coating material and the deposited thickness are controlled through an instruction. Even more general, a production system is capable of producing a three-dimensional titanium structure, however, the details of the structure are stored in a 3D file while the metal powder to be used is set in the handling instruction.

Therefore, in K-RAMP it is assumed that the production system has general abilities to create products. Together with the applied instructions, the specific capabilities of the production system are specified. K-RAMP does not look into the details of the production system, but only a subset of the possible settings of the applied instructions.

The (logical) access path to the equipment is considered on a very abstract level in K-RAMP. With respect to the control input, aforementioned instructions are applied. In terms of process results (process data, inline metrology data, quality assurance data) K-RAMP does not care about the mapping from the production system's sensors to process parameters but assumes that there is a rather product-independent (see previous section) mapping concept between the production system and the process segments already in place. Therefore, K-RAMP only focuses on parameters which are collected on the level of process segments and does not care about specific production system sensors or sensor busses.

As a conclusion of the two previous sections, the control-related features are out of scope of K-RAMP and are considered encapsulated by process segment parameters as output and instruction settings as input.

Although, the assumptions of K-RAMP deviate from the situation as it is described in usage scenario 3, the highlighted needs for semantic support are similar on a more abstract level. Product functions, product design, and process capabilities have to be represented (N1), heterogeneous views with respect to the product, the process and process control have to be represented (N2), or analytics of process data is essential in order to classify capabilities of process segments (N3).

Due to the similarities with the aforementioned usage scenarios, it can be shown that the same needs for semantic support are valid for the ramp-up of new products.

- There is the need for integration of a variety of information models (N1) considering the functional domain, the physical domain, and the process domain.
- The integration of those information models in a consistent way (N2) is the next challenge to be faced in the context of a K-RAMP.
- Querying from an integrated variety of information models during the ramp-up phase (N3) enables K-RAMP to recommend a comprehensive process plan which comprises aspects from the physical and the process domain.
- One part of the information which has to be considered during the ramp-up of a new product is semi-structured data (N4), like machine vendor specific formats for machine setup (equipment recipes). This information has to be combined with structured handling instructions in order to compile the comprehensive setup of process segments (see previous discussion of usage scenario 3).
- Constraints are applied in order to specify quality criteria of (sub-)products. Such constraints have to be met while (sub-)products are created in a production system. Automated matching of process segments with the constraints of new (sub-)products is an essential contribution of K-RAMP to the ramp-up phase. This demand is highlighted by N5, the need for flexible and intelligent engineering applications.
- The ramp-up phase is performed by a multidisciplinary team, which leads to the need N6, the support for multidisciplinary engineering process knowledge.
- Finally, the approach to be discussed shall recommend a strategy for manufacturing the product to be ramped up in the volume production system. Therefore, a consistent visualization of the status of the ramp-up of the new product has to be provided during runtime of a production system as stated by N7.

As a summary, K-RAMP can be seen as another usage scenario with a broad variety of needs for semantic support. All requirements (addressed in Chap. 2 and further extended in this chapter) are shown in Table 9.1.

Requirements R-1 and R-2 are satisfied by capability C2—formal and flexible semantic modeling—as it is discussed in Table 2.2 of Chap. 2 in this book. Requirement R-3 is covered by the Semantic Web through capability C3—linked data. Requirement R-4 is supported by the Semantic Web only to a certain extent. It is possible to import ontology submodels to a knowledge base. However, reasoning capabilities of the Semantic Web are handicapped due to the underlying open world assumption (OWA).

For this reason, it is not possible to deduce mappings between forerunner products and new functional requirements, as the case may be between existing process segments and new (sub-)products. For this reason, requirement R-5 was split from demand R-4 although both requirements are intended to be solved at once. In order to solve requirement R-5, a specific procedural ontology matchmaking algorithm was implemented outside the scope of the Semantic Web core of the K-RAMP knowledge base.

The following requirements are not covered by Semantic Web technologies at all. Requirements R-6 and R-7 need an association rule mining algorithm and

Table 9.1 List of K-RAMP knowledge base requirements

The K-RAMP knowledge base requirements	
R-1	A modular, multidisciplinary engineering model which covers the specification of functional requirements of products, characteristics of the product design, settings of the process setups and control software setups including their composition structures. The latter are reduced to the process setup for individual production machines and metrology devices. Moreover, integrated information models for control of externally implemented procedures (process control software) shall be provided and accessible
R-2	Ability to describe generalization structures (taxonomies) for functional requirements, (sub-)products and process segments
R-3	Means to extract and deliver partial models to other knowledge base instances
R-4	Means to add partial models of other knowledge base instances
R-5	Matching of partial models: Ability of matching functional requirements of two partial models. This requirement is derived from activity A of Fig. 9.6. Ability of matching product design of two partial models. This requirement is encapsulated in activity F of Fig. 9.6
R-6	Classification (assertion of new classes) of (sub-)products and process segments from production data of forerunner products. This requirement is needed because of activity E of Fig. 9.6
R-7	Coverage between design domains: Ability to determine degree of coverage between functional requirements and (sub-)products based on production data of forerunner products (needed because of activity C of Fig. 9.6). Ability to determine degree of coverage between (sub-)products and process segments based on setup data of forerunner production processes (needed because of activity encapsulated in F of Fig. 9.6)
R-8	Correlations between design domains: Ability to determine correlations between characteristics of functional requirements and characteristics of (sub-)products which are needed because of activity A of Fig. 9.6. Ability to determine correlations between (sub-)products and process segments (and parameters of handling instructions, respectively, which are needed because of activity F of Fig. 9.6)
R-9	Deduction of new design domain elements: Ability to deduce new (sub-)product elements due to R6. Ability to deduce new process segments due to R6. Both abilities are needed to establish parts of the knowledge base's content from production data. This happens before the K-RAMP process is performed
R-10	Ability to query a process plan of a product, which includes its sequence, the setup of each process segment for each capable production machine or metrology device and the setup of each control software component. This happens after the K-RAMP process is performed
R-11	Means to visualize the initial status of the ramp-up phase after the first completion of the K-RAMP process and ongoing update of the ramp-up phase status during production system's runtime

statistical functions, respectively. The execution of this algorithm is controlled by a partial ontology model which is supported through R-1. This way, the association rule mining algorithm receives the parameters which have to be used for mining information about a particular product or process segment from production data. Determining useful association rules causes the creation of new classes (restrictions) within the K-RAMP knowledge base. Requirement R-8 involves the

Table 9.2 Requirements and how they are implemented (++ full coverage, + partial coverage) by the Semantic Web or the procedural programming paradigm

The semantic web coverage of K-RAMP requirements			
Requirement	Semantic Web	Procedural	Reason
R-1	++		RDF, RDF(S), OWL2
R-2	++		
R-3	++		SPARQL
R-4	+	+	Import of ontology models, but weakness with respect to matching based on specification ranges of characteristics
R-5		++	Overcome lack of support for R-4
R-6		++	Procedural programming of association mining algorithm and statistical functions
R-7		++	
R-8		++	
R-9	+	++	SWRL, but weakness of metamodel based reasoning rules. Therefore, reasoning of new classes and class members based on procedural programming and SPARQL for reading and writing knowledge base
R-10	++		SPARQL
R-11	+		SPARQL

execution of mathematical and statistical functions. Similar to requirement R-6, there is a procedural library of essential functions available where the execution is controlled through a partial ontology model from requirement R-1.

Table 9.2 rates and explains which requirements can be fully met by Semantic Web, or may require additional, external algorithms.

Requirement R-9 sounds reasonable for Semantic Web technologies in accordance to capability C5—Quality assurance of knowledge with reasoning. However, it turned out that particularly SWRL is limited with respect to reasoning based on metamodels. SWRL does only provide a rule-based reasoning on concrete ontology models. However, this is not sufficient due to requirement R-6, which may cause the introduction of new classes (restrictions) and therefore the need for reasoning of new knowledge based on a metamodel.

Requirements R-10 and R-11 are widely covered by capability C4—Browsing and exploration of distributed data sets. The visualization of information is out of scope of K-RAMP and can be performed by any visualization framework which uses SPARQL, the Semantic Web query language.

In accordance to these requirements and the highlighted support of Semantic Web technologies, a software architecture was developed which implemented the K-RAMP knowledge base. This architecture is described in more detail in the next sections of this chapter.

9.6 Architecture and Ontology Models

Despite of all the advantages which have been enumerated above, Semantic Web technologies also have certain shortcomings as discussed in the previous sections. The partial ontology models and the complementary software procedures, which are needed to compensate these shortcomings are described in this subchapter.

The K-RAMP architecture utilizes Semantic Web technology particularly for knowledge modeling, generalization and specialization of concepts, querying and updating the knowledge base, as well as for reasoning from existing knowledge to generate new knowledge. The latter is used with some limitations as already highlighted. Other functions of K-RAMP were implemented in classic form following the procedural programming paradigm (Fig. 9.9).

The K-RAMP knowledge base leaves it up to the factory which software is used for the purpose of production control or process control. The single requirement is that all software procedures use SPARQL for interacting with the knowledge base.

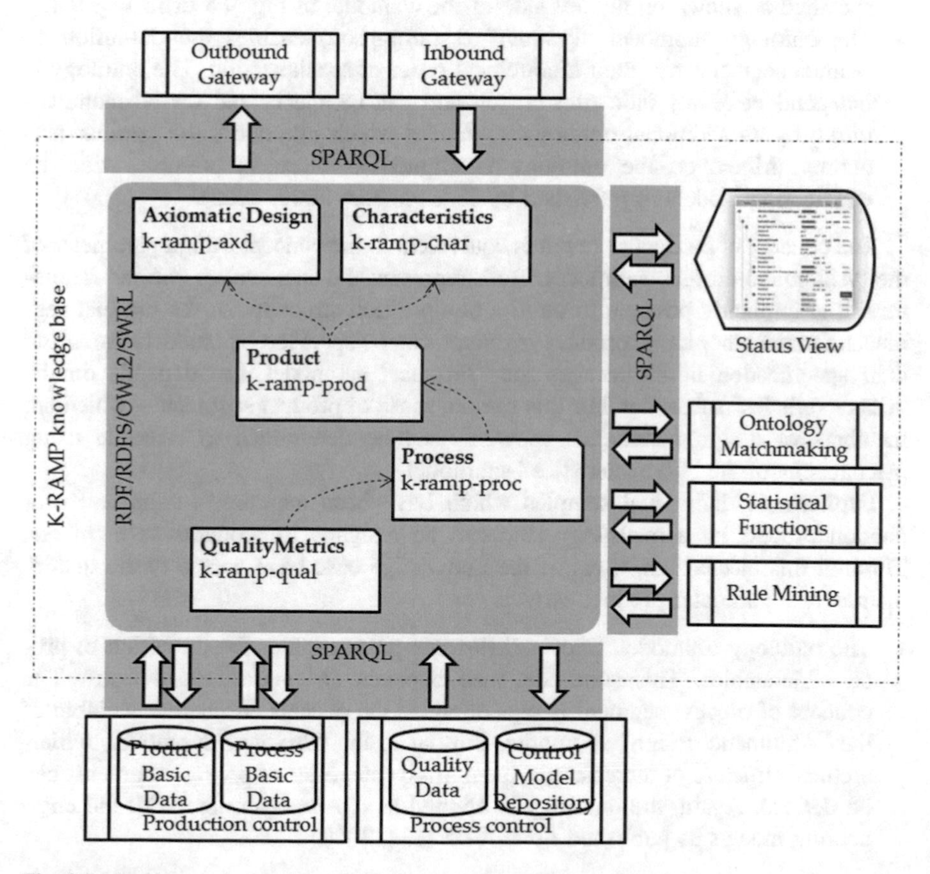

Fig. 9.9 Architectural overview of the K-RAMP knowledge base

The K-RAMP knowledge base stores information which is structured in accordance to specialized ontology submodels in order to fulfill the requirements R-1 to R-3.

- The ontology submodel “Characteristics” (k-ramp-char) supports the definition of specifications with upper and lower specification limits (tolerance ranges of functional requirement specifications or (sub-)product specifications) in general. This submodel makes no restrictions on concepts to be specified—such as, functional requirements, (sub-)products or process segments. It provides reasoning rules for comparing arbitrary concepts (classes) based on the specification limits of their characteristics.
- The ontology submodel “Axiomatic design” (k-ramp-axd) specifies the composition and dependencies of the functional, the physical, and the process domain on the most generic level. With the help of this ontology model, it is possible to specify composition hierarchies of functional requirements, design parameters or process variables, in accordance to the axiomatic design terminology. As a second aspect, mapping associations between all three domains are specified as shown on the left side of the V-model in Fig. 9.5 or in Fig. 9.4.
- The ontology submodel “Product” (k-ramp-prod) enables the definition of product segments and their hierarchical order of specialization. The ontology is independent of any industrial branch and can be specialized for IC manufacturing by an additional ontology submodel which introduces the familiar taxonomy. Moreover, the ontology is aligned as close as possible with the engineering models as published by ISA-95 (ISA 2000, 2005).

The concept of product segment is equivalent to the concept design parameter of the “Axiomatic design” submodel. Both represent the same set of entities. Therefore, it is implicitly possible to build a composition structure for the product segments. On the other hand, product segments can be specified by characteristics and their specification limits because the “Product” submodel also depends on the “Characteristics” submodel. For this reason, reuse of product segments—which are members of design parameters as well—can be determined as reusable using mechanisms of the “Characteristics” submodel.

Furthermore, individual samples which have been measured in the real production process by a metrology step can be assigned to products as members. Through this mechanism, users of the knowledge base have access to the quality metrics of a particular product as well.

- The ontology submodel “Process” (k-ramp-proc) enables the definition of process hierarchies. This submodel also depends on “Axiomatic design”. The concept of process segment is equivalent to the concept of process variable of the “Axiomatic design” submodel. Similar to the “Product” submodel, a hierarchical structure of increasingly specialized categories of process segments can be defined. Again, the ontology is aligned as close as possible with the engineering models as published by ISA-95 (ISA 2000, 2005).

By depending on the "Characteristics" submodel, the characterization of process segment categories and individual process segments can be carried out within the "Process" submodel as well.

The processing and measurement of individual samples in the context of a process segment is referred to as process job. For this reason, each process job, including all relevant production data, is considered as member of a process segment. Moreover, due to its connection to individual workpieces, but not only for this single reason, the "Process" submodel also depends on the "Product" submodel.

- By means of the ontology submodel "Quality Metrics" (k-ramp-qual), it is possible to define settings for the execution control of data mining and quality management functions which are executed outside of the ontology domain in the "Statistical Functions", the "Rule Mining", or the "Ontology Matchmaking" components.

These settings for execution control comprise of the characteristics which have to be used for the data selection of the rule mining or statistical functions execution. The information is immediately connected to specification limits which are needed for the calculation of process capability indexes (c_{pk}), a well-known quality metric for every production process. Thresholds for the c_{pk} and explicit knowledge about process results and process setup parameters are used either as input for performing rule mining or in conjunction with regressions. By mapping this information as ontology, it is possible to determine settings for execution control of data mining and statistical functions automatically by deduction.

In addition to the knowledge base, every part of the common K-RAMP ontology has some complementary software programs, also called components. These components overcome aforementioned shortcomings of Semantic Web technologies and provide interactive access of users to the contents of the particular part of the common K-RAMP knowledge base. These complementary components implement the requirements R-4 to R-9 and R-11.

- The "rule mining" component implements association rule mining as complementary function of the "Characteristics" submodel. Within the submodel, it is possible to hold the information about characteristics and specification ranges. It is also possible to determine overlapping of specification ranges by the use of reasoning rules.
  However, association rules—a special group of machine learning algorithms—are used in order to determine the membership of a single product in a new class of products, using existing production data. Based on the association rule mining results, new classes (restrictions) of (sub-)products are created. In addition, new products are attached to existing classes of (sub-)products.
- By means of the "statistical functions" component, missing features of Semantic Web technologies tailored to the efficient calculation of more complex mathematical functions are compensated. Statistical functions are used, for instance,

while investigating the existing production data. Such investigations are, for example, the determination of statistical distribution of measured values, the calculation of average and standard deviation, or the calculation of the c_{pk} and parts per million (ppm). Also, the results of such statistical functions are stored as new information inside the knowledge base.

- The "ontology matchmaking" component is always involved when fragments of information are gathered from another part of a company's knowledge base, as this is the case during a new product's ramp-up.
  The ontology matchmaking performs an algorithm following the K-RAMP process as described in Sect. 9.4. It uses SPARQL for querying and writing information as well as the previously introduced "rule mining" and "statistical functions" components.
  As a result of ontology matchmaking, the fragment of another knowledge base is stored and matched with content in the targeted production system's knowledge base. Conformities between the existing portion of information and the new portion are determined and linked accordingly.
  Finally, matchmaking allows querying of a recommended process plan which invokes specific instructions for process segments of the targeted production system and appropriate configuration settings of process control procedures.
- Last but not least, the "status view" component is used to visualize the content of the local part of the knowledge base. It provides an interactive insight to the queried process plan. This process plan may also contain gaps if no solution was found. Also assumptions are included if a new process segment was derived from an existing one by the use of regression functions.
  In the sequel of the ramp-up project, the status view may also serve as a tracking tool, as the specific process plan of the new product can be updated as soon as gaps are closed by the ramp-up team.

The ontology submodels and their complementary components together result in a deterministic process which determines the potential reuse of (sub-)products and process segments. The implicit procedural logic of this behavior is called the K-RAMP process.

9.7 Reuse of Process Control Settings

In the previous sections, the process perspective of K-RAMP and the architecture of the K-RAMP knowledge base have been discussed. The current subchapter describes how K-RAMP dissolves Question 2, which is focused on reasoning of settings of the control software components. After answering Question 1, this question can be solved solely by means of reasoning using OWL2 and SWRL.

Figure 9.10 shows an example which is based on a part of the "k-ramp-proc" ontology submodel. The example simplifies the Epitaxy process, which is the first

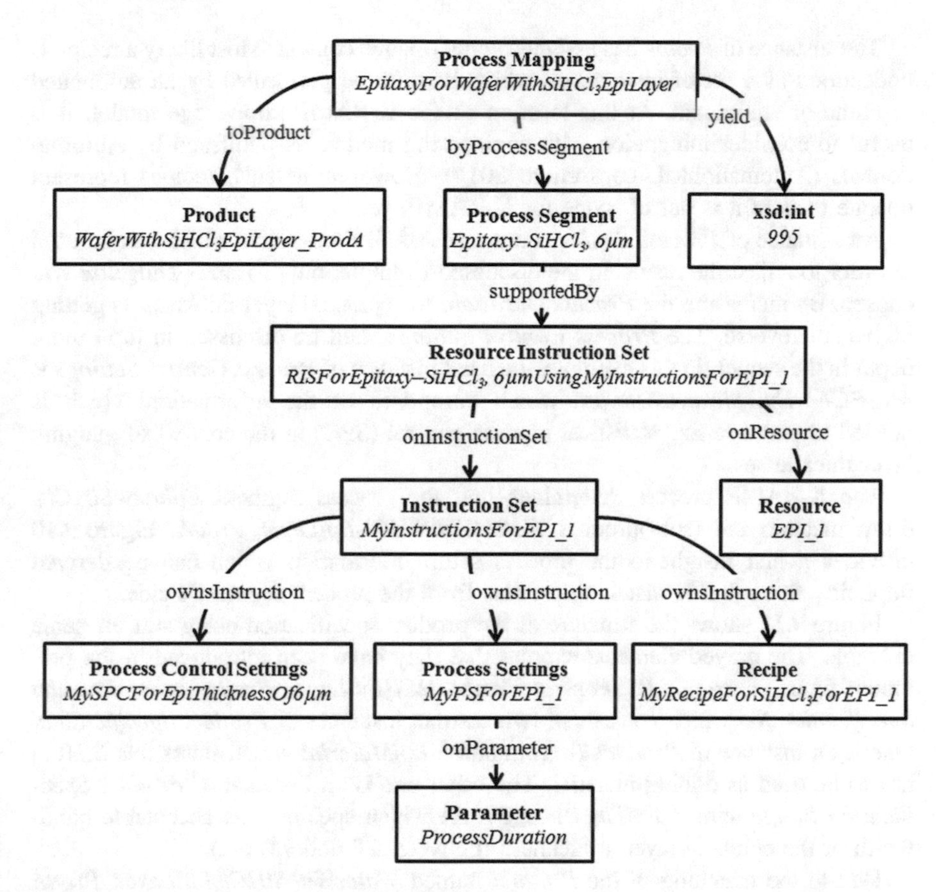

Fig. 9.10 Support of process segments by resources and their instructions

process segment of IC manufacturing (Xiao 2012, pp. 111–116). The *Process Mapping* instance, as shown in this figure, represents the mapping between a *Product* instance and a *Process Segment* instance and thus the mapping between the process domain and the physical domain in terms of the axiomatic design. The yield property says that the *Process Segment* named *Epitaxy–SiHCl₃, 6* µm of the viewed production system covers the quality needs of the *Product* named *WaferWithSiHCl₃EpiLayer_ProdA*—which shall be the result of the process segment—with a FPY of 99.5 %. The name *Epitaxy–SiHCl₃, 6* µm summarizes the essence of the capability of the process segment. This process segment has the target to grow a thin crystalline layer of 6 µm which is doped with $SiHCl_3$ (Trichlorosilane) on a thin silicon disk (which is commonly named “wafer”).

In order to achieve the target (epitaxial layer of 6 µm thickness), for each possible *Resource* (like the epitaxy equipment *EPI_1* in the example) at least one *Instruction Set* instance is needed, which comprises of the applied *Recipe*, some *Process Settings,* and the *Process Control Settings*.

The instance of *Recipe* has assigned some opaque content. Most likely a recipe is understood as a set of instructions which have to be performed by an automated machine or equipment. At this location of the K-RAMP knowledge model, it is useful to consider integration with engineering models as published by AutomationML (AutomationML consortium 2014). However, as such recipes represent opaque content it is out of scope for K-RAMP, yet.

An instance of *Process Settings* comprises of all *Parameters* which can be used in order to adjust the result. In the discussed example, only *Process Duration* was chosen. By increasing the *Process Duration,* the epitaxial layer thickness is getting higher, or reverse. The *Process Control Settings* shall be discussed in little more depth in the sequel. In this example, the used instance of *Process Control Settings* is *MySPCForEpiThicknessOf6* µm which comprises of the information which is needed to parameterize statistical process control (SPC) in the context of gauging layer thicknesses.

The K-RAMP process determines that the process segment *Epitaxy-*$SiHCl_3$*, 6* µm matches the (sub-)product *WaferWith*$SiHCl_3$*EpiLayer_ProdA*. Figure 9.10 provides a first insight to the process setup information which can be derived implicitly through this matching process from the process segment's side.

Figure 9.11 shows the structure of the product specification using still the same example. The grayed elements express that they have been introduced in the previous figure already. *PSSForWaferWith*$SiHCl_3$*EpiLayer_ProdA* is a *Product Specification Set* which consists of two distinct instances of *Product Specification.* One is an instance of *Product Target* named *EpiMaterial* which states that $SiHCl_3$ has to be used as doping material. The other one is an instance of *Product Specification Range* named *EpiThickness_ProdA* which specifies the acceptable bandwidth of the epitaxial layer's thickness (between 5.0 and 7.0 µm).

Due to the matching of the *Product* named *WaferWith*$SiHCl_3$*EpiLayer_ProdA* with the *Process Segment* named *Epitaxy–*$SiHCl_3$*, 6* µm a new instance of *Process Control Setting,* namely *MySPCForEpiThicknessOf6* µm *ForProdA*, was asserted as variant of the existing *MySPCForEpiThicknessOf6* µm. It is the case that K-RAMP matches *Epitaxy–*$SiHCl_3$*, 6* µm and *WaferWith*$SiHCl_3$*EpiLayer_ProdA* because the process target of 6.0 µm is within the specification range of 5.0 to 7.0 µm. However, thereby product-specific constraints may be identified which have to be considered during process control. For instance, in case of a statistical process control model, the set of applied decision rules can be different for a new product. Consequently, asserting a new *Process Control Settings* instance causes the assertion of new variants of the matching *Instruction Set* instances, *Resource Instruction Set* instances, and *Process Segment* instances as well.

In Fig. 9.12, it is shown how one SPC setting instance is represented as a variation of another SPC setting instance. To a certain extent, their object properties and data properties are referring to the same elements, however, there are variations with respect to the set of decision rules and with respect to the applied control limits. If SPC settings are reused in the context of a new *Process Segment* it is rather likely that the same *SPC Chart Type* and even the same *sampling rate* are applicable. With respect to *SPC Decision Rules,* it may be useful to apply a reduced

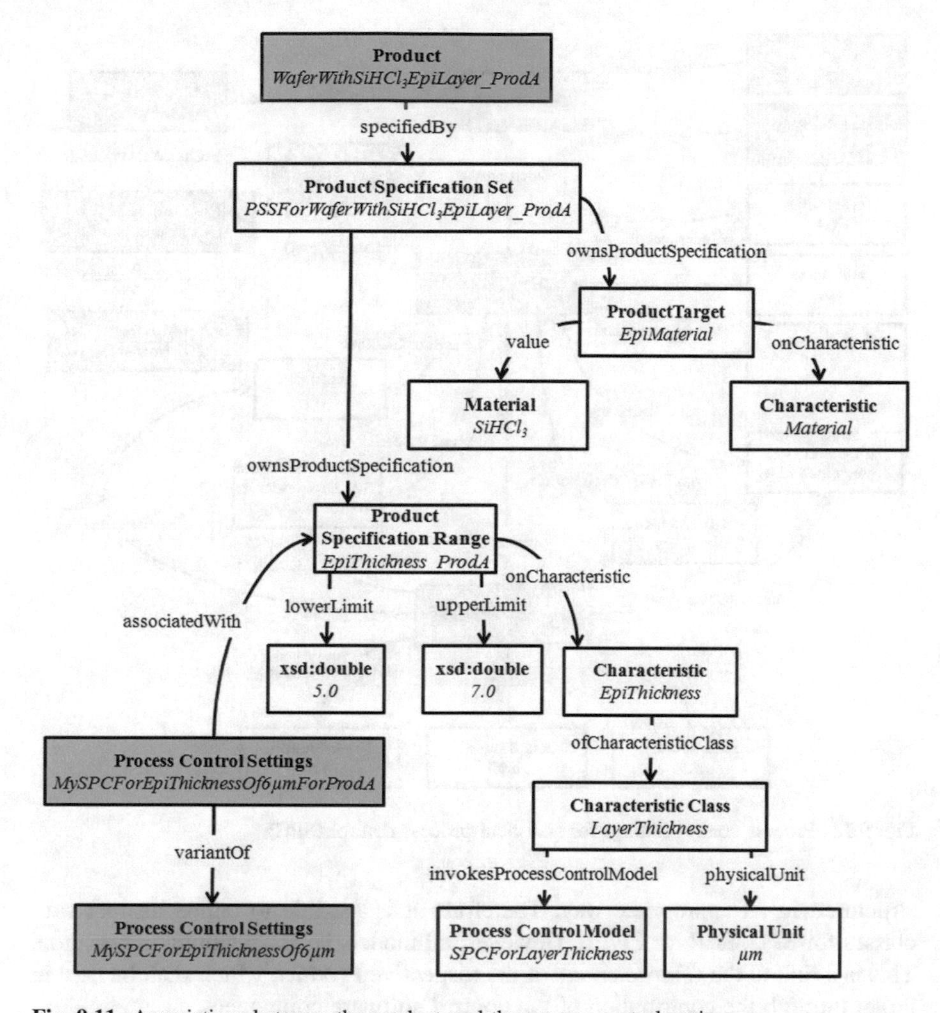

Fig. 9.11 Associations between the product and the process control settings

set during the ramp-up phase of the new product in order to avoid too many notifications. Concerning the *control limits* and *average,* it is either possible to take over the same values as the original—for instance, if a whole (sub-)product can be reused and therefore the associated process segments can be reused without changes as well—or it is necessary to recalculate *average* and *control limits* based on cluster members from activity D in Fig. 9.6.

The concept *Process Control Settings* represents the parameterization of *Process Control Model*. By separating it into two concepts the more stable knowledge is represented by *Process Control Model* instances while the product-specific parameterization is performed through *Process Control Settings* instances.

Depending on the purpose of the control model and the individual needs of the used control software component, occurrences of *Process Control Model* are

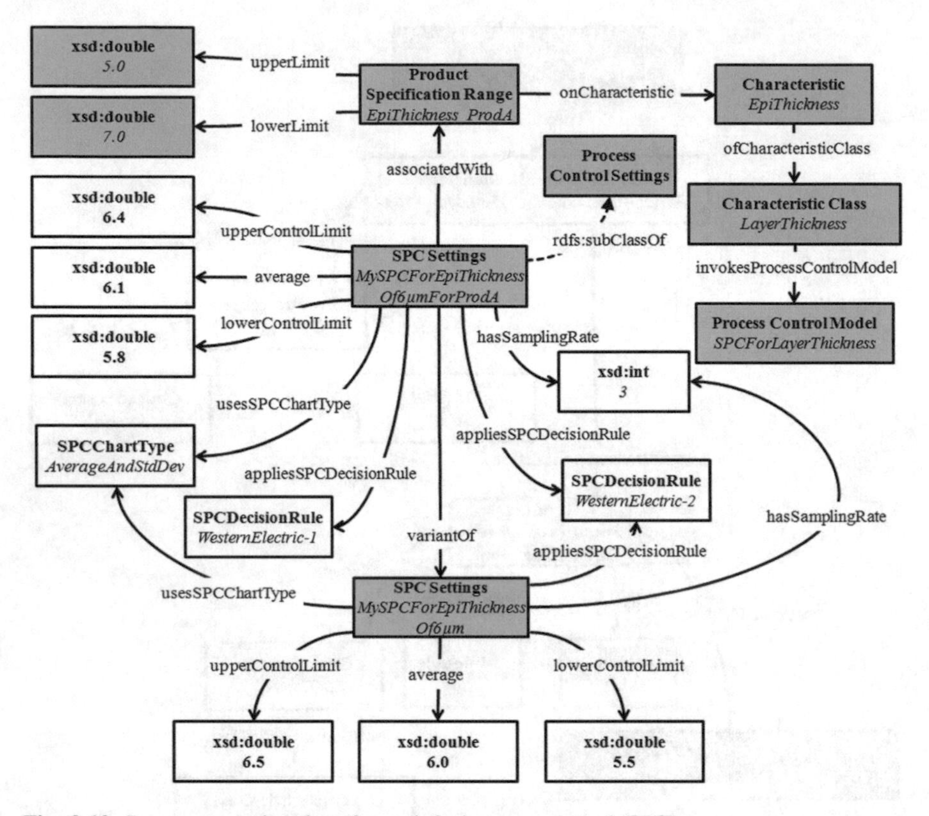

Fig. 9.12 Process control settings for statistical process control (SPC)

structured in an appropriate way. Therefore, it is feasible to define distinct subclasses for SPC, R2R, or FF/FB. However, all models have something in common. This is a link to the *Characteristic* of the respective *Product*, which shall be held in target through the contribution of the control software component.

9.8 Conclusions and Outlook

Particularly, but not exclusively targeted for the semiconductor industry, K-RAMP makes the ramp-up of new products more deterministic, by automated matchmaking between the ontology of the targeted production system and the ontology of a new product.

Using an approach like K-RAMP enables automated or at least semiautomated recommendation of device-specific process plans for the creation of new products in a targeted production system. Knowledge about existing capabilities of a targeted production system is used for generating this device-specific process plan.

The approach particularly determines reusable (sub-)products, as well as process segments. If exact matching of existing (sub-)products or process segments is not possible K-RAMP considers potential adjustment based on correlation functions which are part of the knowledge base. If there is no possibility for reuse of existing knowledge a gap appears in the device-specific process plan, which has to be closed by the product ramp-up. This is performed through development and qualification of a new process segment for the targeted production system. In the future, knowledge about such additional efforts shall be maintained in the K-RAMP knowledge base and will enable K-RAMP to predict efforts for activities which are needed to close such gaps.

Because many kinds of control software settings are linked with the (sub-) product structure, the approach is also able to recommend adapted settings for control software components.

For IT experts, the K-RAMP approach is an example for the integration of Semantic Web-based knowledge bases with existing production databases and production-IT (Willmann et al. 2014). Vulnerabilities and how they can be bridged by a hybrid knowledge base concept are presented in this chapter. Through this hybrid concept, data mining and statistical calculations, but also reasoning based on metamodels enhance the standardized or recommended features of Semantic Web technologies.

From a corporate management perspective, there are prerequisites which enforce the successful adoption of K-RAMP.

- A product technology platform approach for products like in the automotive industry or the concept of process technologies and silicon IP in the semiconductor industry introduce a modularization concept. Such a modularized concept for products and the production process enforces the reusability of manufacturing knowledge between product generations.
- Comprehensive and flexible collection of production data (process data, inline metrology data, quality assurance data) based on single workpieces (single part tracking) provides sufficient data quality in order to derive accurate correlations.

It is mandatory for manufacturers which face the challenges of *Industrie 4.0* to implement those prerequisites in their production systems. In conjunction with decreasing product life cycles and knowledge management for product ramp-up within the production systems and along their supply chains, K-RAMP is then the next consequent step in order to master the predictability of product ramp-up projects.

Acknowledgment This work was partly supported by the Christian Doppler Forschungsgesellschaft, the Federal Ministry of Economy, Family and Youth, and the National Foundation for Research, Technology and Development in Austria.

References

AutomationML consortium. <AutomationML/> The Glue of Seamless Automation Engineering —Whitepaper AutomationML—Part 5—Communication. www.automationml.org Accessed Sept 2014

Breyfogle III, F.W.: Implementing Six Sigma, 2nd edn. John Wiley and Sons Inc, Austin (2003). ISBN 0-471-26572-1

Dietrich, E., Schulze, A.: Statistische Verfahren zur Maschinen- und Prozessqualifikation (Statistical Methods for Machine and Process Qualification). (Hanser, Ed.) Munich, Vienna (2003). ISBN 3-446-22077-1

Fransoo, J.C., de Kok, A.G.: What determines product ramp-up performance? A review of characteristics based on a case study at nokia mobile phones. Beta, Research School for Operations Management and Logistics (2007). ISSN 1386-9213

Hüttenrauch, M., Baum, M.: Effiziente Vielfalt: Die dritte Revolution in der Automobilindustrie (Efficient Variety: The Third Revolution of the Automotive Industry). Springer, Berlin (2008). ISBN 978-3-540-72115-4

ISA: ANSI/ISA-95.00.01-2000. Enterprise Control System Integration—Part I: Models and Terminology. ISA–The Instrumentation, Systems, and Automation Society, Research Triangle Park, North Carolina, USA (2000). ISBN 1-55617-727-5

ISA: ANSI/ISA-95.00.03-2005. Enterprise Control System Integration—Part 3: Activity Models of Manufacturing Operations Management. ISA–The Instrumentation, Systems, and Automation Society, Research Triangle Park, North Carolina, USA (2005). ISBN 1-55617-955-3

Konrad, K., Hoffmeister, M., Zapp, M., Verl, A., Busse, J.: Enabling fast ramp-up of assembly lines through context-mapping of implicit operator knowledge and machine-derived data. Int. Fed. Inf. Process. 163–174 (2012)

Kurttila, P., Shaw, M., Helo, P.: Model factory concept—enabler for quick manufacturing capacity ramp-up. In: European Wind Energy Conference and Exhibition, EWEC, vol. 4 (2010)

Macleod, P.: A review of flexible circuit technology and its applications. In: PRIME Faraday Partnership (2002). ISBN 1-84402-023-1

Moser, T., Mordinyi, R., Winkler, D., Melik-Merkumians, M., Biffl, S.: Efficient automation systems engineering process support based on semantic integration of engineering knowledge. In: 16th IEEE International Conference on Emerging Technologies and Factory Automation (ETFA 2011), pp. 1–8. IEEE Computer Society (2011). doi:10.1109/ETFA.2011.6059098

Moser, T., Biffl, S., Sunindyo, W.D., Winkler, D.: Integrating production automation expert knowledge across engineering stakeholder domains. In: IEEE (ed.) IEEEXplore Digital Library, pp. 352–359, Krakow (2010)

Nenni, D., McLellan, P.: FABLESS: The transformation of the semiconductor industry. SemiWiki. com (2013). ISBN 978-1-4675-9307-6

Ruegsegger, S., Wagner, A., Freudenberg, J., Grimard, D.: Optimal feedforward recipe adjustment for CD control in semiconductor patterning. In: Characterization and Metrology for ULSI Technology: 1998 International Conference, pp. 753–577. The American Institute of Physics (1998). ISBN 1-56396-753-7

Said El-Haik, B.: Axiomatic Quality: Integrating Axiomatic Design with Six-sigma, Reliability, and Quality Engineering. John Wiley & Sons, Inc., New Jersey (2005). 0-471-68273-X

Slamanig, M., Winkler, H.: Management of product change projects: a supply chain perspective. Int. J. Serv. Oper. Manag. **11**(4) (2012)

Slamanig, M., Winkler, H.: Wissensmanagement bei Produktwechselprojekten (Knowledge management in product change projects). ZWF Wissensmanagement. Document number ZW110416 (10), 893–900 (2010). www.zwf-online.de

Smietana, M., Mroczyński, R., Kwietniewski, N.: Effect of sample elevation in radio frequency plasma enhanced chemical vapor deposition (RF PECVD) reactor on optical properties and deposition rate of silicon nitride thin films. Materials **7**(2), 1249–1260 (2014). doi:10.3390/ma7021249

Suh, N.P.: The Principles of Design. Oxford University Press, Oxford (1990). ISBN 0-19-504345-6

Suh, N.P.: Axiomatic Design—Advances and Applications. Oxford University Press, New York (2001). ISBN 978-0-19-513466-7

Terwiesch, C., Bohn, R.E., Chea, K.S.: International product transfer and production ramp-up: a case study from the data storage industry. In: R&D Management, vol. 31, no. 4, pp. 435–451. Blackwell Publishers Ltd, Oxford (2001).

Terwiesch, C., Xu, Y.: The copy exactly ramp-up strategy: trading-off learning with process-change (2003)

Wappis, J., Jung, B.: Taschenbuch - Null-Fehler-Management - Umsetzung von Six Sigma (*Zero-defects management - Implementation of six sigma*), 3rd edn. Carl Hanser Verlag, Munich (2010). ISBN 978-3-446-42262-9

Willmann, R., Biffl, S., Serral Asensio, E.: Determining qualified production processes for new product ramp-up using semantic web. In: Proceedings of the 14th International Conference on Knowledge Technologies and Data-Driven Business (I-KNOW). ACM, Graz (2014). ISBN 978-1-4503-2769-5

Winkler, D., Biffl, S.: Process Automation and Quality Management in Multi-Disciplinary Engineering Environments. Vienna University of Technology (2010). http://cdl.ifs.tuwien.ac.at/files/CDL_M1_03%20Process_Automation_Quality_Management%20Research%20100321.pdf. Accessed 03 Oct 2013

Xiao, H.: Introduction to Semiconductor Manufacturing Technology, 2nd edn. SPIE, Bellingham (2012). ISBN 978-0-8194-9092-6

Chapter 10
Ontology-Based Simulation Design and Integration

Radek Šindelář and Petr Novák

Abstract Strict requirements on the quality of industrial plant operation together with environmental limits and the pursuit of decreasing energy consumption bring more complexity in automation systems. Simulations and models of industrial processes can be utilized in all the phases of an automation system's life cycle and they can be used for process design as well as for optimal plant operation. Present methods of design and integration of simulations tasks are inefficient and error-prone because almost all pieces of information and knowledge are handled manually. In this chapter, we describe a simulation framework where all configurations, simulation tasks, and scenarios are obtained from a common knowledge base. The knowledge base is implemented utilizing an ontology for defining a data model to represent real-world concepts, different engineering knowledge as well as descriptions and relations to other domains. Ontologies allow the capturing of structural changes in simulations and evolving simulation scenarios more easily than using standard relational databases. Natively ontologies are used to represent the knowledge shared between different projects and systems. The simulation framework provides tools for efficient integration of data and simulations by exploiting the advantages of formalized knowledge. Two processes utilizing Semantic Web technologies within the simulation framework are presented at the end of this chapter.

Keywords Simulation design · SCADA · Model-driven design · Complex simulations · Knowledge integration

R. Šindelář (✉)
Institute of Software Technology and Interactive Systems, CDL-Flex,
Vienna University of Technology,
Vienna, Austria
e-mail: radek.sindelar@tuwien.ac.at

P. Novák
Czech Institute of Informatics, Robotics, and Cybernetics, Czech Technical
University in Prague, Prague, Czech Republic
e-mail: novakp46@fel.cvut.cz

S. Biffl and M. Sabou (eds.), *Semantic Web Technologies for Intelligent Engineering Applications*, DOI 10.1007/978-3-319-41490-4_10

10.1 Motivation

Present industrial systems and accompanying automation systems are very complex. In order to develop and operate them, engineers of various disciplines have to cooperate and share their knowledge. Simulations can be very helpful in any phase of the automation system's life cycle and they can be used in a wide range of applications as follows:

- System design
  - operation analysis, testing and verification
  - production planning
  - virtual commissioning
- System operation (run-time phase)
  - operation analysis, planning and optimization
  - decision-making support
  - fault detection
  - advanced process control

To utilize simulations in practice, they have to be integrated into the native working environment of users. *Simulation integration* means finding an architecture and a general methodology for connecting the heterogeneous parts of industrial systems which take part in the simulation life cycle (Šindelář and Novák 2012). Simulation integration can be regarded as the problem of finding a suitable integration platform as well as steps like a definition of data and object models for storing and handling parameters, simulation data, message transformation, etc. The goal of the simulation integration should not be to incorporate only one particular simulation environment or only one group of simulators with a similar object model and purpose but to prepare the simulation environment to be able to deal with event-based, dynamic, or agent-based simulators. Some of the aforementioned tasks will only work successfully when tightly connected to SCADA system (abbrev. for **S**upervisory **C**ontrol **A**nd **D**ata **A**cquisition). In the architecture of industrial control systems, the SCADA system frequently represents a subsystem responsible for system monitoring, data collection, and process control.

In the present approaches, design and run-time phases of a simulation life cycle suffer from several problems. For example, all tools must be configured manually and even small changes affect many files, parameters, and interfaces. Thus, it is very difficult to reconfigure and update the system when needed. Knowledge is available at the very low level of formalization; thus it is very hard to reuse it for different purposes. Also there are no managements providing services related to data processing, user access, or simulation executions.

The basic goal is to provide an environment for the integration of simulations and advanced process control tools. The technical and semantic integration is based on

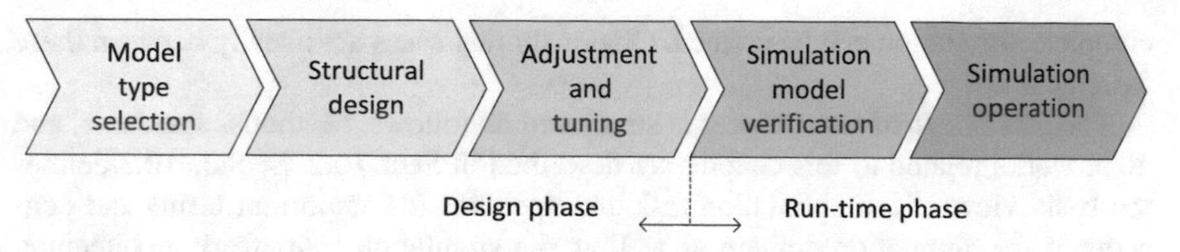

Fig. 10.1 Simulation life cycle

the Open Engineering Service Bus (OpenEngSB)[1] which is an open source, extensible tool integration platform (Biffl et al. 2009). It provides foundations for a flexible technical integration of heterogeneous engineering environments for strengthening engineering process support, improving collaboration of involved stakeholders, and providing a mechanism for capturing process data based on events for project monitoring and quality assurance activities. In this chapter, the environment for simulation integration is called the simulation framework.

The entire engineering process is very complex and cannot be handled at once. Two subtasks have been chosen as the most important ones where Semantic Web technologies will be used:

Model-driven configuration It is a very important part of an integration process ensuring a smooth transition between the design and run-time phase (see Fig. 10.1). It is obvious that the configuration of a complex integration environment, i.e. defining the run-time behavior of all interconnected tools, is a very complex task and cannot be effectively solved by manually written configurations. Thus the goal is to find the methods that support such configuration tasks by utilizing information and knowledge stored in the machine-readable form from different domains. Such an approach is called a model-driven configuration (Dreiling et al. 2005). Its crucial element is a knowledge base where all required information about the different domains, executed simulation tasks, and their results is stored.

Model-based simulation design Large and complex simulation tasks (e.g., virtual commissioning) require cooperation of several simulation tools (e.g., simulation of a mechanical system with simulation of an industrial computer). It is very difficult to maintain consistency between source codes of different simulation tools in the design phase. The knowledge base can be also used to support a simulation design. There are several sorts of cases that can be considered as a model-driven simulation design. Basically two main groups can be distinguished. The first one covers simulation tasks where a simulation library with simulation blocks for a given simulation tool is available. Under the assumption that the simulation design is driven by mappings between devices in a real plant and simulation blocks provided by a library. The second group is comprised from methods where various parts of a simulation structure are designed based on information from the knowledge base. On the one hand only interfaces, input and output points can be designed, on the other hand a

[1] Available online: http://openengsb.org/.

complete simulation can be created. Obviously real cases are usually between these extreme ones.

The remainder of this chapter is structured as follows. Methods, standards, and other works related to this chapter are described in Sect. 10.2. Section 10.3 defines the basic view of the simulation task and specifies the important terms and concepts of the simulation domain as well as the simulation framework architecture. Sect. 10.4 presents the architecture of the simulation domain, which is the one of the most important parts of the knowledge base (KB). The structure of the knowledge base is then described in Sect. 10.5. Section 10.6 shows the usage of the knowledge base for the configuration of the simulation framework, Sect. 10.7 presents an example utilizing our approach for simulation model design. Conclusions in Sect. 10.8 summarize the experience with the prototypical implementation.

10.2 Related Work

For more than 20 years, researchers have worked on managing knowledge for supporting simulation engineering and integration. The extraction and the reuse of knowledge for supporting the engineering of plants as well as their automation and control systems are crucial tasks in the emerging area of knowledge-based engineering (KBE) (Rocca 2012; Verhagen et al. 2012). In this chapter, we apply the KBE in the area of process simulations.

10.2.1 Simulation Model Design

Simulation and mathematical models for industrial processes can be represented in many forms (Franklin et al. 2009). From the modeling point of view, simulations can be distinguished according to modeling approaches—signal-based methods and power-based modeling methods (Karnopp et al. 2006). In this work we focus on systematical creating of a mathematical model of a physical system with the bond-graph theory (Gawthrop and Bevan 2007). This theory deals with creating mathematical models for complex mechatronic systems. The term "bond" is related to device interfaces as the interaction of physical subsystems is influenced by balancing energy leading to the maximal entropy of the energy distribution within the system. The bond graph theory utilizes generalized terms flow and effort, whose product is energy transferred via specific bonds. The main limitation of bond graphs is that they are primarily intended for manual use. Although various software implementations of bond graphs have been proposed and implemented, their typical usage does not satisfy the current requirements of module-based simulation model composition and scalability of models. The resulting set of equations coming from a bond graph modeling method is called the simulation model, which can be simulated by standard numerical methods in various simulation tools.

Since the simulation framework with the KB is oriented on large-scale industrial problems, the engineering and behavior investigation should be solved at an appropriate level of granularity. This assumption leads very often to the block-oriented methods. One of the block-oriented approaches is presented in (Durak et al. 2007). The paper illustrates an approach for generating MATLAB-Simulink blocks and defining them via DAVE-ML[2] according to the Trajectory Simulation Ontology. Connections of these blocks are consequently done manually.

The approach described in (2007) is based on two ontologies; the first one is called domain ontology and it categorizes knowledge including a problem vocabulary in the domain scope. The second one is called modeling ontology and it is used for the simulation model description. Such an approach guarantees a high degree of reconfigurability of the solution and a separation of the whole problem knowledge into appertaining engineering scopes. The discrete event model represented by the DeMO (e.g., Discrete Event Modeling Ontology, see Silver et al. 2009). The presented approach is based on mapping concepts from domain ontologies to a modeling ontology, translating ontology instances to an intermediate XML markup language, and generating an executable simulation. The approach presented in this chapter is based on a similar idea, but it addresses other engineering tools and reflects on the features of large-scale industrial systems.

10.2.2 Simulation Model Integration

Data and tool integration is a standard task in current business IT systems. However, integration of simulation models within the rest of automation systems is not common (Novák and Šindelář 2012; Novák et al. 2014). Therefore, benefits of simulation models are not achieved decently. Existing integration approaches in the simulation domain are focused on reusing the simulation code. In order to provide the results of simulations to all groups of users, a complex generic integration environment is required, but does not exist yet.

The CO-LaN (the CAPE-OPEN Laboratories Network), formerly CAPE-open, is *a neutral industry and academic association promoting open interface standards in process simulation software*[3] This very detailed standard defines single parts of a process simulator. The concept of CAPE-open respects the structure of industrial processes, such that the representation of material and energy flows is possible. The main goal of these standards is to ensure simulation module reuse, their cooperation, and data exchange, but it does not involve the integration with other tools such as process data sources, knowledge bases, etc.

---

[2] Available online: http://daveml.org/.

[3] http://www.colan.org/.

The process of integration in the industrial automation area is standardized by ISA-95. ISA-95 defines vertical (also called hierarchical) integration of automation tools and it systemizes this type of integration (e.g., ERP systems). The data models defined by this standard and the applications satisfying it are described in (Unver 2012).

ISA-88[4] defines standards for design and specifications of batch process control systems. The approach presented in this article adopts basic principles of ISA-88 and one of the basic scenarios used for testing of the proposed infrastructure is the simulation of batch processes in hydraulics.

The standard ISO 15926 (Ind 2009) is at the borderline between technical and semantic integration. Although it has been originally intended to oil industry, its ideas are general and usable in other domains as well. In the original version of the standard, the description of data models and reference data was based on EXPRESS language (International Organization for Standardization 2004). It was difficult to use, hence the standard was enhanced with supporting OWL ontologies (Kim et al. 2011).

Among existing ontologies, the most relevant is OntoCAPE (Morbach et al. 2009), which is a set of (possibly intermapped) ontologies covering a description of large-scale physical systems, signals, etc. The OntoCAPE has a layered modular structure and it is implemented in OWL-DL. In a broad research context, the Onto-CAPE is described in (Morbach 2009) that also summarizes other existing ontologies in related domains.

Many UML-based and XML-based representations of plant topologies can be found in literature. The Systems Modeling Language (SysML) is a general-purpose modeling language for model-based systems engineering (OMG 2012), which is built on top of UML. This language can be used not only for the description of systems, but also to support the design of simulation models typically in Modelica language (Kapos et al. 2014). Examples of transformers from the SysML to simulation models are SysML2Modelica or SysML4Modelica.

A widely cited XML-based language is AutomationML,[5] which is intended for the description of industrial plants and software tools from the automation, instrumentation, and control perspectives. AutomationML is based on four hierarchies that can be intermapped. They describe system components including their parameters, system units, roles, and interfaces. The common weak point of such XML formats is the absence of semantics as well as XML files are difficult to query and perform knowledge reasoning, hence their application for the configuration of the integration platform is limited nowadays.

Industrial data integration and acquisition is very often based on OPC (OLE for Process Control). OPC Unified Architecture (OPC UA) is an industrial specification developed on the basis of the classic OPC specification and it combines all of the following classic OPC standards: OPC Data Access, OPC Historical Data Access, and OPC Alarms and Events into one unifying specification (Lange et al. 2010).

[4]https://www.isa.org/isa88.

[5]https://www.automationml.org.

10.3 Simulation Process

The goals mentioned in Sect. 10.1 require a general definition of simulation tasks applicable in a wide variety of usage. In this section, basic concepts and terms describing a simulation process and building the knowledge base are defined. At the end of this section a basic description of the simulation framework is given.

A simulation task or *simulation scenario* is a set of calculations. From the user's perspective it represents a final task like "operator training", "job planning", or "production line testing". A simulation task consists of *simulation modules*. *A simulation module* is one of the basic objects in the simulation framework. It represents any calculation or simulation executable in integrated tools. The module concept (abstraction) is defined by the logical name and lists of input and output variables, and it is independent from its implementation. From the mathematical point of view, it can be seen as a general function $(O) = f(I)$ where I stands for all input data and O for all output data. This abstraction is independent from the implementation and one module concept can be implemented in more tools. The general view of the simulation module and its relation to other objects in the simulation framework is depicted in Fig. 10.2. There are input data (input variables), output data (output variables) and several sets of parameters. *A plant or process model* is a mathematical description of the system/plant's behavior. It can be implemented by a single module as well as by by many simulation modules in different simulation tools.

A simulation task is described by a *simulation workflow*. This workflow consists of simulation modules, signal connections, and simulation task inputs and outputs. Simulation workflows can be seen as a graph where nodes represent modules and edges represent data flows between modules and simulation task inputs and outputs. A simulation workflow contains all pieces of information required to obtain data

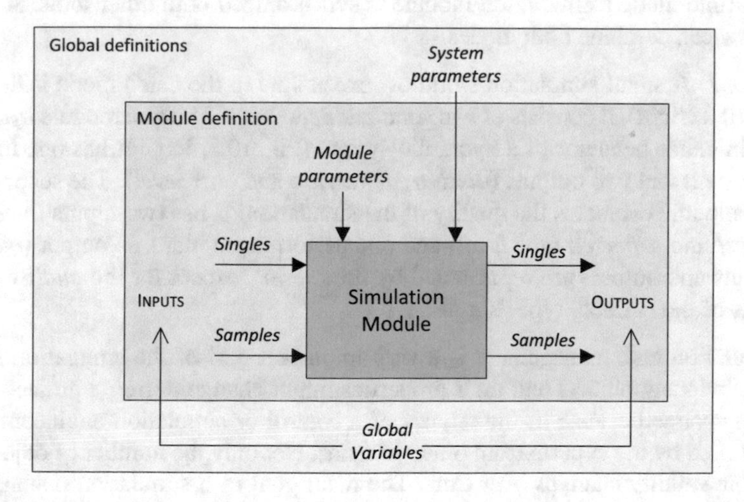

Fig. 10.2 Description of a simulation module

from external data sources and executes all simulation modules. *A simulation run* is a single execution of a simulation task described by a simulation workflow.

In order to execute a simulation task, data from various data sources is required. Basically two kinds of data are distinguished in simulation tasks – parameters and variables.

Parameters represent the values used for parametrization of simulation modules or the whole modeled plant/system. Therefore, two groups of parameters are implemented in the simulation framework—model and system parameters. From the workflow point of view both the groups of parameters are constant and cannot be calculated or changed by any module in the simulation workflow. *Model parameters* are variables allowing a parametrization of simulation source codes. Time constants, filter parameters, or switch settings are defined within the model parameters. It is obvious that model parameters must be defined for the particular implementation of the simulation module. They can be used to control or define the module operation and behavior. *System parameters* define parameters of a real system and they are defined globally—pipe diameters, tank overflow and bottom altitudes, pump power, etc. Parameters remain constant during the simulation execution.

Variables are the data entering simulation modules or produced by simulation modules. Two basic types of variables can be distinguished. A *sample* is a variable whose value is changing with the simulation time. A sample is a triple of the variable name, variable value, and time stamp. A collection of samples is called a time series. "Tank level" or "Valve pressure" are the examples of variables in the hydraulic system. On the contrary, a *single* is a variable, whose value is constant with the simulation time. The single is a couple of the variable name and value. Statistics of input time series—means or variations—are examples of single values.

From the execution point of view, two basic modes can be used. While in the batch mode a simulation module runs without any interaction with other tools, in the run-time mode a simulation module is synchronized with other tools. A single workflow can combine both modes.

Example 1 A small simulation workflow executable in the batch mode is depicted in Fig. 10.3 (right). It consists of two simulation modules. The first module *hydraulic plant* simulates behavior of a hydraulic system (Fig. 10.3, left). It has one input (*a pump speed*) and two outputs (*discharge pressure* and *tank level*). The second simulation module evaluates the quality of the simulation. It has two inputs (*measured tank level* and *expected tank level*) and one performance index as output (*quality*). All inputs and outputs are represented by time series, except for the *quality* output which is of the variable type "single".

Simulation task management is a very important part of the simulation framework. Obviously objects and their properties might change during a project. Some of them change because of the request of a control or simulation engineer, others are modified by the experts from other domains. Not only the number of objects but also their validity changes over time. The main goal of a simulation management is to maintain mappings between objects in different domains. It must also support assembling simulation workflows and ensuring a validity and feasibility.

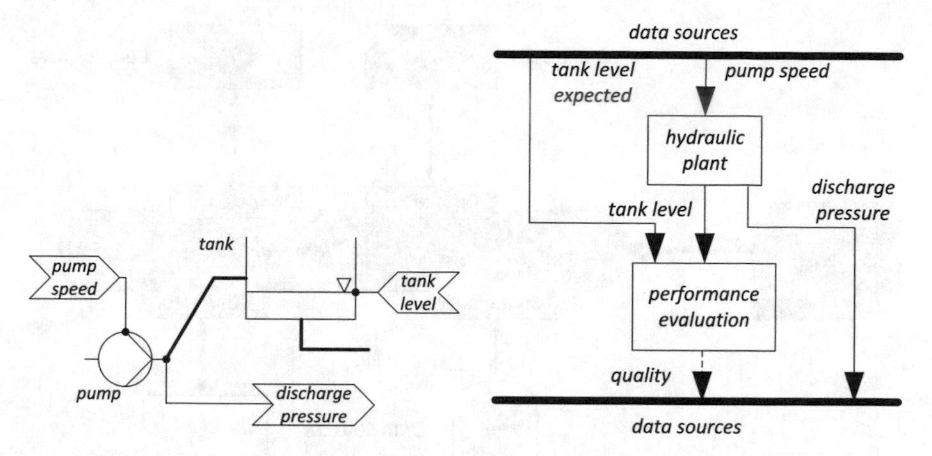

Fig. 10.3 A small hydraulic system (*left*) and a workflow evaluating the simulation quality (*right*)

10.4 Simulation Domain Architecture

In this section, the concept of the simulation framework and simulation domains is presented. The goal is to define concepts and their relations based on definitions from the previous section. The object model and architecture is defined independently of its implementation because ontologies described in Sect. 10.5 represent only one way of machine-readable form.

10.4.1 Simulation Framework

Present solutions of simulation integration are usually based on ad-hoc approaches utilizing direct interconnections using available interfaces and manual configurations. The main goal of the simulation framework is to provide a configurable environment for more flexible integration of simulators, data sources, and other tools used in applications utilizing simulations. The basic architecture of the simulation framework is depicted in Fig. 10.4.

The simulation framework is formed by the hierarchical structure of tool domains. *Tool domain* is an abstract description of interfaces of a group of tools. A domain defines the common methods for this set of tools. *Tool connectors* are specific implementations exposing services for a particular tool. The examples of tools connectors in the data source domain are a JDBC connector for databases or a connector for csv files. Domains are defined hierarchically to provide concepts for various groups of users, such as control engineers, operation managers, and others. In the next sections two basic domains are described—simulation domain and data source domain. The more detailed description of the simulation framework can be found in (Šindelář and Novák 2011) or on (Novák et al. 2014).

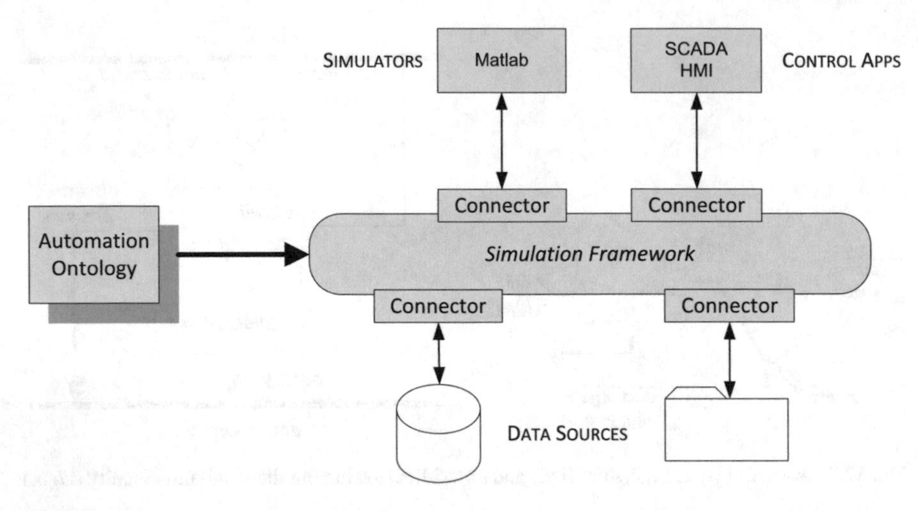

Fig. 10.4 Simulation framework overview

10.4.2 Data Sources and Data

In contrast to automation systems in industrial environments, the simulation framework has a much more complex view of data. A simple occurrence of each variable is expected in industrial automation systems, whereas multiple occurrences are required in the simulation framework. Typical industrial automation systems provide an actual value as well as historical records for each variable. Other occurrences of same variables may be present in the simulation framework. Another occurrence may originate in a simulation as a result or can be prepared by a control engineer manually to describe and represent expected behavior or status.

Regardless of the origin of the values, all of them can be inputs for simulations. For that reason, input and output of simulation modules cannot be directly related to one of these occurrences but can be related to a more general object. For that reason, in the simulation domain a special layer is created providing an abstract interface to other domains (like data domain or real plant domain). The main goal of this layer is to provide data for all kinds of users or tools in the form they are used to work with. It must also allow to define data in an abstract name workspace which is independent from physical data sources.

The most abstract object in this view is a *variable*. It is related to a particular place in the system or plant and describes the general features of data connected to this point. The variable object has a name which is unique in the project workspace. It is the basic object representing data in the simulation framework (Sect. 10.4.2). Any variable can be an input or output of a simulation module, particular occurrence of this variable will be used based on the context of simulation tasks. There are more layers between the general variable and the physical values stored at the data server. An example of this structure is a flowmeter in the plant, which produces the variable *input flow* depicted in Fig. 10.5. There are two different time series stored in different data sources (i.e., in a database and in a file storage).

Listing 1: The variable discharge pressure from Example 1 could be defined as:

Listing 10.1 XML variable definition

```
<tns:variable name=''discharge pressure''
              variableType=''pressure''
              dataType=''double''
              unit=''kPa''>
  <tns:lowLimit>0</tns:lowLimit>
  <tns:highLimit>21</tns:highLimit>
  <tns:userAttributes>
    <tns:attribute name=''facility location''
                   type=''string''>zoneA3652</tns:attribute>
  </tns:userAttributes>
</tns:variable>
```

Variables can be grouped into a *dataset*. The goal of the dataset is to group data related to one simulation task and to simplify the handling of values. It is also very important to make data accessible and usable for nonexperts. For each dataset more profiles can be assigned. Each profile represents a particular combination of physical data sources but it must contain exactly the same group of variables as defined in the related dataset. Datasets and profiles create an abstract layer which is built on top of a physical layer. The overview of the data architecture is illustrated in Fig. 10.5.

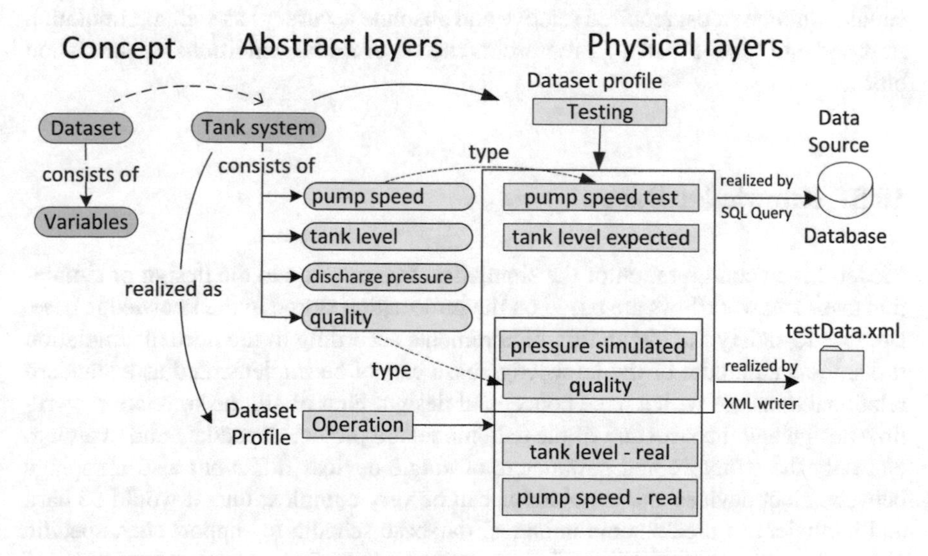

Fig. 10.5 Data architecture (not all relations between objects are depicted)

10.4.3 *Simulation Modules*

The architecture of simulations is very similar to the data architecture described in the previous section. As defined in Sect. 10.3, a simulation module can be seen as a black box with an interface. It is an abstract layer in the architecture. The physical layer contains information about implementations of simulation tools interfaces and various configurations.

In order to define a complex simulation tasks, the abstract layer of the simulation module must be defined in a very general form. Because this work is focused on the simulations of processes or mechatronic systems, a description of continuous-time finite dimensional dynamical systems (Antsaklis and Michel 2006) has been chosen as a basis of simulation module interface. It can be simplified and used also for a description of more simple systems. All datasets related to a simulation module can be considered as 6-tuple:

$$M = (I_i; I_o; P_s; P_m; P_r; P_c)$$

where I_i and I_o represent inputs and outputs of a module, P_s denotes system parameters. It is a set of parameters related to the real system, for example lengths and diameters of pipes, volumes of tanks, etc. P_m are parameters for simulation purposes related to the particular model representation like time constants of first-order models, steady-state gains, or transfer functions. P_r are parameters of single simulation runs. This set of parameters involves solver settings (e.g., minimum and maximum simulation time steps, required relative and absolute accuracy) as well as simulation start and end time. Finally, P_c represents the set of initial conditions of simulation blocks.

10.5 Knowledge Base

Model-driven configuration of the simulation framework and the design of simulation tasks and workflows are based on the information stored in the knowledge base. Because to satisfy the integration requirements according to the needed simulation tasks, the data model of the knowledge base cannot be implemented as a standard relational database with a fixed conceptual design. First of all, the simulation workflow design and the structure of the data model are project-dependent and evolving. Not only the structure and parameters of single devices differ but also mappings between plant devices and simulations can be very complex; thus it would be hard and inefficient to create some universal database schema to support such specific tasks. The primary goal of simulation engineers is to find an optimal structure of simulation modules. The internal structure of simulation models is thus dynamically evolving during a simulation engineering project.

An effective and efficient solution to this problem is the usage of ontologies. An ontology allows to specify concepts and relationships that collectively characterize a domain in a machine-understandable form. Ontologies are widely utilized in modern artificial intelligence approaches for representing knowledge. The utilization of ontologies in a role of a knowledge base for simulation model design and integration is beneficial as ontologies offer a high level of flexibility, reuse, and inference of new pieces of knowledge that are not explicitly specified in source data. However, the shortcoming of ontologies is the need of more time for querying compared to indexed relational databases. But as ontologies are used for design and configuration phases of the simulation model life cycle in the presented case, this feature does not pose a crucial restriction.

The data model of the knowledge base is implemented by the classes and properties of the automation ontology, which was designed by the authors of this chapter (Novák et al. 2015). It consists of several domain subontologies and their mappings. The most important domains represented in the automation ontology are the real plant domain, simulation domain, variable domain, and parameter domain. A crucial part of the ontology are the mappings, which are used to interrelate corresponding entities. The most important concepts of the automation ontology are depicted in Fig. 10.6 and they are addressed in the following description in detail.

The foundation for the structure of a simulation model is the topology of the real industrial plant. It is assumed that each industrial plant consists of devices. The real plant domain of the automation ontology represents the structure of the real system, i.e., it includes physical devices and their connections. The topology of the system means how the devices are interconnected, which is represented in the ontology with

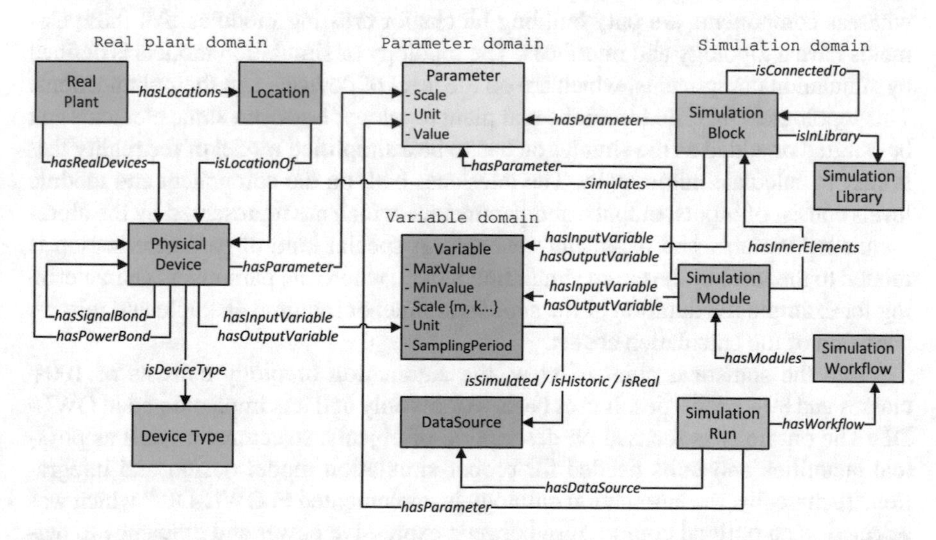

Fig. 10.6 Engineering KB—overview of the automation ontology (not all details are depicted). Adapted from (Novák et al. 2015)

the object properties "hasPowerBond". As the real plant description is considered as a foundation for engineering simulation models, the level of details is typically more coarse-grained and only the very basic characteristics are needed in comparison to the AutomationML data format or to the AutomationML ontology described in Chap. 5.

Both the real plant and the simulation domains have assigned variables and parameters. Variables are measured within the plant or the setup by control system to influence the industrial plant. Variables have a crucial role at the borderline between the physical system and the control or monitoring algorithms at the automation software level. On the contrary, parameters are properties of devices that characterize behavior and features of devices, models, etc. They are considered as parameters in the mathematical sense. The difference between variables and parameters is that variables are time series (i.e., sets of samples), whereas parameters are constant values (i.e., single values). The variable domain formalizes variables in the system and interrelates diverse variable representations in different domains. Since parameters are common concepts for both real plant description and simulation description, they pose an independent domain that is referenced where needed. The parameter domain is a taxonomy of existing parameters that formalizes parameters in the system and interrelates diverse parameter representations in different domains. Both variables and parameters have assigned physical quantities and units.

The simulation domain is focused on simulation models, which approximate the behavior of the real plant. This is the reason why the simulation domain and the real plant domain are connected, expressing which parts of the real plant are simulated by simulation modules. The entire simulation models consist of simulation modules and those consist of simulation components. The models and modules can be executed, whereas components are only building blocks for creating modules. All these elements have a topology and interfaces. The topology of simulation models is defined by simulation components, which are on the level of devices, and their connections. This topology is inherited from the real plant topology; however, some elements can be omitted or added as the simulation has to be a simplified model of the reality that is easy to calculate numerically. The interfaces both on the component and module levels consist of inputs, outputs, and parameters, which are represented by the aforementioned variable and parameter concepts. A special kind of parameterization is related to the level of the entire simulation model, where the parameters characterizing for example the duration of the simulated time, or required absolute and relative precision of the calculation are set.

From the statistical point of view, the automation ontology consists of 100+ classes and 80+ properties. It uses basic axioms only and it is implemented in OWL-DL. The ontology is focused on description of objects, structures as well as physical quantities and units needed for proper simulation model design and integration. Technically, the automation ontology is implemented in OWL-DL,[6] which was selected as an optimal compromise between expressive power and efficiency to perform reasoning. In the technical infrastructure, the ontology is embodied within the

[6] https://www.w3.org/TR/owl-guide/.

Engineering Knowledge Base (EKB) framework (Moser and Biffl 2012), which is an ontology-based data-modeling storage. The EKB enables querying and inferring new pieces of knowledge efficiently, as it has been explained in (Novák et al. 2014) for the case of technical integration of simulations. Required pieces of knowledge are retrieved with SPARQL queries.

On top of the EKB, we implemented the ontology tool that enables to access ontologies in such a form that is comprehensible for simulation experts and that is integrable with tools utilized in the simulation model engineering frequently. The ontology tool encapsulates the ontology API called Apache Jena,[7] which provides the access to the automation ontology technically. Since ontologies are considered as a back-end part of the simulation framework, users are not intended to access them directly. Direct access would be complicated because process engineers are usually not skilled in editing ontologies. In addition, it would be a safety and security risk, hence additional integrity and plausibility checks would be required. Therefore, the ontology is encapsulated by an ontology tool, accessing the ontology and providing input and output interfaces for populating the ontology with new pieces of knowledge or querying and retrieving required knowledge from the ontology in an appropriate form. While the ontology tool provides several interfaces, the most important one serves for populating the automation ontology with individuals representing the real plant structure. Other interfaces are used for retrieving knowledge out of the ontology. In particular, they support creating configuration files for the simulation framework and generating simulation models.

10.6 Model-Driven Configurations

Configuration of the technical level, including that one defining the run-time behavior of the service bus, is a very complex task, which would be inefficient and error-prone when carried out manually. Thus it is important to support configuration of the technical integration level of the simulation framework. All tool connectors and services in the EngSB will be configured using the knowledge base. A configuration of the technical integration level for run-time tool integration is based on a specification of interfaces of tools and mapping between tool interfaces declaring which tool is a consumer and which is the producer of a particular variable.

A target domain expert does not work directly with the knowledge base and the simulation framework. A set of tools is provided to assemble and execute the simulation workflow based on run-time parameters P_r (see Sect. 10.4.3) provided by the user. The process of the model-driven configuration is depicted in Fig. 10.7. The main goal is to bridge the gap between required implementation of simulation modules and specific data sources and parameters. Any resulting configuration must be consistent and executable in the simulation framework.

[7] https://jena.apache.org/.

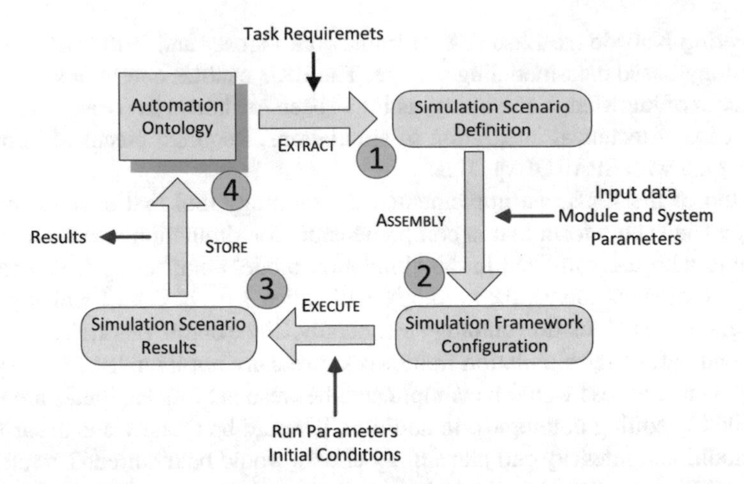

Fig. 10.7 A life cycle of a single simulation scenario

First, information and objects related to solved simulation tasks are extracted from the knowledge base using SPARQL (step 1 in Fig. 10.1). This step allows to detect inconsistencies between intermapped domain ontologies are detected by utilizing information and relations in the KB. An OWL reasoner can then, e.g., detect different physical units assigned to various objects. Some of these inconsistencies can be solved later (like psi to kPa conversion) or others lead to a simulation task redefinition. Unfortunately, not all potential problems can be solved directly during this phase, e.g., a detection of uneven numbers of variables in datasets and simulation modules (see Chap. 15). These problems must be solved in the assembly phase (2) when an XML configuration file is created. This XML configuration file is provided to the simulation framework and executed (3). Note that there is no cooperation between the ontology and the running simulation task in the run-time phase, since the entire process runs in the batch mode. After a simulation workflow is finished, results are stored and intermapped with other objects (4).

Listing 2: The part of the configuration file for the workflow from Example 1:

Listing 10.2 XML configuration file

```
+ <tns:ParameterList> + <tns:TagList> - <tns:Simulation>
  + <tns:Header>
  + <tns:Parameters>
  - <tns:Workflow>
    - <tns:Module name=''hydraulic_plant'' description)>
    + <tns:Parameters />
    <tns:Initialize />
     - <tns:Input>
        <tns:Tag name=''pump_speed'' />
       </tns:Input>
     - <tns:Output>
        <tns:Tag name=''tank_level'' />
        <tns:Tag name=''discharge_pressure'' />
```

```
        </tns:Output>
    - <tns:SimulationTool>
      - <tns:Service>
          <tns:Matlab type=''MATLAB_API_C'' workspace= ...
              ...''C:\demo\simulation\data''>LoadFcn</tns:Matlab>
        </tns:Service>
      </tns:SimulationTool>
```

10.7 Simulation Model Design

The design of simulation models is a complex task including various process steps. The two basic scenarios of simulation design can be distinguished:

1. Design of a simulation model without any prior artifacts (i.e., design from scratch)
2. Creating a simulation model by assembling available simulation blocks

The individual process steps of these two basic scenarios are compared in Fig. 10.8. In both cases, a general knowledge about the type of the real plant and its topology must be captured in the knowledge base. For example, such general knowledge involves that water distribution networks can contain pumps, pipes, tanks, water wells, consumers, or disturbances; pumps have flow and pressure as their inputs and flow and pressure as their outputs, real parameters can be length, diameter, or elevation, see Sect. 10.3 for details about parameters. The general knowledge is a kind of a knowledge skeleton, which can be filled with real values when describing a specific plant. If needed, it can be extended with other parameters which are device-specific. The further process steps differ in case of available and unavailable simulation libraries.

When a simulation library is not available, the method handles the process of gathering the following sequence of knowledge: decomposition of the real plant into simulation modules, selection of simulation blocks and their interfaces, declaration of simulation parameters, entering values of the parameters and finally uploading the new simulation model into the simulation integration framework environment and registering it as one of the available simulations. One of the most crucial parts of this process is its post-analysis and retrieval of a simulation library as a partial result of this process. Therefore, the simulation library depicted as a part of the process on the left-hand of Fig. 10.8 is not an input, but the output obtained with reverse engineering from existing simulations designed manually by simulation experts.

The simulation design scenario, which is easier to be used, utilizes predefined atomic components considered as templates for simulation design. These components can be designed based on physical description coming from various monographs such as (Halliday et al. 1997), or they can be in some cases measured in the real system and considered as black-box characteristics.

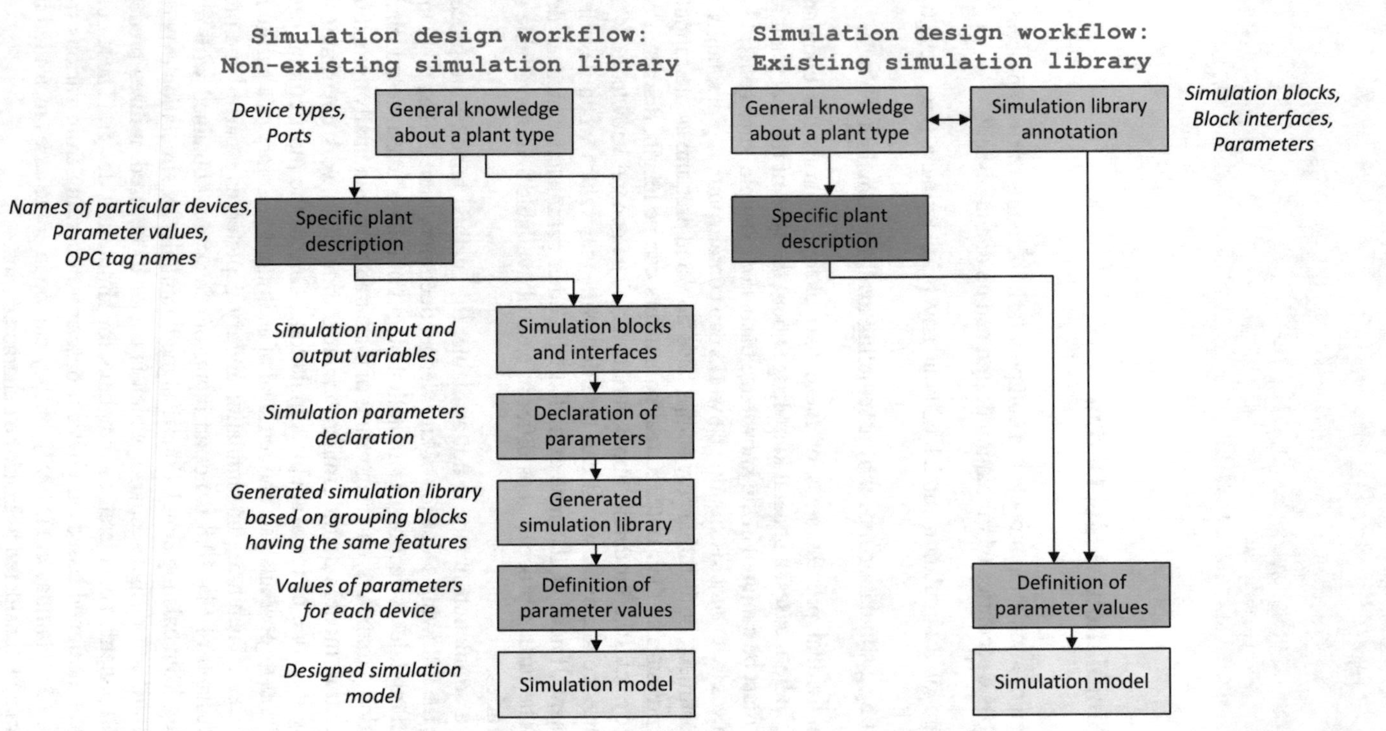

Fig. 10.8 Simulation design workflow: Comparison of scenarios starting from scratch and utilizing a simulation library. Adapted from (Novák and Šindelář 2012)

In the case, when a simulation library involving simulation components is available (see the right-hand side of Fig. 10.8), the library blocks are annotated in the automation ontology incorporating their inputs, outputs, simulation parameters, and initial conditions. Consequently, the plant description has to be formalized, which means that a real plant topology and real parameters are stored in the ontology. Technically, this step means populating the automation ontology with individuals denoting real devices, their interconnections, variables/tags, and others. Based on those pieces of knowledge, an ontology tool is able to generate the simulation model structure automatically. Finally, the simulation expert is required to insert the simulation parameter values for simulation blocks, such as diameters or lengths of pipes.

The mapping between the real component (e.g., a pipe) and the simulation block approximating the pump does not have to be a simple mapping (i.e., $1 : 1$), but may also be defined as complex mapping (i.e., $1 : n$). For example, one simulation block representing a pipe can calculate pressure loss for a given flow through the pipe, or vice versa, based on the flow given as input to the block, the pressure loss can be calculated.

To tackle the issue of various simulation blocks having different interfaces, the well-proven bond graph theory was extended. A bond graph can be considered as a graphical support for engineers to describe the behavior of a given system mathematically and to transform this mathematical behavior description to the form required by simulators. As the bond graph theory introduces generalized signals flow and effort, bond graphs can be used for mechatronic systems combining components of diverse engineering disciplines as well. Following the semiautomated simulation design process proposed in (Novák and Šindelář 2014), simulation models can be designed with extended bond graphs in s semiautomatic manner from available simulation components.

However, the designed simulations still need to be configured with a set of system and plant parameters to justify the geometrical positions and shapes of components as well as required simulation time and precision. In addition, the generated simulations typically require a set of variables representing boundary conditions for the system under simulation (see Sect. 10.3). Furthermore, their output variables (i.e., simulated results) have to be saved and visualized in a form that is easy to be read and comprehended by human domain expert. These tasks are solved by the simulation integration approach, which is proposed in the first part of this chapter.

10.8 Conclusions and Future Work

We proposed the knowledge base for the simulation integration environment in the present chapter. This knowledge base is connected to the simulation framework and used for its configuration. Following this approach the testing and operating simulations can more effectively reflect changes from the design phase. The proposed approach eliminates manual work of domain experts and reduces possible errors. Moreover the knowledge base keeps also mappings between objects participating in a single simulation run. Future work will focus on better utilization of model-driven techniques in the simulation framework configuration, i.e., while in the present version it is possible to detect inconsistencies in physical units, future work should allow to detect inconsistencies between variable features and properties, an automatic conversion, resampling and filtration of input and out time series.

In the area of the simulation design, we have compared two scenarios for developing simulation models. If the simulation library with simulation components is available, the simulation model can be semiautomatically assembled from these components. As future work in the simulation model design area, we will tackle the transformation of energy between various physical disciplines, such as between electrical and mechanical systems. The common problem of simulation design and integration that we will address in the future work is the validation of designed simulation models, especially in the relationship to the used components, from which the simulation is assembled.

Acknowledgments This work was supported by the Christian Doppler Forschungsgesellschaft, the Federal Ministry of Economy, Family and Youth, and the National Foundation for Research, Technology and Development in Austria

References

Antsaklis, P.J., Michel, A.N.: Linear Systems, 2nd Corrected printing edn. Birkhäuser, Boston (2006)

Biffl, S., Schatten, A., Zoitl, A.: Integration of heterogeneous engineering environments for the automation systems lifecycle. In: 2009 7th IEEE International Conference on Industrial Informatics, pp. 576–581. IEEE (2009)

Dreiling, A., Rosemann, M., van der Aalst, W., Sadiq, W., Khan, S.: Model-Driven Process Configuration of Enterprise Systems, pp. 687–706. Wirtschaftsinformatik, Physica-Verlag (2005)

Durak, U., Güler, S., Oğuztüzün, H., İder, S.K.: An exercise in ontology driven trajectory simulation with MATLAB SIMULINK(R). In: Proceedings of the 21th European Conference on Modelling and Simulation (ECMS) 2007, pp. 1–6 (2007)

Franklin, G.F., Powell, J.D., Emami-Naeini, A.: Feedback Control of Dynamic Systems. Prentice Hall, London (2009)

Gawthrop, P., Bevan, G.: Bond-graph modeling. IEEE Control Syst. Mag. **27**(2), 24–45 (2007)

Halliday, D., Resnick, R., Walker, J.: Fundamentals of Physics, 5th edn. extended edn. Wiley (1997)

Ind: Industrial automation systems and integration—integration of life-cycle data for process plants including oil and gas production facilities (ISO 15926), International Organization for Standardization (2009)

International Organization for Standardization: Industrial automation systems and integration—Product data representation and exchange—Part 11: Description methods: The EXPRESS language reference manual (2004)

Kapos, G.-D., Dalakas, V., Tsadimas, A., Nikolaidou, M., Anagnostopoulos, D.: Model-based system engineering using sysML: deriving executable simulation models with QVT. In: 8th Annual IEEE Systems Conference (SysCon 2014), pp. 531–538 (2014)

Karnopp, D.C., Margolis, D.L., Rosenberg, R.C.: System Dynamics: Modeling and Simulation of Mechatronic Systems. Wiley (2006)

Kim, B.C., Teijgeler, H., Munc, D., Han, S.: Integration of distributed plant lifecycle data using ISO 15926 and Web services. Ann. Nucl. Energy **38**, 2309–2318 (2011)

Lange, J., Iwanitz, F., Burke, T.J.: OPC—From Data Access to Unified Architecture. VDE Verlag 2010

Morbach, J.: A reusable ontology for computer-aided process engineering. Ph.D. thesis, RWTH Aachen University (2009)

Morbach, J., Wiesner, A., Marquardt, W.: OntoCAPE—a (re)usable ontology for computer-aided process engineering. Comput. Chem. Eng. **33**, 1546–1556 (2009)

Moser, T., Biffl, S.: Semantic integration of software and systems engineering environments. IEEE Trans. Syst. Man Cybern. Part C: Appl. Rev. **42**(1), 38–50 (2012)

Novák, P., Serral, E., Mordinyi, R., Šindelář, R.: Integrating heterogeneous engineering knowledge and tools for efficient industrial simulation model support. Adv. Eng. Inform. (2015)

Novák, P., Šindelář, R.: Semantic design and integration of simulation models in the industrial automation area. In: Proceedings of the 17th IEEE International Conference on Emerging Technologies and Factory Automation (ETFA 2012)—2nd Workshop on Industrial Automation Tool Integration for Engineering Project Automation (iATPA 2012) (2012)

Novák, P., Šindelář, R.: Component-based design of simulation models utilizing bond-graph theory. In: Proceedings of the 19th IFAC World Congress (IFAC 2014), Cape Town, pp. 1–6 (2014)

Novák, P., Šindelář, R., Mordinyi, R.: Integration framework for simulations and SCADA systems. Simul. Model. Pract. Theory **47**, 121–140 (2014)

OMG: 2012, OMG Systems Modeling Language (OMG SysML(TM)) [online]. Version 1.3. http://www.sysml.org/docs/specs/OMGSysML-v1.3-12-06-02.pdf

Rocca, G.L.: Knowledge based engineering: between AI and CAD. Review of a language based technology to support engineering design, Adv. Eng. Inform. **26**(2), 159–179 (2012). Knowledge based engineering to support complex product design

Silver, G., Bellipady, K., Miller, J., Kochut, K., York, W.: Supporting interoperability using the Discrete-event Modeling Ontology (DeMO). In: Proceedings of the 2009 Winter Simulation Conference (WSC), pp. 1399–1410 (2009)

Silver, G., Hassan, O.-H. Miller, J.: From domain ontologies to modeling ontologies to executable simulation models. In: Proceedings of the 2007 Winter Simulation Conference, pp. 1108–1117 (2007)

Unver, H.O.: An ISA-95-based manufacturing intelligence system in support of lean initiatives. Int. J. Adv. Manuf. Technol. 853–866 (2012)

Verhagen, W.J., Bermell-Garcia, P., van Dijk, R.E., Curran, R.: A critical review of knowledge-based engineering: an identification of research challenges. Adv. Eng. Inform. **26**(1), 5–15 (2012). Network and Supply Chain System Integration for Mass Customization and Sustainable Behavior

Šindelář, R., Novák, P.: Framework for simulation integration. In: Proceedings of the 18th IFAC World Congress 2011, vol. 18, pp. 3569–3574. IFAC, Bologna (2011)

Šindelář, R. and Novák, P.: Simulation integration framework. In: 10th IEEE International Conference on Industrial Informatics (INDIN) 2012, pp. 80–85 (2012)

Part IV
Related and Emerging Trends in the Use of Semantic Web in Engineering

Chapter 11
Semantic Web Solutions in Engineering

Marta Sabou, Olga Kovalenko, Fajar Juang Ekaputra and Stefan Biffl

Abstract The *Industrie 4.0* vision highlights the need for more flexible and adaptable production systems. This requires making the process of engineering production systems faster and intends to lead to higher quality, but also more complex plants. A key issue in improving engineering processes in this direction is providing mechanisms that can efficiently and intelligently handle large-scale and heterogeneous engineering data sets thus shortening engineering processes while ensuring a higher quality of the engineered system, for example, by enabling improved cross-disciplinary defect detection mechanisms. Semantic Web technologies (SWTs) have been widely used for the development of a range of Intelligent Engineering Applications (IEAs) that exhibit an intelligent behavior when processing large and heterogeneous data sets. This chapter identifies key technical tasks performed by IEAs, provides example IEAs and discusses the connection between Semantic Web capabilities and IEA tasks.

Keywords Intelligent engineering applications · Model integration · Model consistency management · Flexible comparison

---

M. Sabou (✉) · O. Kovalenko · F.J. Ekaputra · S. Biffl
Institute of Software Technology and Interactive Systems, CDL-Flex,
Vienna University of Technology, Vienna, Austria
e-mail: Marta.Sabou@ifs.tuwien.ac.at

O. Kovalenko
e-mail: Olga.Kovalenko@tuwien.ac.at

F.J. Ekaputra
e-mail: Fajar.Ekaputra@tuwien.ac.at

S. Biffl
e-mail: Stefan.Biffl@tuwien.ac.at

S. Biffl and M. Sabou (eds.), *Semantic Web Technologies for Intelligent Engineering Applications*, DOI 10.1007/978-3-319-41490-4_11

11.1 Introduction

The *Industrie 4.0* trend (Bauernhansl et al. 2014) envisions increased flexibility of production systems as well as an improved vertical and horizontal integration of production system components (Lüder and Schmidt 2016) as also explained in Chap. 1. Essential elements for realizing these goals are: self-ware and self-adaptable production components as well as flexible production systems relying on adaptation and plug-and-work capabilities. For example, future factories should be able to flexibly respond to changing business conditions and to handle disruptions in the production process (Legat et al. 2013). To achieve such capabilities, it is essential to modernize factory engineering processes in particular by ensuring an integrated exchange of engineering-component information during the engineering design and run-time phases of the production system life cycle (Legat et al. 2013; Lüder and Schmidt 2016).

The need for more flexible and adaptable production systems, which are redesigned more often than before (Bauernhansl et al. 2014), requires, on its turn, that the process of engineering production systems becomes faster; likely leading to higher quality, more complex plants. However, the optimization of engineering processes is often hampered by their heterogeneous and collaborative nature. Heterogeneity is a key characteristic because a large and diverse set of stakeholders is involved in the process of engineering production systems, often bridging the boundaries of several organizations. Usually, the engineering of a production system, for example, a hydro power plant, involves: the production system owner, a main contractor, between 10 and 500 subcontractors, and up to 1,000 component vendors. These stakeholders span diverse engineering disciplines (including mechanical, electrical, and software engineering), make use of a diverse set of (engineering) tools and use terminologies with limited overlap. Indeed, Feldmann et al. (2015) identified heterogeneous and semantically overlapping models as a key characteristic and challenge of engineering settings.

Despite their heterogeneity, the involved stakeholders need to collaborate toward designing and building a complex production system. Indeed, they all provide data and engineering effort to the engineering process. Based on these inputs, many engineering decisions are taken that shape the detailed engineering and implementation of the intended production system (Schmidt et al. 2014).

The study of Lüder and Schmidt (2016) states that engineering of complex mechatronic objects, especially production systems, is increasingly driven by information models that enable representing different aspects of the produced system. This opens up the need for model-driven technologies, such as Semantic Web technologies (SWTs) where *ontologies* (Gruber 1993; Studer et al. 1998) are used as models (Berners-Lee et al. 2001; Shadbolt et al. 2006). Indeed, knowledge-based approaches in general have been observed to be particularly suitable to support the process of engineering production systems as well as to enable advanced functionalities of such systems (e.g., handling disturbances, adapting to new business requirements) (Legat et al. 2013). Knowledge-based systems support "(1) the explicit representation of knowledge in a domain of interest and (2) the exploitation of such knowledge through appropriate reasoning mechanisms in order to provide high-level problem solving performance" (Tasso and Arantes e Oliveira 1998). SWTs, explained in Chap. 3, extend the principles of knowledge-based approaches to Web-scale settings which introduce novel challenges in terms of data size, heterogeneity, and level of distribution. In such setting, SWTs focus on large-scale (i.e., Web-scale) data integration and intelligent reasoning-based methods to support advanced data analytics. Important data analytics tasks include project monitoring, defect detection, and control.

While the use of SWTs to create intelligent engineering applications (IEAs) has seen considerable uptake, there is a lack of understanding of:

- *Q1: What are the key technical tasks that should be solved by an IEA?* and
- *Q2: How are typical IEA tasks enabled by SWT capabilities?*

As an answer to Q1, clear requirements emerge from the area of mechatronic engineering for the *technical tasks* that should be solved by model-driven technologies such as the Semantic Web. Lüder and Schmidt (2016) identify a set of concrete technical tasks that are still challenging to perform in mechatronic engineering, which should be better supported in *Industrie 4.0* settings. They identify that techniques are needed for *model generation, model transformation, model integration,* and *model consistency managemen*t. These tasks are made more difficult by the fact that engineering models are created by engineers from different disciplines. In addition, to support adaptation and plug-and-work capabilities, approaches are needed that allow *flexible comparison*. Flexible comparison can occur, for example, during the selection of an appropriate mechatronic unit, where matchmaking is performed between the required functionalities and those offered by the mechatronic unit to be selected. We consider the tasks put forward by Lüder and Schmidt (2016) as good indicators for typical technical tasks that Semantic Web-based IEAs should solve. Approaches to model generation and model transformation that rely on SWTs are discussed at length in Chap. 5. In this chapter, we focus, in particular, on *model integration, model consistency management,* and *flexible comparison*.

To answer Q2 we perform the following two analysis tasks. First, for each of the three technical tasks we discuss in Sects. 11.2–11.4 a set of IEAs that achieve one of these tasks. The goal is to provide example applications and therefore the material is not meant as an extensive survey of the domain. For each application, we also discuss how SWT capabilities, presented in Chap. 3, support achieving the technical tasks at hand. As such we go beyond the initial version of this work published in (Sabou et al. 2015). We recall that the SWT capabilities identified in Chap. 3 are: (C1) formal and flexible semantic modeling; (C2) intelligent, Web-scale knowledge integration; (C3) browsing and exploration of distributed data sets; (C4) knowledge quality assurance with reasoning, and (C5) knowledge reuse. Section 11.5 concludes our analysis. Second, in Sect. 11.6 we provide an outlook to the remainder of this book, namely Chaps. 12, 13 and 14, and perform a similar analysis of which technical tasks are addressed and which SWTs are used in these three chapters.

11.2 Semantic Web Solutions for Model Integration

Model integration aims to bridge semantic gaps in engineering environments between project participants (and their tools), who use different local terminologies (Aldred et al. 2006; Hohpe and Woolf 2003; Moser et al. 2009; Moser et al. 2010), thus ultimately supporting the analysis, automation, and improvement of multi-disciplinary engineering processes. Semantic model integration is defined as solving problems originating from the intent to share information across disparate and semantically heterogeneous data (Halevy 2005). These problems include the matching of data schemas, the detection of duplicate entries, the reconciliation of inconsistencies, and the modeling of complex relations in different data sources (Noy et al. 2005). Noy (2004) identified three major dimensions of the application of ontologies for supporting semantic model integration: the task of finding cross-source mappings (semi-)automatically, the declarative formal representation of these mappings, and reasoning using these mappings.

Engineering setups introduce important constraints for the semantic integration of engineering knowledge, namely: (1) the high number of involved engineering disciplines with a limited terminological overlap between them, thus further hampering data integration possibilities; (2) the variety of software tools and tool data models in these engineering disciplines; (3) the requirement of domain experts to continue using their well-established tools and processes; (4) the use of domain-specific jargon to represent a (large) part of the engineering knowledge; (5) the distributed and concurrent nature of engineering projects, with geographically dispersed experts working on the project at the same time. Such constraints make semantic integration challenging in engineering environments.

In the remainder of this section, we exemplify some approaches to SWT-based model integration and we discuss which SWT capabilities are used by each approach. Table 11.1 sums up the dependencies between SWT capabilities and the task of model integration for the examples given in this section.

Terkaj and Urgo (2014) present an ontology-based solution to achieve interoperability between systems and tools that may be developed according to different data structures and by employing heterogeneous technologies. Concretely, by leveraging the benefits of conceptual modeling *(C1)*, an ontology represented in the Web Ontology Language[1] (OWL) is used to integrate partial design information from the stakeholders with different competences and expertise (i.e., plant planner or PLC programmer). For the ontology representation, the authors extend the

[1]OWL Reference: https://www.w3.org/TR/owl2-overview/.

Table 11.1 SWT capabilities used by example approaches for model integration

C1: Semantic modeling	C2: Knowledge integration	C5: Knowledge reuse
(Grünwald et al. 2014; Terkaj and Urgo 2014)	(Grünwald et al. 2014; Kovalenko et al. 2013)	(Terkaj and Urgo 2014)

Virtual Factory Data Model (VFDM), a standard data model for representing factory objects related to production systems, resources, processes, and products (Terkaj and Urgo 2012). The provided infrastructure is tested in the context of supporting the design of production systems and generating simulations to check the performance and other aspects of the production system. For that, a simulation model is generated in a semiautomatic way from the obtained virtual representation of the designed production system. This allows faster verification of the current design (i.e., current model). The produced simulation is used to generate the 3D visualization of the dynamic behavior of the production system, based on which the various parameters of the current design of production system (e.g., performance) can be analyzed. The provided solution therefore facilitates the production system reconfiguration in the design phase. The created ontologies exploit the state-of-the-art technical standards and were designed in an application-oriented fashion in order to ensure their future reuse in similar scenarios *(C5)—knowledge reuse*.

Driven primarily by the context of project management in multidisciplinary engineering settings, the *Semantic Model Editor* (SMEd) provides an intuitive way to model and integrate ontologies describing engineering knowledge (Grünwald et al. 2014). Thanks to the formal representation in *conceptual modeling* (*C1*) of the integrated data, a set of SPARQL queries can be executed to extract information relevant for project management (e.g., the number of open features, the number of collaborations between project participants, and the project status). SWTs that support data integration in this tool are formal and conceptual modeling (*C1*) as well as knowledge integration through model mapping, *knowledge integration* (*C2*). An interesting feature of this tool is the use of UML[2] (Unified Modeling Language) class diagram notations for modeling ontological knowledge, thus making ontology development more intuitive to nonexperts in SWTs.

Kovalenko et al. (2013) analyze the types of relations that have to be modeled to integrate heterogeneous data sets in multidisciplinary engineering environments, such as those specific for mechatronic engineering projects. Based on these relations authors derive the requirements for mapping types, which will be needed while integrating heterogeneous data across engineering disciplines and tools with SWTs. Different technologies for mapping definitions and representation between ontologies are then analyzed with respect to their capabilities and limitations to support the identified mappings types. Authors focus on SPARQL[3] CONSTRUCT,

[2]UML: http://www.uml.org/.

[3]SPARQL Overview: https://www.w3.org/TR/sparql11-overview/.

SPIN[4], SWRL[5], and Jena[6] rules as the most widely used alternatives that enable the Semantic Web capability of mapping-based model integration for *knowledge integration* (*C2*). This analysis of mapping representation languages and their strengths could be helpful while choosing an appropriate mapping language for a specific application scenario. More details on this line of work are available in Chap. 6.

11.3 Semantic Web Solutions for Model Consistency Management

In multidisciplinary engineering projects, defects in artifacts of individual disciplines can be propagated to artifacts in other disciplines, causing a major impact on product and process quality in terms of additional risk and time for defect repair. For instance, the sensor type specified in the physical topology model (mechanical engineering) of the automation system has to match the information in the corresponding electrical plan (electrical engineering) and the value range for control variables (software engineering) to describe a correct system. Defects may also come from inconsistencies between disciplines, which are not defects in any of the single discipline views. Because these interdisciplinary relations are not represented in a machine-understandable way, they cannot be checked and managed easily with standard tool support.

Model consistency management refers to the task of detecting defects and inconsistencies in models of individual engineering disciplines as well as across interrelated models from diverse engineering disciplines. The following works are examples of addressing model consistency management with SWTs. Table 11.2 sums up how approaches to consistency management discussed in this section make use of diverse SWT capabilities.

Feldmann et al. (2015) focus on providing a solution for identifying inconsistencies that may arise among diverse engineering models created during the engineering process of automated production systems. Such inconsistency detection contributes to the increased productivity of the engineering process as it supports the detection of potentially severe errors early in the engineering process. To that end, the authors propose the use of SWTs, in particular: (1) RDF[7] (Resource Description Framework) is used as a means to represent knowledge from various engineering models and its simple, triple-based data model acts as a common formalism to represent a variety of models with *conceptual modeling* (*C1*);

[4]SPIN Overview: https://www.w3.org/Submission/spin-overview/.

[5]SWRL Overview: https://www.w3.org/Submission/SWRL/.

[6]Jena: https://jena.apache.org/.

[7]RDF: https://www.w3.org/RDF/.

Table 11.2 SWT capabilities used by example approaches for model consistency management

C1: Semantic modeling	C2: Knowledge integration	C3: Browsing and exploration of distributed data sets	C4: Quality assurance with reasoning	C5: knowledge reuse
(Feldmann et al. 2015; Kovalenko et al. 2014; Abele et al. 2013; Feldmann et al. 2014b; Sabou et al. 2016)	Feldmann et al. 2015; Kovalenko et al. 2014)	(Sabou et al. 2016;	(Feldmann et al. 2015; Kovalenko et al. 2014; Abele et al. 2013; Feldmann et al. 2014b; Sabou et al. 2016)	(Kovalenko et al. 2014; Abele et al. 2013)

(2) representing all models using the same formalism enables specifying explicit links between the elements of those models for *knowledge integration* (*C2*); (3) the SPARQL query language is chosen as a mechanism to explicitly define and verify various inconsistency checks—this benefits from the formality of Semantic Web languages as well as employs reasoning (*C4*). Feldmann et al. (2015) identify that the use of ontologies for capturing knowledge that is shared among various models should be beneficial as a future extension of their work. Such an ontology-based approach to inconsistency detection is already taken by (Kovalenko et al. 2014), as discussed next.

Kovalenko et al. (2014) present an ontology-based approach to automatically detect inconsistencies across heterogeneous data sets produced by different engineering disciplines and tools in multidisciplinary engineering projects. OWL ontologies are used to explicitly represent the discipline/tool-specific knowledge and data in a machine-understandable form with *conceptual modeling* (*C1*). Mappings are then defined between the ontologies to make cross-disciplinary (or cross-tool) relations between the data models and data sets explicit for *knowledge integration* (*C2*). SPARQL queries are executed over the discipline/tool ontologies regarding the defined mappings in order to perform inconsistency detection across discipline/tool boundaries, thus taking advantage of the reasoning capabilities of SWTs (*C4*). Another advantage of using SWTs lies in the possibility to reuse the developed ontologies and formulated checks (SPARQL queries) in subsequent projects for *knowledge reuse* (*C5*).

An approach for the automated validation of plant models is presented in (Abele et al. 2013). Inconsistencies in plant models may arise when integrating different engineering views on the plant created by different experts (e.g., mechanical and electrical engineers), who work concurrently on developing models for the same plant. The approach relies on representing plant models in terms of ontologies and subsequently applying reasoning techniques for validation purposes. In particular, the focus is on models conforming to CAEX data format (Schleipen et al. 2008), a meta-model for the storage and exchange of engineering models defined by IEC 62424 (IEC 2008). An OWL representation of the CAEX plant models is obtained through an automated transformation, which relies on a set of mappings that the authors defined between CAEX and OWL constructs. A modular ontology design is adopted: first, taking advantage of *conceptual modeling* (*C1*), a CAEX base ontology captures in OWL the basic design decisions of representing CAEX models; second, a plant ontology imports the base CAEX ontology and extends it with vendor-specific information and instance data from the specific CAEX file. The base ontology can be *reused* across projects (*C5*). Then SPARQL queries and (for some checks) reasoning are used to perform consistency checks on the obtained ontology. *Reasoning* and querying mechanisms enabled by OWL allow validation and retrieval of implicit knowledge from ontologies (*C4*). The final decision of whether the identified inconsistencies are indeed a problem is left to domain experts, i.e., there is no automatic correction/fixing.

In the area of requirements and test case management, Feldmann et al. (2014b) propose a Semantic Web solution for verifying the consistency of requirements as well as of requirements and test cases. A *conceptual model* that describes the main elements for this use case is developed and then formalized as an OWL ontology (*C1*). *Reasoning mechanisms* (*C4*) such as satisfiability checking, instance classification, and model consistency checking are applied to support various consistency-related use cases in the management of requirements and their associated test cases.

The AutomationML Analyzer[8] (Sabou et al. 2016) uses Semantic Web and Linked Data technologies to provide an interface for analyzing data from integrated AutomationML files. AutomationML (Drath 2010) is an emerging format for exchanging engineering data. While the concerted use of AutomationML in an engineering project makes the integration between data from different engineering disciplines easier, tools are still needed for more easily navigating and analyzing integrated AutomationML data. To that end, the AutomationML Analyzer uses ontology-based technologies to integrate AutomationML data; to provide easy navigation support within the AutomationML data as well as to detect project level inconsistencies and defects through SPARQL querying of the integrated data. The main SWT capabilities used are: semantic modeling (*C1*) of an AutomationML ontology with which the input data is semantically enriched (see also Chap. 5); browsing and exploration of the semantic data through Linked Data based mechanisms (*C3*) and the use of reasoning mechanisms as part of the SPARQL querying activities (*C4*).

[8]AutomationML Analyzer: http://data.ifs.tuwien.ac.at/aml/analyzer.

11.4 Semantic Web Solutions for Flexible Comparison

In engineering settings, comparisons are often performed between engineering objects that should be replaced or interchanged. For example, it is often requested that a comparison be made between the capabilities of an engineering unit to be replaced (e.g., a device) and a new unit. *Flexible comparison* refers to performing such comparison among descriptions of engineering objects, as exemplified in the works described in this section. Table 11.3 depicts how approaches to flexible comparison discussed in this section make use of diverse SWT capabilities.

Feldmann et al. (2014a) consider the problem of parts exchange in an evolving manufacturing system, in particular, checking the compatibility of the old part and the new part. This is a complex problem since insights from multiple contributing disciplines must be taken into account. Therefore, support for such operations leads to an increased productivity of the engineering process. The authors offer a solution where model-based and Semantic Web approaches are combined: the SysML language provides a means for modeling interdisciplinary manufacturing systems using graphical means but lacks the formal foundations to allow automated compatibility checks between various components. This shortcoming is compensated by translating SysML models into OWL ontologies and exploring the formality of OWL for checking compatibility constraints expressed in terms of SPARQL queries. The Semantic Web capabilities of *conceptual modeling* (*C1*) and *reasoning enabled quality assurance* (*C4*) play a key role in supporting this application.

The use of SWTs for creating simulation models is discussed in (Novák and Šindelár 2011). Simulation models are widely used in industrial engineering in general, and for engineering production systems in particular, to perform experiments that would be too dangerous or expensive if performed in vivo. A major task is the generation of a simulation model, which consists of the selection of a set of suitable simulation blocks that accurately represent the state of the real plant. To improve this manual model creation process, the authors propose the use of three ontologies which allow the explicit representation of knowledge with *conceptual modeling* (*C1*) about simulation blocks (simulation ontology), the real industrial plant (plant ontology), and the signals of each simulation block (signal ontology). Mappings are created between these ontologies for *knowledge integration* (*C2*). This formalized knowledge enables the creation of a semantic engine that performs *reasoning*-based,

Table 11.3 SWT capabilities used by example approaches flexible comparison

C1: Semantic modeling	C2: Knowledge integration	C4: Quality assurance with reasoning	C5: Knowledge reuse
(Feldmann et al. 2014a; Novák and Šindelár 2011; Willmann et al. 2014)	(Novák and Šindelár 2011)	Feldmann et al. 2014; Novák and Šindelár 2011; Willmann et al. 2014)	(Novák and Šindelár 2011; Willmann et al. 2014)

flexible comparisons between the available simulation blocks and the industrial plant status, thus automating the creation of the simulation model (*C4*). The created ontologies and engine can be *reused* across diverse simulation events (*C5*).

Flexible comparison among products and production processes is an important aspect of any product ramp-up activity, which aims at identifying a suitable production process at a target site in order to produce a certain product with the same quality as at a source site. Willmann et al. (2014) propose K-RAMP, a knowledge-based production ramp-up process, where a core task is flexibly finding matches between semifinished products (production processes) available at the source and target production sites. SWTs were used to define, design, and evaluate the ramp-up process knowledge-base. SWT capabilities such as *conceptual modeling* (*C1*), *reasoning* for matchmaking (*C4*) as well as *reuse of knowledge* (*C5*) between various ramp-up activities are the most useful features in this setting. Chapter 9 provides more insights into this line of work.

11.5 Conclusions

A key goal of this chapter was to better understand how typical IEA tasks can be enabled by the SWT capabilities introduced earlier in Chap. 3. To that end, the chapter exemplified typical technical tasks supported by IEAs, in particular, a set of tasks that, according to (Lüder and Schmidt 2016), require support in the context of the engineering of production systems (i.e., *model generation, model transformation, model integration, consistency management, and flexible comparison*). Approaches to model generation and model transformation that rely on SWTs are discussed at length in Chap. 5. In Sects. 11.2–11.4 a set of current approaches was analyzed that make use of SWTs for supporting the engineering of production systems and address the tasks of model integration, consistency management, and flexible comparison. Table 11.4 sums up the analysis of the earlier chapters and

Table 11.4 Number of example approaches that solve a given technical task with a certain SWT capability

	C1: Semantic modeling	C2: Knowledge integration	C3: Browsing, exploration of distributed data sets	C4: Quality assurance with reasoning	C5: Knowledge reuse
Model integration	2	2			1
Model consistency management	5	2	1	5	2
Flexible comparison	3	1		3	2

identifies the number of approaches that address certain technical tasks by using the SWT capabilities.

The analysis leads to the following conclusions:

- *SWTs support various aspects of production system's engineering*. Although our focus was restricted to the engineering phase of production systems, it was observed that SWTs have been applied to support various aspects of this process ranging from requirements management (Feldmann et al. 2014b), to simulation (Novák and Šindelár 2011) and project management (Grünwald et al. 2014). Although diverse, use cases form these various life cycle stages or production systems, are enabled at a technical level by a few individual tasks. This chapter focuses on the tasks of model integration, model consistency checking, and flexible comparison.
- *Model consistency management* tasks are of major importance. They have applications in a wide range of settings. The task of *flexible comparison* is more complex and therefore less explored to date. *Model integration* is often not a goal *per se*, but rather an enabler for the other tasks, especially in the settings that consider engineering models from multiple disciplines.
- *Formal and flexible conceptual modeling (C1) is the most used SWT capability.* Often performed by using modeling approaches of wide adoption in engineering such as SysML (Feldmann et al. 2014a) or UML (Grünwald et al. 2014), this feature is essential for attaining all three technical tasks we analyzed. The *reasoning* SWT capability (*C4*) is used by all approaches dealing with consistency management and flexible comparison. *Knowledge integration* techniques (C2), such as model mapping, play an important role for model integration being applicable in all scenarios, in which multiple models are involved that need to be integrated before advanced reasoning-driven analytics can be performed. The possibility to easily *reuse* formally represented conceptual knowledge (C5) was perceived as a clear benefit across the various usage scenarios. This is somewhat conflicting with other evidence from the literature which alerts to the difficulty of performing ontology reuse both in general (Oberle 2014; Simperl 2009) and in engineering contexts in particular (Legat et al. 2014; Terkaj and Urgo 2014).
- *Web-based SWT capabilities are less frequently used.* Although one of the strengths of SWTs is the combination of traditional knowledge representation and reasoning techniques with Web compliance features, there is a clear tendency, at least in the papers we reviewed, to primarily explore the semantic features of these technologies as opposed to those related to Web compliance, in particular, C3 related to *Browsing and Exploration of Distributed Data Sets*. This could be a consequence of the fact that SWTs are primarily used as an enterprise data integration and management solution, where Web-oriented features (unique URIs, reuse from other Web data sets) are of less importance. An interesting future research question is therefore the investigation of how the Web-compliance-related features of SWTs could be of use in engineering of mechatronic systems, and more broadly in *Industrie 4.0*.

11.6 Outlook on Part IV

Part IV showcases three works that use SWTs in engineering settings to solve all three basic tasks discussed above: model integration, consistency management, and flexible comparison. We provide a summary of these chapters and discuss how they use SWT capabilities to address the three technical tasks.

Chapter 12 (*Semantic Web Solutions in the Automotive Industry*) reports on two use cases in the context of the automotive industry solved with SWTs. The first use case aims to support the engineer in deriving an optimized design layout starting from a system specification, which is refined in iterative design steps. To enable these tasks, ontologies are used to represent requirements, to allocate them to parts of the systems, to attach constraints to requirements and parts of the system, and to keep track of different versions of requirements during the subsequent processing (i.e., design phases). The main advantage of using ontologies in this case was representing requirements explicitly and in a machine processable way (*C1*). This allowed the versioning of the requirements and attaching constraints which were then verified with a *Relational Constraint Solver* (RCS), a constraint engine specialized to check numeric equations. Other benefits of this solution were: (1) enhanced *reuse* (C5) of previous knowledge (e.g., requirement templates and system structure information could be reused across design problems); (2) the *formal representation* of requirements and associated constraints was successfully used to guide the engineers during the design phase and prevent them from entering incorrect values; (3) thanks to version control over the iterative steps, this solution proved more manageable than the baseline, an *Excel*-based approach.

The second use case (UC2) focuses on supporting the collaborative development process of an automatic transmission gearbox by distributed engineering groups from different engineering disciplines needed to realize this complex mechatronic object, namely: mechanical, hydraulic, electrical, and software engineering. This use case illustrates the task of *consistency management* of several overlapping models of the same system (the authors refer to these as views, similarly to the terminology in Chap. 13). An additional need was to enable change propagation among the interrelated models, meaning that a change in one model would lead to changes in all the related models whenever necessary (i.e., according to the interdependencies specified between models). SWTs were used to address these needs in particular with the capabilities *C1, C2, C4*.

As the reported work was performed in the early years of Semantic Web research when the representation languages were still under development, the authors made use of classical knowledge representation techniques such as a Frame logic-based knowledge representation and the *Flora2*[9] reasoning framework. This use case also demonstrates solving a *model integration* task. Domain-specific knowledge is integrated by reference to so-called *Engineering ontologies*, which capture common concepts shared across engineering disciplines. The authors observed several

[9]Flora2 reasoning framework: http://flora.sourceforge.net/.

benefits to engineers such as enabling automatic consistency checking and change propagation among engineering models, which was not possible before. In addition, engineers also benefitted from the fact that the dependencies between their models were made explicit—as such they achieved a better understanding of how changes in their model might affect other models.

Chapter 13 (*Leveraging Semantic Web Technologies for Consistency Management in Multi-Viewpoint Systems Engineering*) focuses on *consistency management* among different, overlapping views (or models) of the same complex systems. As such, this chapter captures a typical problem since such overlapping models are often created by engineering tool networks, each model representing one engineering discipline's view on the system. The authors propose a solution where RDF is used to encode different system views in a uniform manner and the emerging *Shapes Constraint Language* (SHACL) is employed to define the inter-viewpoint dependencies (*C1, C2*). These dependencies can be automatically checked during modeling time to uncover potential inconsistencies between the various models. The *Reasoning* SWT capability supports this task (*C4*). Going beyond the case when dependencies are specified between a set of views, the authors also consider a use case which requires the *semantic integration* of multiple viewpoints prior to checking the consistency among these viewpoints. As a technical solution for solving the engineering data integration, they choose the *hybrid ontology integration* approach, where local models are mapped to a global ontology, which contains shared knowledge (such as common concepts discussed in Chap. 5). SHACL expressions are used to define the correspondences between local and global views in line with the SWT capability *C2* on *data integration*. The authors note that the Semantic Web solution described in the chapter is highly compatible with and complementary to Model-Driven Engineering approaches, as models can be easily transformed into Semantic Web-specific languages.

Chapter 14 (*Applications of Semantic Web Technologies for the Engineering of Automated Production Systems—Three Use Cases*) details three use cases from the process of engineering automated production systems where SWTs are used. The first use case (UC1) focuses on ensuring compatibility between mechatronic modules that need to be replaced in a given system configuration, illustrating a setting that needs *flexible comparison* approaches. This use case requires means for (1) identifying modules compatible with a module that needs to be replaced and (2) identifying and resolving conflicts in a given system configuration as a follow-up of a module change. There is a need for *knowledge representation* and for intelligent access mechanisms that can accomplish such comparisons. For this, the authors propose using an ontology for representing compatibility information (*C1*) and encoding and checking compatibility through SPARQL queries.

The second use case (UC2) focuses on the task of *ensuring consistency* between requirements specified for a production system and test cases corresponding to checking that requirement. The authors use semantic modeling to formally *represent domain knowledge* (C1) and reasoning services offered by Semantic Web *reasoners* to check whether test cases are compatible with requirements (C4).

The third use case (UC3) aims to *detect consistency* between different engineering models of the same system. Because the engineering models describe the same system, they overlap to some extent and as such these overlapping parts should be kept consistent. To achieve this task, there is also a need to define which parts of the models correspond to each other as a basis for compatibility checks. The authors propose achieving such *engineering data integration* by (1) defining a base vocabulary that contains the common concepts used by the various models considered (C1) and (2) using a common data representation language (namely RDF) to encode the various models in a uniform way and describe equivalent *mappings* between their corresponding elements (C2).

All use cases rely on a combination of Semantic Web and *Model-based Engineering* (MBE). Concretely, the authors leverage the widespread adoption and good tool support for MBE, especially the SysML4Mechatronics language to collect relevant system models from engineers. These models are then translated into Semantic Web formats that allow specifying correspondences between, querying and reasoning on the various engineering models.

Table 11.5 provides an overview of the discussion in this chapter by depicting the task(s) addressed by each chapter and the SWT capabilities used for solving those tasks. Several of the conclusions drawn from the analysis of example applications (Sect. 11.5) can also be made when considering the chapters of Part IV. We observe that these works cover various aspects of production system's engineering and that model consistency management is a task addressed by all chapters. It is also remarkable that all chapters showcase the need for model integration prior to enabling model consistency management and that they rely on similar data integration approaches. As with the example approaches, for the chapters in Part IV, the conceptual modeling SWT (C1) is most frequently used while capabilities C3 and C5 are not used at all. Another interesting observation is that Chaps. 13 and 14 aim to establish synergies with system engineering languages that are more widely spread among engineers as a way to facilitate the acquisition of domain models.

Table 11.5 Overview of technical tasks addressed by chapters in Part IV and the SWT capabilities used to address those tasks

	C1: Semantic modeling	C2: Knowledge integration	C4: Quality assurance with reasoning
Model integration	Ch12 UC2, Ch13, Ch4 UC3	Ch12 UC2, Ch13, Ch4 UC3	
Model consistency management	Ch12 UC2, Ch13, Ch14 UC2		Ch12 UC2, Ch13, Ch14 UC2
Flexible comparison	Ch14 UC1		

Acknowledgments This work was supported by the Christian Doppler Forschungsgesellschaft, the Federal Ministry of Economy, Family and Youth, and the National Foundation for Research, Technology and Development in Austria.

References

Abele, L., Legat, C., Grimm, S., Müller, A.W.: Ontology-based validation of plant models. In: 11th IEEE International Conference on Industrial Informatics (INDIN), pp. 236–241. IEEE (2013)

Aldred, L., van der Aalst, W., Dumas, M., Hofstede, A.: Understanding the challenges in getting together: the semantics of decoupling in middleware. In: BPM Center Report BPM-06-19, BPMCenter.org, p. 36 (2006)

Bauernhansl, T., ten Hompel, M., Vogel-Heuser, B. (eds.): Industrie 4.0 in Produktion, Automatisierung und Logistik. Springer (2014)

Berners-Lee, T., Hendler, J., Lassila, O.: The Semantic Web. Sci. Am., 29–37 (2001)

Drath, R.: Datenaustausch in der Anlagenplanung mit AutomationML: Integration von CAEX. PLCopen XML und COLLADA. Springer, DE (2010)

IEC: IEC 62424. Representation of process control engineering. Requests in P&I diagrams and data exchange between P&ID tools and PCE-CAE tools (2008)

Feldmann, S., Kernschmidt, K., Vogel-Heuser, B.: Combining a SysML-based modeling approach and semantic technologies for analyzing change influences in manufacturing plant models. In: Proceedings of the 47th CIRP Conference on Manufacturing Systems, pp. 451–456 (2014)

Feldmann, S., Rösch, S., Legat, C., Vogel-Heuser, B.: Keeping requirements and test cases consistent: towards an ontology-based approach. In: 12th IEEE International Conference on Industrial Informatics, INDIN, pp. 726–732 (2014)

Feldmann, S, Herzig, S.J.I., Kernschmidt, K., Wolfenstetter, T., Kammerl, D., Qamar, A., Lindemann, U., Krcmar, H., Paredis, C.J.J., Vogel-Heuser, B.: Towards effective management of inconsistencies in model-based engineering of automated production systems. In: Proceedings of IFAC Symposium on Information Control in Manufacturing (INCOM 2015) (2015)

Gruber, T.R.: A translation approach to portable ontology specifications. Knowl. Acquis. **5**(2), 199–220 (1993)

Grünwald, A., Winkler, D., Sabou, M., Biffl, S.: The semantic model editor: efficient data modeling and integration based on OWL ontologies. In: Proceedings of the 10th International Conference on Semantic Systems, SEM '14, pp. 116–123. ACM (2014)

Halevy, A.: Why your data won't mix. In: Queue, vol. 3, no. 8, pp. 50–58 (2005). ISSN 1542-7730

Hohpe, G., Woolf, B.: Enterprise Integration Patterns: Designing, Building, and Deploying Messaging Solutions. Addison-Wesley Longman, pp. 480 (2003). ISBN 978-0321200686

Kovalenko, O., Debruyne, C., Serral, E., Biffl, S.: Evaluation of technologies for mapping representation in ontologies. In: On the Move to Meaningful Internet Systems: OTM 2013 Conferences, pp. 564–571. Springer (2013)

Kovalenko, O., Serral, E., Sabou, M., Ekaputra, F.J., Winkler, D,, Biffl, S.: Automating cross-disciplinary defect detection in multi-disciplinary engineering environments. In: Knowledge Engineering and Knowledge Management, pp. 238–249. Springer (2014)

Legat, C., Lamparter, S., Vogel-Heuser, B.: Knowledge-based technologies for future factory engineering and control. In: Borangiu, T., Thomas, A., Trentesaux, D. (eds.) Service Orientation in Holonic and Multi Agent Manufacturing and Robotics, Studies in Computational Intelligence, vol. 472, pp. 355–374. Springer, Berlin (2013)

Legat, C., Seitz, C., Lamparter, S., Feldmann, S.: Semantics to the shop floor: towards ontology modularization and reuse in the automation domain. In: Proceedings of 19th IFAC World Congress, Cape Town, South Africa, pp. 355–374 (2014)

Lüder, A., Schmidt, N.: Challenges of mechatronical engineering of production systems—an automation system engineering view. In: Ghezzi, L., Hömberg, D., Landry, L. (eds.) Math for the Digital Factory. Springer, Heidelberg (2016)

Moser, T., Mordinyi, R., Mikula, A., Biffl, S.: Making expert knowledge explicit to facilitate tool support for integrating complex information systems in the ATM domain. In: Proceedings of the International Conference on Complex, Intelligent and Software Intensive Systems (CISIS), pp. 90–97. IEEE (2009). ISBN 978-1-4244-3569-2

Moser, T., Mordinyi, R., Mikula, A., Biffl, S.: Efficient integration of complex information systems in the ATM domain with explicit expert knowledge models. In: Xhafa, F., Barolli, L., Papajorgji, P.J. (eds.) Complex Intelligent Systems and Their Applications, vol. 41, pp. 1–19. Springer (2010). ISBN 978-1-4419-1635-8

Novák, P., Šindelár, R.: Applications of ontologies for assembling simulation models of industrial systems. In: Proceedings of the Confederated International Conference on the Move to Meaningful Internet Systems, pp. 148–157. Springer (2011)

Noy, N.F.: Semantic integration: a survey of ontology-based approaches. ACM SIGMOD Record **33**(4), 65–70 (2004)

Noy, N.F., Doan, A.H., Halevy, A.Y.: Semantic integration. AI Mag. **26**(1), 7 (2005)

Oberle, D.: How ontologies benefit enterprise applications. Semant Web J. **5**(6), 473–491 (2014)

Sabou, M., Kovalenko, O., Ekaputra, F.J., Biffl, S.: Beiträge des Semantic Web zum Engineering für Industrie 4.0. In: Vogel-Heuser, B., Bauernhansl, T., ten Hompel, M. (eds.) Handbuch Industrie 4.0. Springer (2015). doi:10.1007/978-3-662-45537-1_90-1

Sabou, M., Ekaputra, F.J., Kovalenko, O.: Supporting the engineering of cyber-physical production systems with the AutomationML analyzer. In: Proceedings of the CPPS Workshop, at the Cyber-Physical Systems Week, Vienna (2016)

Schleipen, M., Drath, R., Sauer, O.: The system-independent data exchange format CAEX for supporting an automatic configuration of a production monitoring and control system. In: IEEE International Symposium on Industrial Electronics, pp. 1786–1791. IEEE (2008)

Schmidt, N., Lüder, A., Biffl, S., Steininger, H.: Analyzing requirements on software tools according to functional engineering phase in the technical systems engineering process. In: Proceedings of the 19th IEEE International Conference on Emerging Technologies and Factory Automation (ETFA). IEEE (2014)

Shadbolt, N., Berners-Lee, T., Hall, W.: The semantic web revisited. IEEE Intell. Syst. **21**(3), 96–101 (2006)

Simperl, E.: Reusing ontologies on the semantic web: a feasibility study. Data Knowl. Eng. **68** (10), 905–925 (2009)

Studer, R., Benjamins, V., Fensel, D.: Knowledge engineering: principles and methods. Data Knowl. Eng. **25**(1–2), 161–197 (1998)

Tasso, C., Arantes e Oliveira, E.D. (eds.) Development of Knowledge-Based Systems for Engineering. Springer, Vienna (1998)

Terkaj, W., Urgo, M.: Virtual factory data model to support performance evaluation of production systems. In: Proceedings of OSEMA 2012 Workshop, 7th International Conference on Formal Ontology in Information Systems, pp. 24–27. Graz, Austria (2012)

Terkaj, W., Urgo, M.: Ontology-based modeling of production systems for design and performance evaluation. In: 12th IEEE International Conference on Industrial Informatics (INDIN), pp. 748–753. IEEE (2014)

Willmann, R., Biffl, S., Serral, E.: Determining qualified production processes for new product ramp-up using semantic web technologies. In: Proceedings of the 14th International Conference on Knowledge Technologies and Data-Driven Business. ACM (2014)

Chapter 12
Semantic Web Solutions in the Automotive Industry

Tania Tudorache and Luna Alani

Abstract This chapter describes how we employed ontologies to solve and optimize different design tasks at an automotive company. We first introduce five core engineering ontologies that provides the formal grounding for the described use cases. We used these ontology to represent engineering systems, and to perform change propagation and consistency checking of the design models. The first use case presents an approach that helps engineers derive an optimized design starting from a system specification, whose parameters are refined in iterative design steps. This use case will demonstrate how we represented requirements and their iterative refinements using ontologies, and how we used a constraint solver in conjunction with an ontology to derive an optimized design with respect to different criteria. The second use case comes from the collaborative development process of an automatic gearbox, in which distributed engineering teams developed different models (i.e., geometric and functional) of the same product in a parallel way. We used ontologies to represent the different models of the same engineering product, and used formal mappings to define the correspondences between the different models. We could then use reasoning to check the consistency of the two models with each other, and to propagate changes from one model to the other.

Keywords Engineering ontologies · Modular and generic Systems modeling · SysML · Real-world use cases · Collaborative development · Constraint solving · Change propagation · Specification-driven design · Viewpoints · Mappings · Modeling requirements · Automotive industry

---

T. Tudorache (✉)
Stanford Center for Biomedical Informatics Research, 1265 Welch Rd,
Stanford, CA 94040, USA
e-mail: tudorache@stanford.edu

L. Alani
Kropbacher Weg 10, 35398 Giessen, Germany
e-mail: dr.luna.alani@gmail.com

S. Biffl and M. Sabou (eds.), *Semantic Web Technologies for Intelligent Engineering Applications*, DOI 10.1007/978-3-319-41490-4_12

12.1 Introduction: Models in the Engineering Domain

Engineering modeling is the activity of building models of engineering systems to be used in solving different tasks in product development, such as design, analysis, diagnosis, etc. Engineering modeling can be seen as a form of design, which is composed of several subtasks (Top and Akkermans 1994). Building an engineering model became a difficult process with the increase of product complexity. Engineers rely more and more on the support of model editors to help them build high quality design models.

Most models that are built in the modern development process are simplified representations of a system in a computer language. They are used to investigate the properties of the system, to understand better the system behavior, or they may even serve as a specification for building physical prototypes of the product.

Very often, modeling is done in a compositional fashion (Falkenhainer and Forbus 1991): General model fragments are assembled together to form the structure of a new model. In the case that the internal structure of the general components is not changed, the modeling process can be seen as a configuration task (Breuker and Van de Velde 1994).

Besides the model management aspects, other types of knowledge support the engineer in the modeling task. For example, consistency checking of a design model can be performed in an automated fashion, if general rules or constraints of the domain are specified explicitly in a formal manner.

All these modeling support activities are possible if the underlying representation of models is rich enough to capture all necessary types of knowledge. Ontologies offer a rich and computer–interpretable representation of a domain that can be used to support an improved modeling process.

In this chapter, we will investigate how engineering ontologies have been used to improve product design. We have performed the work described here as part of two research projects at a major automotive company. First, we will describe briefly a well–accepted way of modeling engineering systems using a compositional approach with SysML (Sect. 12.2), then we will introduce five generic engineering ontologies that provide the formal basis of this work (Sect. 12.3), and finally we will detail two real–world use cases of applying ontologies to engineering problems. The first use case (Sect. 12.4) is in the context of the iterative refining of product requirements, and will show how we used ontologies and a relational constraint solver to reach an optimal product design that fulfills all requirements. The second use case (Sect. 12.5) shows how we were able to check the consistency and propagate changes across two different models (geometric and functional) of an automatic gearbox. After each use case, we also discuss the ontology–specific benefits of that use case. More details about the work presented in this chapter can be found in (Alani 2007; Tudorache 2006b, 2008).

12.2 Systems Engineering and SysML

Systems engineering deals with the methods necessary for developing and implementing complex systems (Stevens et al. 1998). Systems engineering has been very successfully applied in software engineering. Following this success, systems engineering has also been adopted as the basis for the mechatronic development process (VDI 2004) to respond to the increasing product complexity in a challenging market situation. It is current practice, that activities in the system engineering process—for example, requirements specification, analysis, design, verification, validation, etc.—use different modeling languages, techniques and tools.

The Object Management Group (OMG[1]) recognized the need for a general-purpose modeling language for describing systems that also provides a unified graphical representation. To that end, SysML (SysML 2006) was created as a general modeling language for systems engineering. SysML has a graphical notation based on a UML 2 profile (UML 2006), and provides a simple, yet powerful language for specification of complex systems. For example, it has constructs for representing requirements, systems, composition and interconnection of systems, allocation of requirements to systems, behavioral diagrams, etc. However, even if SysML provides a basic language and a graphical notation for systems engineering, it lacks a formal semantics.

The center piece of engineering design are the systems. According to the ISO Standard (ISO/IEC 2002), a system is a combination of interacting elements organized to achieve one or more goals. In engineering, computer models of systems are built for different purposes: better understanding a system, investigating its properties, or even to building it.

A system is defined by its parts and connections. Connections between components can be made only in the interaction points called *ports* or *connectors*. SysML denotes the connectors as *ports* as they are derived from the UML 2 notion of *ports*. They support building modular systems from reusable components with clearly defined interfaces. SysML also makes an important distinction between the definition of a component and its usage. The definition of a component is given by the declaration of its parts and their connections, whereas its usage is given by the role that the component plays as part of another component.

An example of a system is given in Fig. 12.1, which depicts the components and connections of a powertrain subsystem of a car. Only two components of the powertrain system are shown for the sake of simplicity. In this example, a powertrain system is composed of two components—an engine and a transmission—, and contains three connections.

We will use the following notation to explain the example.

`Component`—to denote a component. *Example*: `Powertrain` or `Engine`.
`Component.port`—to denote a port of a component. *Example*: `Engine.in`.

[1] http://www.omg.org.

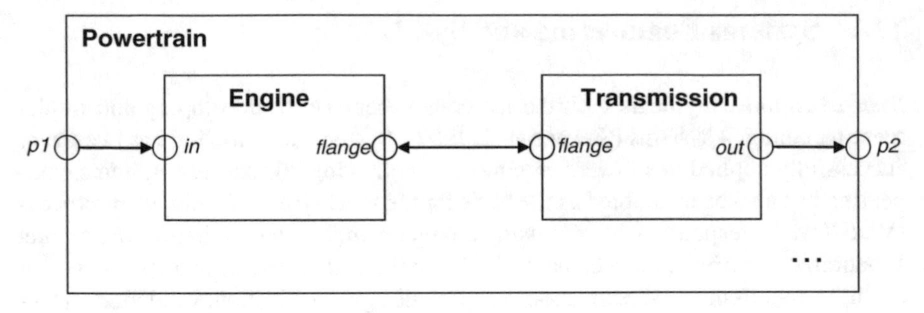

Fig. 12.1 An example of a simple system

`connected(Component1.port1, Component2.port2)`—to denote that `port1` of `Component1` is connected to `port2` of `Component2`. *Example*: `connected(Engine.flange, Transmission.flange)`.

The system `Powertrain` is defined by its parts: `Engine` and `Transmission` and the set of connection between them:

```
{connected(Powertrain.p1,Engine.in),
 connected(Engine.flange,Transmission.flange),
 connected(Transmission.out,Powertrain.p2)}
```

The connection notation has been inspired from Modelica (Modelica Specification 2005)—an object-oriented language for modeling physical systems. The semantics of the connection is related to the type of ports that it connects. Modelica supports two ways of handling connected ports. If the variables in the ports are *across variables*, then they are set equal in the connected ports (for example, voltages in two connected electrical pins). If the variables are of type *flow* (like in the case of electrical currents), then Kirchhoff's law is applied, i.e., the sum of the variables in a node is equal to zero.

We have used SysML as an inspiration for building a set of five modular ontologies for modeling engineering systems, which we describe in the following section.

12.3 The Engineering Ontologies

There are many definitions of the term ontology. Gruber gave one of the first definition to gain acceptance in the community: "*An ontology is a formal specification of a conceptualization*" (Gruber 1993).

Ontologies have long been used to model engineering systems. For example, a type of design that is especially of interest for the engineering domain is the *parametric design task*, which is formally defined by Wielinga (Wielinga et al. 1995) and applied in different frameworks (Motta 1998; Motta et al. 1999). The parametric design task aims at finding of a structure composed of parameterized building blocks that are constructible and that satisfy all the requirements. Borst (1997), who investigated the reuse of engineering ontologies, extends the definition given by Gruber by emphasizing that in order to be useful, an ontology should be reusable and shared across several applications.

To support different applications scenarios at an automotive company, we have developed a set of **engineering ontologies** using SysML (briefly introduced in Sect. 12.2) as inspiration: the *Components*, *Connections*, *Systems*, *Requirements* and *Constraints* ontologies. The engineering ontologies are general and abstract ontologies, to facilitate their reuse in several engineering applications. The ontological commitments are kept to a minimum, which is a desiderata in building ontologies (Gruber 1993). The engineering ontologies form building blocks that can be assembled together to serve the purposes of different applications.

The rest of this section is structured as follows: Sect. 12.3.1 describes the knowledge representation formalism that we chose for the engineering ontologies, Sect. 12.3.2 provides some motivation for the formalism that we chose, and the rest of the section is dedicated to describing briefly the five engineering ontologies. For a detailed description of the ontologies, we refer the reader (Tudorache 2006b). The engineering ontologies can be downloaded from (Tudorache 2006a).

12.3.1 Representing the Engineering Ontologies

There are several knowledge representation formalisms which may be used to specify an ontology, such as frame–based systems, description logics, rule systems, etc. They differ in the expressiveness of the language (for example, some may allow the representation of defaults, while others will not); in the supported inference mechanisms (for example, subsumption reasoning, constraints checking); in the assumptions about the world (closed world vs. open world assumption); they have different tool support, etc. A description of the different ontology representation languages and ontology editors is given in (Gómez-Pérez et al. 2004).

The knowledge representation formalism that we chose for specifying the engineering ontologies in this work is a frame–based formalism, Protégé (Noy et al. 2000), which is compatible with the Open Knowledge–Base Connectivity (OKBC) protocol (Chaudhri et al. 1998). OKBC is a common communication protocol for frame–based systems. Protégé[2] is a free, open source ontology editor and knowledge–base framework developed at the Stanford Center for Biomedical Informatics Research.

In the OKBC and Protégé knowledge model, *frames* are the principal building blocks. A *class* is a set of entities. An entity is said to be an *instance* of a class, and the class is known as the *type* of the instance. An instance may have multiple types.

[2] http://protege.stanford.edu/.

Classes are organized in taxonomic hierarchies formed by the *subclass-of* relationship. Classes are used as templates for building instances.

Slots represent the properties of the entities in the domain. For example, a slot "*weight*" may describe the weight of a person. Slots are first order entities: They are themselves frames and can exist independently of their attachment to classes. In this case they are called *top–level slots*. *Constraints* are attached to the definition of slots that limit the values that the slot may take. The properties of the slots that can be constrained are: the value type (e.g., String, float, boolean, Instance, etc.), cardinality (e.g., single, multiple), minimum and maximum value for numerical slot, and so on. When a top–level slot is attached to a class, it becomes a *template slot* of the class, which is inherited to subclasses. When an instance of a class is created, its template slots are also instantiated and become own slots of the instance and may take values specific for that instance.

Facets describe properties of slots. They are used to define constraints on allowed slot values. Examples of facets are the cardinality of a slot that constrains how many values the slot may have, the range of a slot that constrains the valid type of values for the slot, the minimum and maximum values for a numeric slot, and so on.

The reasoner used in our work is *F*LORA-*2*. *F*LORA-*2*[3] (F-Logic translator) is a declarative object oriented language used for building knowledge intensive applications. It is also an application development platform (Yang et al. 2005). *F*LORA-*2* may be used for ontology management, information integration or software engineering.

*F*LORA-*2* is implemented as a set of XSB[4] libraries that translate a language created by the unification of F-logic, HiLog (Chen et al. 1993) and Transaction Logic into Prolog code. *F*LORA-*2* programs may also include Prolog programs. It provides strong support for modular software through dynamic modules. Other characteristics make *F*LORA-*2* appealing for using it in reasoning with frame–based ontologies, such as support for objects with complex internal structure, class hierarchies and inheritance, typing, and encapsulation.

12.3.2 Why Frames and Not OWL

The current knowledge representation standard used by most ontologies is the Web Ontology Language (OWL), which "is an ontology language for the Semantic Web with formally defined meaning" (World Wide Web Consortium 2012). There are several reasons why we chose to use a Frame formalism and not OWL, which we enumerate below.

[3]http://flora.sourceforge.net/.

[4]http://xsb.sourceforge.net/.

1. **Open World versus Closed World Assumption**. OWL was developed for the Semantic Web and uses the Open World Assumption (OWA), in which, if something is not asserted, then it is considered to be unknown. On the contrary, in the Closed World Assumption (CWA), if something is not known to be true, then it must be false. OWL uses the OWA, which makes sense for the Web. However, for the engineering domain, which operates mostly with models of well–defined structures (usually stored in databases), it is more natural to use the CWA. For example, if an engineer queries whether a certain part is used as a component of a system, she would expect the system to answer *Yes* or *No*, rather than *I don't know*. Due to the OWA, it is also difficult in OWL to query for counts of objects (for example, how many parts are contained in this system). Also, because OWL uses the OWA, it does not make the *Unique Name Assumption* (UNA),[5] i.e., different names always refer to different entities in the world, which is something that the engineers (coming from a database perspective) would expect. For example, if two parts of a system are named differently, an engineer would expect that the system would treat them as different entities. However, in OWL, we would have to explicitly model this assertion, as it is not implicit in the formalism. The Frames formalism uses both the CWA and makes the UNA.
2. **Integrity Constraints** are well known in the database domain. They are used to check for errors, rather than to infer new knowledge. They can be used, for example, to check whether the entries made by engineers conform to certain rules of the domain. For instance, in a system only certain types of ports can be connected together. In OWL, due to its use of the OWA and of it not making the UNA, it is quite difficult to model integrity constraints. There are approaches in the literature that try to bring together OWL with CWA and UNA (Motik and Rosati 2007; Motik et al. 2009; Sirin et al. 2008), although they are not widely adopted. The current work on the Shapes Constraint Language (SHACL) (World Wide Web Consortium 2015) for describing and constraining the contents of RDF graphs is also promising in this direction.
3. **Timing and existing tooling**. Last, but not least, at the time that we performed the work described in this chapter, OWL was just in its infancy and was not a standard, yet. Also, there were barely any tools that would support the editing of OWL ontologies, as they were just being implemented. Also, the company where this work was performed had already invested in a Frames representation and had several other tools that would work with this representation, and plugins to the Protégé editor that would support the modeling patterns used in the engineering ontologies.

In the rest of the section, we describe the five engineering ontologies—the *Components*, *Connections*, *Systems*, *Requirements* and *Constraints* ontologies—that we have built using SysML as inspiration, and that serve as the formal basis of our work.

---

[5]Although there are ways of enforcing the UNA in OWL by using the owl:sameAs and owl:differentFrom constructs.

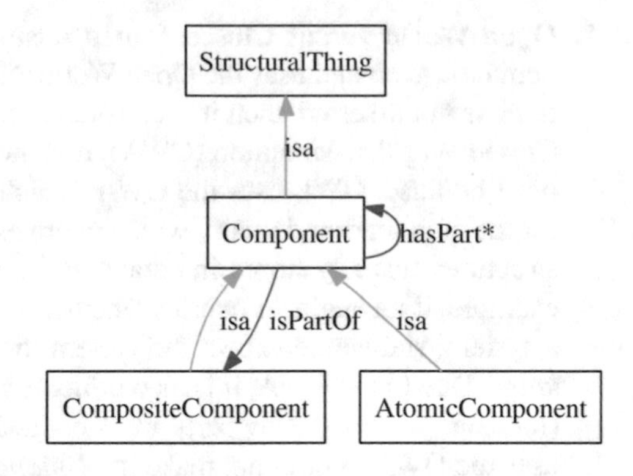

Fig. 12.2 The components taxonomy. A component may be composite or atomic. Composite components may contain other components, whereas atomic components may not

12.3.3 The Components Ontology

The *Components* ontology represents the components and their part–whole decomposition. A component is a fundamental concept that may be used to represent objects from different domains. The ontology contains abstract concepts and is intended to be very general in nature, so that it can be specialized for different domains. In this work, the *Components* ontology has been used to represent engineering systems and their decompositions. However, components do not necessarily represent physical systems. They may also represent abstract things, such as software modules, functional components, or requirements.

A component is a *StructuralThing* that is a general concept encompassing all concepts used to describe the structure of something (Fig. 12.2). A component may be atomic, meaning that it does not have any parts. In this case, we will denote it as *AtomicComponent*. If a component is composed of other parts, it is a *CompositeComponent*. A *CompositeComponent* may contain other *CompositeComponent*-s or *AtomicComponent*-s.

In the example, given in Sect. 12.2, the *Powertrain* is a system that contains two other components, *Engine* and *Transmission*, which may themselves be systems.

hasPart is used to represent the direct parts of a component. For this reason, it is not transitive. However, another transitive predicate *hasSubpart(x,y)* may be used to compute the transitive closure of the parts of a component.

Other properties of the *hasPart* predicate is that it is irreflexive and antisymmetric. An *AtomicComponent* is a component that is not further decomposed, meaning it cannot contain other components. A *CompositeComponent* is a *Component* that has at least one part, i.e., it is not atomic.

In many cases it is desirable to represent also the inverse relationship of a relationship. Inverse relationships are important in the modeling process, for example, they support an easier navigation along relationships (going back and forth); they may be used in consistency checking of the knowledge base, or even in making certain kind of inference more efficient. The inverse relationship for *hasPart* is *isPartOf*.

There are many cases, in which more information about the part relationship needs to be expressed. An example is the case when we want to restrict the cardinality of a certain part of a composite component, such as specifying that a car must have exactly four wheels. The general *hasPart* relationship does not allow representing cardinalities or other constraints for specific classes from its range. We need a more precise representation of the relationships to capture this restriction.

This can be realized in the Protégé knowledge model with subslots. A subslot *hasWheel* of *hasPart* (with domain *Car* and range *Wheel*) can be restricted to have cardinality four at class *Car*. This ensures that an instance of a car must have for the *hasWheel* slot four instances of *Wheel* as values (which will become also values of *hasPart*). Subslots are a mean to represent qualified cardinality restrictions (QCR) in the ontology. We may also place other types of constraints on specific subslots. For example, we may impose a range (min and max) for a value slot, or even default values.

12.3.4 The Connections Ontology

A system is not only defined by the parts it contains, but also by its behavior. The behavior of a system is determined by the interactions of its parts. The interactions are abstracted as connections between the components of a system and form the topology of the system. In engineering systems, the connections between parts represent usually flows of stuff (energy, matter or signals) between the components of the system (Pahl and Beitz 1996).

The components own their ports or connectors and therefore they are part of the specification of the component. Connectors play an important role in the design of systems, since they enable a modular and reusable design.

The *Connections* ontology describes the topology of a system, that is, the way in which components are interconnected with each other. Although it would have been possible to develop the *Connections* ontology independently from the *Components* ontology, we have chosen to include the *Components* ontology in the *Connections* ontology, because the part–whole modeling pattern is essential in the representation of the context of a connection. The main concepts in the ontology are the *Connector*-s, which are the only points through which components can be interconnected, and the *Connection*-s which is a reified concept for describing the actual connection between components. The topological individual which represents anything that can be connected is represented in the ontology by the concept *TopologicalIndividual*.

The ontology contains a general class *TopologicalThing*, that is a superclass of all the classes which are used to describe the topology of a system.
A *TopologicalIndividual* is a subclass of *TopologicalThing* and is the superclass of the classes used for modeling topological entities and relationships.

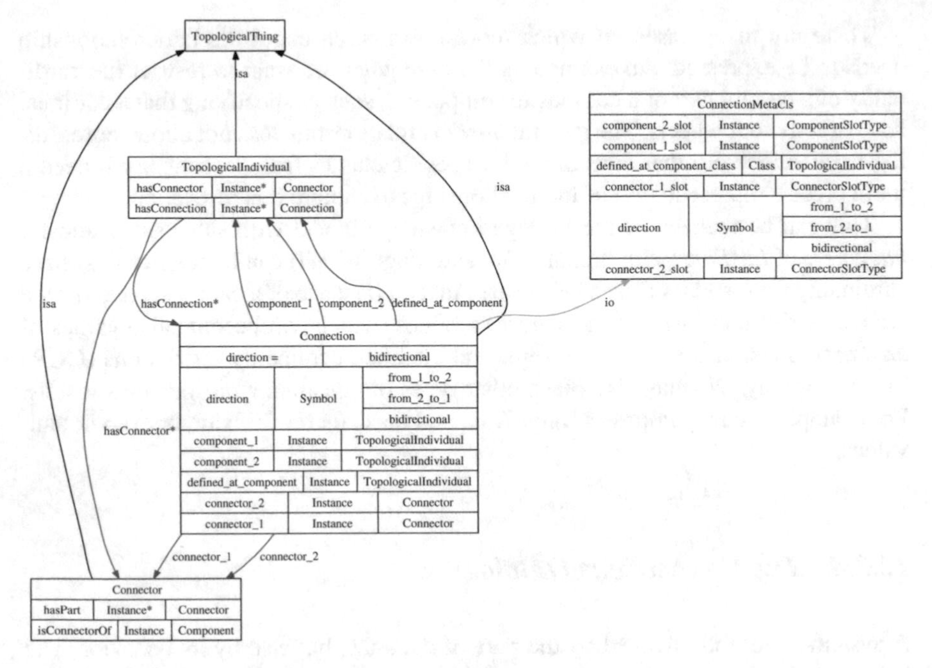

Fig. 12.3 The *Connections* Ontology

The *TopologicalIndividual* is an abstract class (i.e., it cannot have any direct instances), and it is intended to be refined in ontologies that include the *Connections* ontology. For example, the *Engine* class is also a *TopologicalIndividual* (modeled as subclass), since it may have connections to other systems or parts. A graphical representation of the *Connections* ontology is shown in Fig. 12.3.

12.3.5 The Systems Ontology

The *Systems* ontology gives a formal representation of systems and combines concepts from the *Components* and the *Connections* ontologies, shown in Fig. 12.4.

Systems and their parts are topological individuals since they can have connections. This is modeled in the ontology with the subclass relationship, i.e., *System* and *AtomicPart* are subclasses of *TopogicalIndividual*.

The connection between the *Systems* and the *Components* ontologies are given by the fact that systems are composite components, while atomic parts are atomic components.

By combining the *Components* and the *Connections* ontology, systems are defined as composite objects that can have connections to other components, while atomic parts are atomic components that may also be connected to other components.

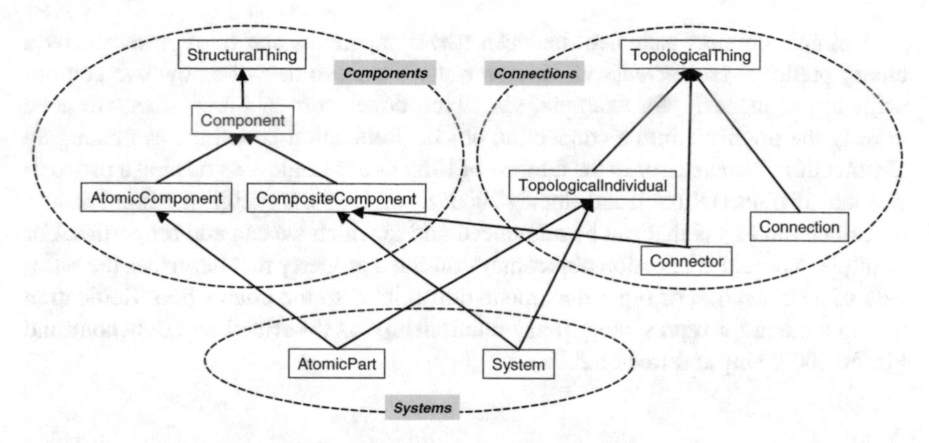

Fig. 12.4 The relationship between the *Components*, *Connections* and *Systems* ontologies

Connectors represent points of interaction between a component and the environment. Connectors are also known as ports in UML2 and SysML. Connectors are owned by the component and are part of the definition of a component. A component can be seen as a black box that exposes a certain behavior and communicates with the outer world (its environment) through the connectors. This type of modeling supports the reuse and modularization of system models: A component can be replaced with another one if it has the same interfaces (types of connectors) and has the same behavior. The internal implementation of the component is of no importance to its environment. The way a component communicates with the environment is only through the connectors.

Connectors are composite components: They may contain other connectors. In this way it is possible to represent more complex connectors, like for example, a serial port that contains several pins. This is realized by making the *Connector* a subclass of the *CompositeComponent* class from the *Components* ontology.

Connectors are attached to a *TopologicalIndividual* through the *hasConnector* slot or one of its subslots. The same modeling pattern as the one used for modeling parts may be used for modeling the attachment of connectors to components. This will bring several advantages: It will be possible to address or reference a certain connector of a component, and also to impose additional constraints on the connector (for example, refine the allowed types of connected components). The domain of the *hasConnector* slot is the *TopologicalIndividual* and the range is *Connector*.

The connectors of components are themselves parts of the components. Hence, the *hasConnector*—that is used to attach a connector to a component—is modeled as a subslot of the *hasPart* slot.

Connectors might be organized in hierarchies of connectors by means of the *is-a* relationship to describe different classification of the connectors. For the engineering domain, the connectors can be classified in mechanical, electrical/electronical and hydraulic connectors. This can be modeled in the ontology by subclassing the *Connector* class with more specific types of connectors.

A simple connect statement between two components can be represented by a binary predicate *connected(x,y)*. However, if we want to describe how two components are connected—for example, via which connectors or ports—then we need to reify the predicate into a connection object. Reification is defined as making an abstract thing concrete, or in the context of First Order Logic—as turning a proposition into an object (Russell and Norvig 2003; Sowa 1999). By reifying a relationship, it becomes an object that can be referenced and to which we can add properties. For example, a reified connection object may contain a property that describes the rationale of a connection or other documentation related to the connection. Reification is also a mean for representing n-ary relationships as described by (Dahchour and Pirotte 2002; Noy and Rector 2006).

12.3.6 The Requirements Ontology

A requirement specifies a capability or condition that the system to be designed must satisfy (SysML 2006). Requirements are the driving force of the development process. They are acquired at the beginning of the development process and are used throughout the entire process in different development stages. Requirements are evolving from the initial phases, when they are very vague, ambiguous, informal to become more precise and formal at the later stages of the development.

Being able to trace the requirements evolution from the initial specification to the design is crucial for an optimal development process (Grabowski et al. 1996; Kim et al. 1999; Lin et al. 1996; Stevens et al. 1998). Another important aspect is the attachment of requirements to systems and their parts.

The requirements are also composite objects, meaning that they may contain other requirements. In order to represent the part–whole relationship between the requirements, the ontology includes the *Components* ontology. The class *Requirement* is modeled as a subclass of the *CompositeComponent* class from the *Components* ontology. In this way it inherits the *hasPart* and *isPartOf* template slots, which are refined at class level to have as range the class *Requirement*. This means that a requirement may be composed of other requirements. For example, a top level requirement object, which might represent the customer specification, may be decomposed in requirements related to performance, ergonomics, safety, etc.

There are different ways in which requirements can be classified. One way to classify requirements is according to their origin, and they may be initial or detailed requirements. Another classification of requirements is according to the aspect of the product that they are constraining. They may be classified in cost, functional, safety, technological, ergonomical, etc. requirements. Different classifications of requirements have been proposed in (Pahl and Beitz 1996; Rzehorz 1998). These categories of requirements are modeled in the ontology as subclasses of the *Requirement* class. Multiple inheritance may be used if a requirement belongs to different categories.

The *is_derived_into* relationship is used to trace the evolution of the requirements. At the first stages of the development, only the customer requirements are specified. They are then further refined in different types of requirements. For this case, it is said that the latter requirements are derived from the source requirements. Usually, the *is_derived_into* relationship links different levels of abstraction between the requirements. The *is_derived_from* is the inverse relationship of *is_derived_into*.

An essential relationship is the *is_fulfilled_by*, which is used to link a requirement to an object that fulfills it. In case of systems design, the relationship has as range a system or part of a system that fulfills the requirement. For example, the minimal braking requirement is fulfilled by the braking subsystem of a car.

12.3.7 The Constraints Ontology

Constraints are an important part of design models. They may be used to represent mathematical relationships between model elements. For example, a constraint might specify that the weight of a car should be less than 1500 kg. Another type of constraint can be used to model the physical behavior of a system, for example, Newton's second law of motion, $F = m \cdot a$. They may also be employed to model costs, or for optimization purposes. Constrains restrict the solution space and are related to the quality of the product.

In our case, constraints are attached at class level and applied at instance level. They may be used to constrain the properties of different type of classes. For example, they are attached to requirements and represent the mathematical translation of the requirement (if possible).

The constraints are modeled in the *Constraints* ontology. A constraint instance contains a constraint statement given as a mathematical relation, e.g., $w < 1500$. It also contains variable declarations that link the variables used in the constraint statement to elements of the design model. For instance, the variable w corresponds to the weight property of the car class. Variables are represented using paths of roles. The declaration of the variable w would be: *Car.weight*. There is no restriction on the length of the variable paths, which are made along the relationships (i.e., slots) defined in the model.

A constraint has also attached to it a documentation that describes in natural language the content of the constraint. It might also contain a relationship to the element to which the constraint has been attached. This is important for the interpretation of the variable paths. For example, if a constraint is directly attached to a *Car* class that has a *weight* property, then the declaration of the w variable can be just *weight* because the constraint is interpreted in the context of the *Car* class and the *weight* property is local to the *Car* class.

The constraint statement has been modeled as a simple String (e.g., "$w < 1500$"), which may leave room for mistakes in creating the constraints. Gruber (Gruber and Olsen 1994) developed an ontology for mathematical modeling in engineering that could be used to overcome this shortcoming.

The variables of a constraint are declared as paths along the relationships in the model. A general declaration of a variable has the form:

$$v := Class.Rel_1[Restr_1].Rel_2[Restr_2] \ldots Rel_n[Restr_n].property \quad (12.1)$$

The definition of the path starts with a *Class* that specifies the context for the constraint. For example, the *Class* might be *Car*. Rel_i represent the relationships (i.e., slots) along which the path is defined. And $Restr_i$ represents a class restriction on the path that is applied to Rel_i. The definition of the path ends with a property to which the variable is bound.

As we have defined the five engineering ontologies, we will show in the following two sections how we used them in two real–world use cases.

12.4 Use Case 1: Stepwise Refinement of Design Requirements

This section describes the first use case that shows how we used ontologies to improve the requirements engineering process to reach an optimal design. This work has been implemented in a research project, Specification Driven Design (SDD), in an automotive company over a period of 3 years. The goal of the project was to develop methods and tools that support the engineer in deriving an optimized design layout starting from a system specification, which is refined in iterative design steps. Checking and maintaining the consistency of the design layout throughout the design iterations was one of the main tasks in the iterative design process.

Requirements are the driving force of the development process and play an essential role in all development phases. After analyzing the customer requirements, a system specification is generated that is more detailed and structured than the user requirements. Out of the systems specification, a design layout is developed, which is optimized in successive refinements steps until a design solution is reached that fulfills all requirements.

One of the greatest challenges that must be confronted in this process is that requirements are usually stored as documents without any kind of formal representation. Design model parameters are constrained based on the requirements. It is a very hard task for the engineers to keep track of the dependencies between different parameters of the design model and to asses their consistency. The typical way of handling complex requirements and model parameters is to use Excel sheets, which contain formulas over the parameters. In this way, it is possible to compute the values of some model parameters. However, this approach has serious limitations. If the parameters involved in different equations have circular dependencies, they cannot be computed in Excel. Another limitation is that it is impossible to track the history of design which is given by the successive refinement of parameter values.

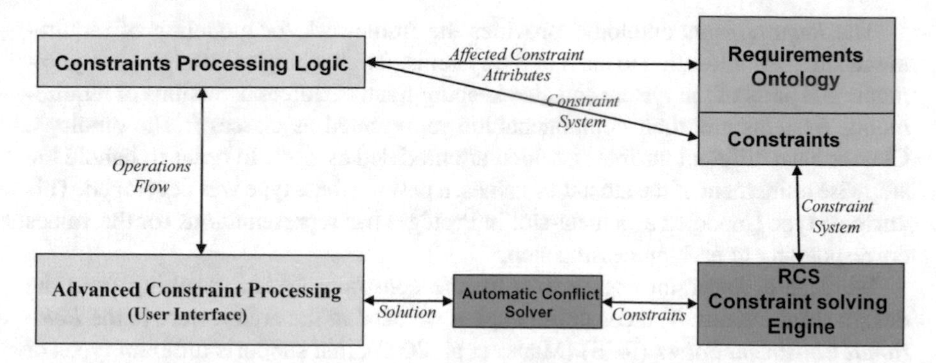

Fig. 12.5 The architecture of the requirement management system

The scenario that we will use to explain the application that we built is in the context of the layout phase in the development of an engine. In the layout phase, the structure of the engine is already known, but optimizations of the parameters of the components are still performed. The goal is to obtain a design that is optimal with respect to some cost function and that does not violate any constraints.

In the layout scenario, the geometry of the motor has to be improved such that the compression ratio[6] of the engine will be increased. The compression ratio depends on a multitude of geometrical parameters of the engine.

Our task was to develop a requirement management system that will support:

- The stepwise refinement of the design parameters,
- The consistency checking of the design based on the defined constraints,
- In case that the design is inconsistent, meaning that at least two constraints are in conflict, the application should support the automatic solving of the conflict based on some cost function of the constraints,
- Support the allocation of requirements to parts and the tracking the requirements' history, and,
- Support for reuse of parts and requirements based on libraries.

12.4.1 The Requirements Management System

The main components of the Requirement Management System and their interdependencies are shown in Fig. 12.5.

[6]The compression ratio is a very important parameter of the engine that describes the efficiency of the engine. The greater the value, the better the engine.

The *Requirements* ontology[7] provides the framework for modeling of requirements, for their allocation to parts of the systems, for attaching constraints to requirements and parts of the system and also keeping track of different versions of requirements. Systems and their components are represented as classes in the ontology. Classes have different attributes, which are modeled as slots. In order to handle the stepwise refinement of the attributes values, a new attribute type was developed. This attribute type (modeled as a meta-slot in Protégé) has representations for the values corresponding to each processing step.

We used a constraint engine to solve the constraint system resulting from the design requirements. The constraint engine we used in the architecture is the *Relational Constraint Solver* (RCS) (Mauss et al. 2002), that supports different types of constraints:

- linear equations and inequations, such as $2x + 3y < 0$. The coefficients may also be intervals with open or closed boundaries, e.g., $[2.8, 3.2)$.
- multivariate polynomial, trigonometric, and other non–linear interval equations, such as $x = sin(y)$, $x = sqrt(y)$.
- assignments of symbols to variables, such as $x = blue$
- disjunctions (*or*) and conjunctions (*and*) of the above relations, such as $x = red \vee x = blue$

The *Constraint Processing Logic* is responsible for the stepwise refinement of the design parameters. The *Automatic Conflict Solver* takes a conflict between two or more constraints as an input and generates a list of possible solution ordered by a cost function.

12.4.2 The User Interface

The graphical interface of the requirement management system (Fig. 12.6) has been implemented as a plug-in for Protégé[8] and provides support for:

- Viewing the attributes that are involved in a constraint system,
- Viewing the constraint system,
- Viewing the conflicting constraints,
- Editing the input values for any processing step and attribute, and,
- Inconsistency detection and conflict solving.

[7]The *Requirements* ontology presented in Sect. 12.3.6 has been developed based on the experiences gained from this project. The ontology used here contains also the definitions of components and constraints.

[8]http://protege.stanford.edu.

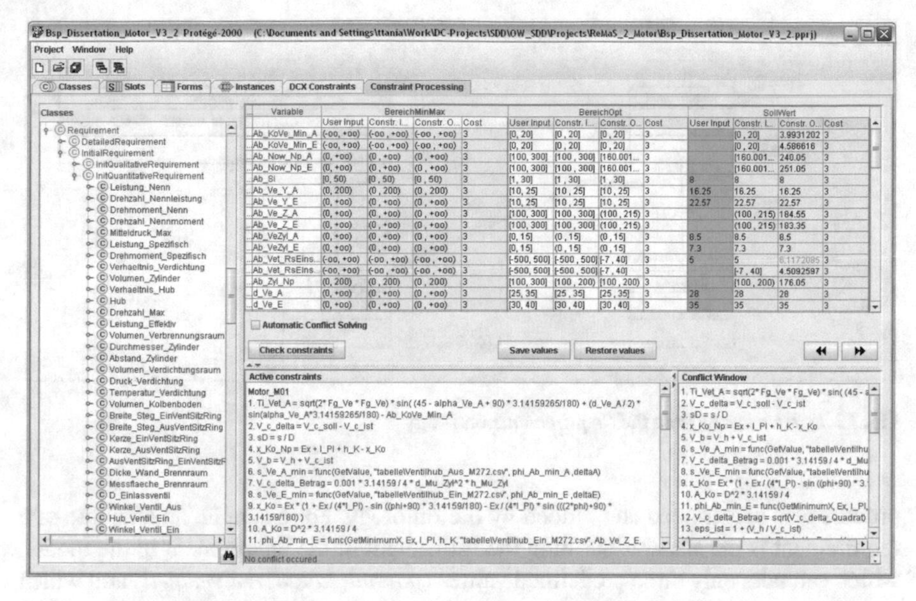

Fig. 12.6 The GUI of the requirement management system. The *left* part of the display shows the Requirements class hierarchy. The *upper* part shows a list of all variables involved in the constraint system. Each variable is associated to an attribute of a class. The *bottom–left* window shows a list of the constraints in the constraint system. The *bottom–right* panel shows a list of the constraints which are in conflict after executing the consistency checking

12.4.3 The Requirements Ontology in SDD

The *Requirements* ontology is used to model the requirements and their interrelationships, such as *hasPart*—for the representation of the requirements decomposition—or *is_derived_into*—to represent the evolution of requirements. An excerpt from the *Requirements* ontology is shown in Fig. 12.7. Constraints may be attached to requirements or to specific parts of a system, modeled as *Component*s.

The ontology supports the modeling of requirements and component templates by representing them as classes. A specific requirement can be modeled as an instance of a requirement class or as a subclass. The *stepValue* attribute attached to the *Component* class is a complex attribute that stores the values of a simple attribute in different processing steps. In this project, each attribute (represented as *stepValue*) was described by three values: *BereichMinMax*—the minimum and maximum values for this attribute, *BereichOpt*—the optimal interval of this attribute, and *CatiaInput*—the actual value (not interval) that goes into the CAD system (in this case, Catia was used). The constraint system would first try to solve the system using the more generic values and would refine the values in consequent refinement steps.

By modeling requirements, components, constraints and their relationships explicitly, the ontology improved substantially the correctness of models. The engineers were not allowed to enter incorrect information, because the knowledge acquisition

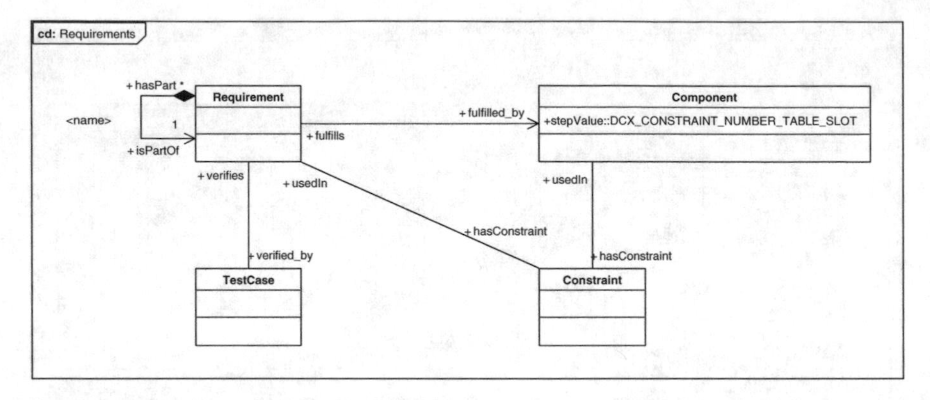

Fig. 12.7 An excerpt from the *Requirements* ontology

process was constrained and guided by the ontology. For example, an attribute of a requirement is the *status* describing the fulfillment of the requirement in the model, which can take only three predefined values (*passed*, *failed*, *not verified*) and which must have a value. When editing the requirements of a component, the engineer had to enter one of the allowed values for the *status* attribute. He was prevented from entering an invalid value for that attribute.

Managing the complex network of requirements and model parameters using this approach has proved to be much easier than it was with the existing approaches. Once the constraints were represented explicitly, the consistency of the model could be checked in an automated way by the constraint engine.

12.4.4 The Constraint Processing Logic

The constraint processing logic defines the flow of operations and data and describes how the information in one processing step affects the information in the next processing step. A rule is that each processing step is more restrictive than its predecessor. Even more, the *actual* input value of a step is the intersection of the user input value for that step and the output value of the predecessor step.

This can be formally described by:

$$constrInputValue_i = constrOutputValue_{i-1} \cap userInputValue_i$$

In this way, it is ensured that a processing step is more restrictive than its predecessor.

Each constraint attribute has a state that depends on which processing step is processed at a certain time. This is implemented by a counter that shows what processing step is currently active. For example, if the user is working in the processing step *BereichOpt*, then the current counter is 2, because *BereichOpt* is the second step of processing. After completing step two, the processing logic activates the next processing step (the one with *orderNr = 3*).

The representation of the constraint attribute type is very flexible, so that it is straightforward to add a new processing step, a processing step property, or to change the name of a processing step.

12.4.5 The Automatic Conflict Solving

If the constraint system defined in the ontology is consistent, then a design solution for a particular processing step is obtained. However, if the resulting system of equations is inconsistent, then we have to relax (i.e., remove) some constraints in order to obtain a consistent state of the system.

The process of choosing what constraint to relax can be done manually by trying different alternatives until a solution is found. However, this process is very tiring and complex. Even if a constraint is relaxed, it is not guaranteed that the new constraint system is consistent.

In order to solve this problem, we developed an algorithm for the automatic conflict solving. This algorithm provides the user with several conflict solving alternatives. It returns a list of constraints that can be relaxed in order to obtain a consistent system, sorted by the values of a cost function. The algorithm may also be optimized by different factors. The full algorithm is presented the appendix of (Tudorache 2006b).

12.4.6 The SDD Application at Runtime

We developed an ontology with the structural representation of an engine, which also contained all the qualitative requirements represented as mathematical constraints (see Sect. 12.4.1). The engine was composed of 137 parts, which were constrained by 71 constraints containing 279 variables. The variables were linked to specific parts using the path representation mechanism described in Sect. 12.3.7. These constraints varied from linear equations and inequations to multivariate polynomial, trigonometric, and other non–linear interval equations, as well as custom functions to read-in values from Excel spreadsheets. We used three steps for the iterative constraint processing: we started with interval values for most variables, which were then iteratively optimized, and in the thrid step the optimal concrete value was chosen. The constraint solver (RCS) was efficient and solved the constraint system for one processing step in 2–3 s.

12.4.7 *Benefits of an Ontology–Based Approach*

We applied the ontology–based Requirements Management System in different scenarios, and we were able to successfully find optimal design solutions in each of them. One of the advantages of using an ontology–based approach is that it **enabled reusing previous knowledge**. We defined a library of components and requirement templates that we were able to instantiate in different design problems.

We saw another benefit coming from the structured information acquisition that **prevented the engineer from entering incorrect information** in a design model. The definitions of component templates are represented as a classes in the ontology. The classes serve as templates for building instances that represent a concrete design model. Therefore, all constraints defined in a template component (at class level) are checked for a concrete design model (at instance level). An intelligent model editing environment can interpret the constraints defined for the template components and prevent invalid input in a concrete design model.

Enabling consistency checking in iterative steps and documenting the history of parameter values was a huge improvement over the existing Excel–based solutions. The constraint system contained several complex constraints involving the parameters of the design model, which were impossible to manage in Excel. By using the ontology–based approach, we were able to **reduce the complexity** of the product, and to **make it more manageable**.

12.5 Use Case 2: Mapping and Change Propagation between Engineering Models

The second use case comes from a research project at an automotive company related to the development of an automatic transmission gearbox of a car. The automatic transmission is a complex mechatronic component that contains mechanical, hydraulic, electrical and software parts, that need to interoperate in order to achieve the overall functionality of the product.

The development of the automatic transmission is made in a parallel fashion by different engineering teams. One team of engineers develops the geometrical model of the gearbox and another team develops the functional model of the same gearbox. The developed models are different in several ways. First, they are represented in different modeling languages. Second, they have different conceptualizations of the gearbox: One models the geometrical characteristics of the product, while the other represents the gearbox as a composition of functional blocks coupled together. However, there are certain correspondences that interrelate the two design models. For instance, a parameter from the functional model, *ratio* of the gearbox is computed as a function of two parameters in the geometrical model. In the following, we will use the keyword *viewpoint* to refer to the geometrical or functional development branch.

The project required that a framework should be implemented that will support the consistency checking between the two design models, as well as, the change propagation from one model to another.

The full details about the implementation of this project can be found in Chaps. 5 and 6 of (Tudorache 2006b).

12.5.1 Mapping Between Libraries of Components

The engineers build the functional and geometrical models using model libraries from their respective tools. Therefore, a model is composed by instantiating template components from the model libraries and by interrelating them. Figure 12.8 shows this situation.

The libraries and the design models are represented in different modeling languages. The names used for corresponding components in the two model libraries are different. The CAD library uses German names, whereas the functional library uses English names for the components.

However, the design models built from these component libraries still have one thing in common: They represent the same product from different perspectives. As a consequence, there will be correspondences between the structures and parameters of the models in the two libraries, shown in Fig. 12.8. We defined the correspondences, also known as, mappings, between components of the library. In this way it is possible to reuse them between all instances of corresponding templates. To map an entire system, we need to be able to compose the mappings between components.

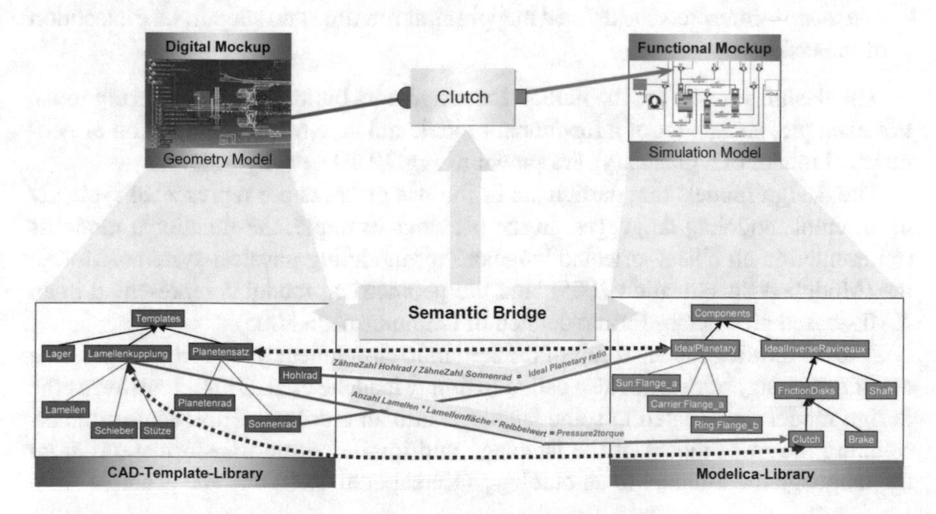

Fig. 12.8 Mappings between component libraries in two viewpoints

We developed a mapping framework, described in the following section, which supports the following operations:

- The definition of mappings between library components in different viewpoints,
- Reuse of template mappings for different instantiations of the design models,
- Computing the mapping of systems based on the mappings of their parts,
- Consistency checking between design models from different viewpoints based on the predefined mappings.

12.5.2 The Mapping Framework

The mapping framework is used to map between different representations of the viewpoints of a product. The mappings can be used to check the consistency of the different viewpoints of a product, or to propagate changes from one viewpoint to the other.

The components of the mapping framework are (Fig. 12.9):

- *Design models*—developed by engineers in different engineering tools (e.g., Catia, Modelica, etc.),
- *Local ontologies*—represent the semantically enriched models of the design models,
- *Engineering ontologies*—used as a common upper ontology for the local ontologies,
- *Mapping ontology*—used to represent declaratively the mappings (correspondences) between the local ontologies,
- *Reasoner*—interprets the defined mappings at run time and supports the execution of the tasks.

The **design models** are the models that engineers build in the engineering tools. For example, an excerpt of a functional model, and an XML representation of geometrical model of a planetary[9] are shown in Fig. 12.10 side–by–side.

The design models that participate in the design tasks are represented typically in different modeling languages. In the previous example, the functional model is represented in an object–oriented language for modeling physical systems, Modelica (Modelica Specification 2005), and the geometrical model is represented in an XML–based proprietary format defined in (Zimmermann 2005).

A **local ontology** is the result of the semantic enrichment of a design model. The act of enriching, sometimes also called lifting (Maedche et al. 2002), transforms the design model represented in some language into an ontology. This requires understanding the meta-model of the language and implies some transformations from the language meta-model to an ontology (Karsai et al. 2003). In the example from

[9]Planetaries are gear–wheels used in gearboxes to change the gears in a car.

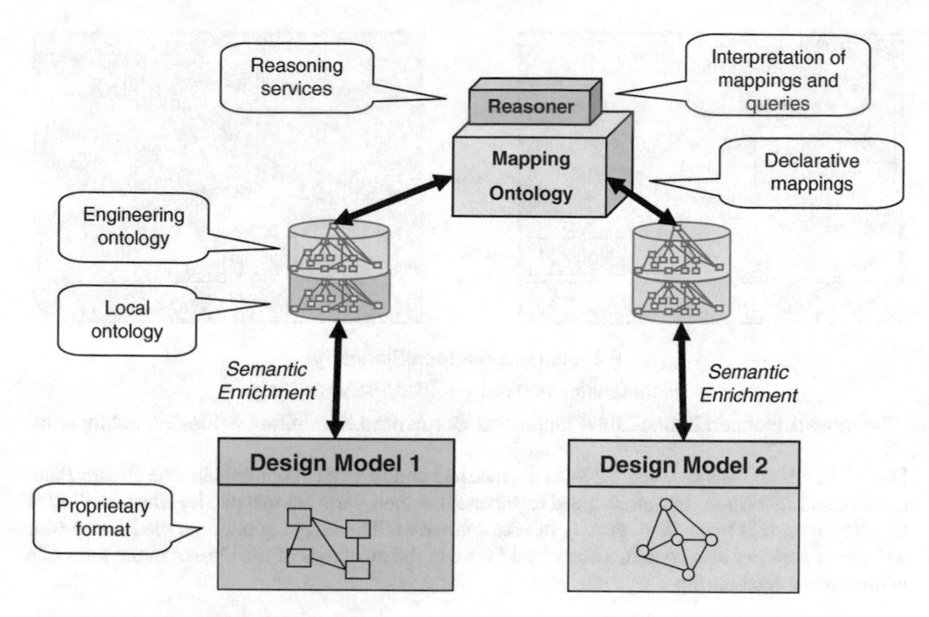

Fig. 12.9 The mapping framework

```
<Class>
   <ClassType>EOclass</ClassType>
        <Identity>
            <FullID>Planetensatz</FullID>
        </Identity>
        <SimpleParameter>
            <Type>float</Type>
            <Identity>
                 <FullID>traegheit</FullID>
            </Identity>
            <PhysicalUnit>kg*mm2</PhysicalUnit>
        </SimpleParameter>
...
</Class>
```

Geometrical model in proprietary format

```
model IdealPlanetary "Ideal planetary gear box"
  parameter Real ratio=100/50
...
  Flange_a sun "sun flange"
  Flange_a carrier "carrier flange"
  Flange_b ring "ring flange"

equation
  (1 + ratio)*carrier.phi = sun.phi + ratio*ring.phi;
...
end IdealPlanetary;
```

Functional model in Modelica language

Fig. 12.10 Excerpts from two design models in different modeling languages side–by–side

Fig. 12.10, the geometrical representation of the planetary uses a XML element—*SimpleParameter*—to represent the definition of a parameter of type float of the planetary.[10] This would be represented in the local ontology using a template slot of type float. The semantic enrichment can be done in different ways: either by annotating the design model elements with the semantic concepts from the ontology, similar to the approach taken by (Wache 2003), or by transforming parts of the design model in classes and instances in the local ontology. We have chosen the second approach in order to allow logical reasoning to be preformed on the concepts in the ontology.

The ***Engineering*** **Ontologies** are used as a common upper level vocabulary for the local ontologies. As in the hybrid integration approach, described in (Wache et al. 2001), the upper level ontologies ensure that the local ontologies share a set

[10] "Planetensatz" (in German) means planetary.

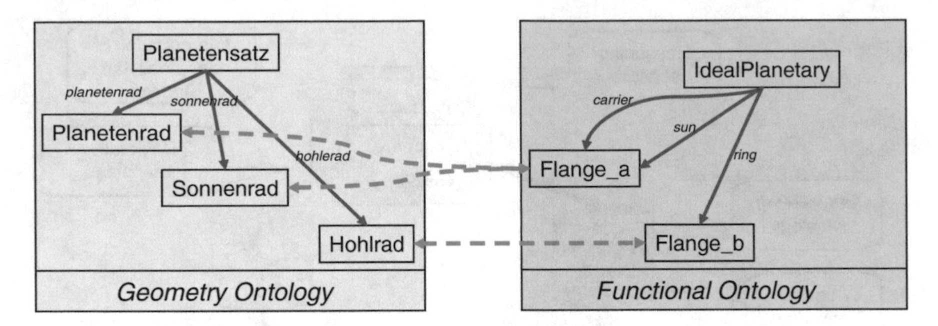

Planetensatz <-> IdealPlanetary

Planetensatz.hohlrad <-> IdealPlanetary.ring

Plantensatz.Hohlrad.ZähneZahl /Plantensatz.Sonnenrad.ZähneZahl = IdealPlanetary.ratio

Fig. 12.11 Mappings between the local geometrical and functional ontologies. The classes *Planetensatz* and *IdealPlanetary* are mapped together. Also their parts are mapped together, *hohlrad* of the *Planetensatz* is mapped to the *ring* of *IdealPlanetary*. The mapping between the *Planetensatz* and *IdealPlanetary* also contain a constraint between the attributes of the classes in the form of a mathematical relationship

of concepts with the same semantics. The local ontologies specialize the upper level concepts. This also ensures that the local ontologies will be comparable with each other. This brings many benefits if mappings between the local ontologies need to be discovered automatically.

The **Mapping Ontology** is used to represent in a declarative way the correspondences between the local ontologies. Having an explicit and well–defined representation of mappings enable reasoning about them, such as, checking whether two mappings are equivalent or mapping composition (Madhavan et al. 2002).

An explicit representation of mappings makes also possible to model different types of mappings, such as, renaming mappings, lexical mappings, recursive mappings, and so on. Park et al. (1998) propose a classification of mapping types. However, the mapping ontology must be designed in such a way that it supports the tasks that must be solved. For instance, mappings between ontologies of engineering systems have to take into consideration the part–whole decomposition of systems and hence they have to support the representation of paths in the part–whole hierarchy.

An example of the mappings between the functional and geometrical ontologies in the previous example is shown in Fig. 12.11.

The **reasoner** is an important part of the mapping framework. The reasoner is used both at design–time of the mappings and at runtime to support the execution of the design tasks. At design time, the reasoner may be used to verify the mappings and to see their effect in a testing environment. It can also provide suggestions for other mappings based on the already defined ones, or warn about missing mappings. At runtime, the reasoner executes the mappings and other operations, such as consistency checking, that are relevant for a particular task.

The reasoner used in the framework is $\mathcal{F}$LORA-2, described briefly in Sect. 12.3.1.

The mapping approach described in this section shares similarities with the Engineering Knowledge Base approach, introduced in Chap. 4 of this book. We also refer the reader to Chap. 6 for an in–depth discussion of mapping approaches, as well as to the following papers (Biffl et al. 2014; Kovalenko et al. 2013; Moser et al. 2011).

12.5.3 Defining the Mappings

In order to define the mappings, the design models in the two viewpoints have to be semantically enriched, as described in the previous section. The enrichment is done by exporting the data from the design models to instances of the local ontologies. The local ontologies already contain the class definitions corresponding to the template components in the library of models in the two viewpoints.

The local ontologies corresponding to the template libraries both include the *Components* ontology presented in Sect. 12.3. In this way, a common vocabulary for the two viewpoints is defined, which already provides a starting point for finding the correspondences between the ontologies. For example, a *Component* in one ontology is typically mapped to another *Component* in the other ontology. The *Plantensatz*[11] class in the geometrical viewpoint is mapped to the *IdealPlanetary* in the Functional Ontology.

The mappings are interrelating components from the two ontologies. Some examples are shown in the previous section. The full mapping algorithm used to compose the mappings between the components in order to map the top–level systems, such as the *Getriebe* and *Gearbox*, is described in detail in Chap. 5 of (Tudorache 2006b).

12.5.4 Consistency Checking and Change Propagation

There are different types of mappings that can hold between different design models. Some of them are simply structural (one component maps to another), but others carry more information. It is common that there are mathematical constraints that interrelate the parameters of the different models.

We modeled the mathematical relationships between the parameters as constraints attached to the mappings between concepts in the ontologies. For example, the *ratio* of the *IdealPlanetary* of the functional model is computed out of two parameters of the parts of the corresponding class *Planetensatz* of the geometrical model. The *Planetensatz* has as parts a *Sonnenrad* and a *Hohlrad* which have each defined a number of teeth for the gears (in German, *ZaehnenZahl*). The constraint between the ratio and the number of teeth of the gears is:

$$IdealPlanetary.ratio = \frac{Planetensatz.hohlrad.Zaehnenzahl}{Planetensatz.sonnenrad.Zaehnenzahl} \tag{12.2}$$

[11] *Planetensatz* in German means *Planetary* in English.

The constraint is defined in the context of a mapping and it is checked whenever the mapping is used. For example, this constraint is checked three times in the mapping between the *Getriebe* and *Gearbox* systems, because they contain three corresponding planetaries.

In order to check the consistency of two design models (e.g., geometrical and functional), we have performed automatically the following steps:

- Convert the local models and their instances (containing the actual parameters of a design model) into a Frame–logic representation,
- Convert the mapping ontology and its instantiation (containing the actual mappings between two design models) into a Frame–logic representation, and,
- Execute in $\mathcal{F}$LORA-*2* a Frame–logic query giving as input the two conversions from above and the top–down mapping–composition algorithm, which will return whether the two design models are consistent with each other.

In a similar way, we can use other predicates we have defined to do the change propagation. In the example from above, if the *IdealPlanetary.ratio* is not set in the functional model, $\mathcal{F}$LORA-*2* is able to compute it based on the mappings and the attached mathematical constraints.

12.5.5 Benefits of an Ontology–Based Approach

One of the benefits we have found by using ontologies in this project was the fact that the correspondences between the design models were described explicitly and formally, which enabled us to **automatically check the consistency between the two models**. Once the framework was implemented, the consistency checking and change propagation could happen at the push of a button, which was previously not possible. Also, the explicit representation of the mappings would make clear to the engineers from both teams what are the **dependencies in the other models**, and **how a change in their model might affect the corresponding model**. Such understanding is not to be taken for granted, as teams are often distributed, and are specialized in certain domains, and do not have a full understanding of other aspects of the design.

The fact that we defined the mappings at the library level also allowed a **better management and building of the models**: the mappings are defined once, and can be used in all the instantiations of the library components. Even though, there was a lot of effort involved in the initial phases, the subsequent mappings of other systems has been much eased, as we could reuse a big part of the existing mappings.

The ontology–based approach for modeling the functional and geometrical design models allowed to **define meta-rules that constrain correct models, and which can be checked at model building time**. For example, certain types of components cannot be connected together in a functional model. This type of constraint can be easily formulated as an axiom in the functional ontology.

The ontology–based representation also **improved the model management process**. For instance, it was straightforward to search for components in a model library that have a certain type of connectors by using a simple query. The design models can also be enriched with other types of semantic relationships that improve the documentation of the model. Different types of meta-data can be attached to components in the ontologies, such as, provenance, version, release status, etc., that plan an important role in the development process.

12.6 Conclusion

We have presented five generic engineering ontologies (*Components*, *Connections*, *Systems*, *Requirements* and *Constraints*), and we have shown how we have used them in two real–world projects from the automotive industry. The first use case focused on the stepwise refinement of requirements, and how an optimal design solution was achieved by using ontologies and a relational constraint solver. The second use case showed how we solved an integration problem, how we were able to check the consistency of geometrical and functional models developed in specific tools using ontologies and a Frame–logic reasoner. Following the description of each use case, we have also talked about the perceived benefits of using an ontology–based approach.

To summarize, we found several benefits of using ontologies in the engineering domain. (1) By building generic and modular ontologies that we could reuse, we were able to reduce the effort necessary to build new engineering models. (2) The formal representation of the structure and constraints of these models allowed us to automatically perform checks, both at model–authoring time, as well as post–priori. This capability allowed us to capture modeling errors earlier in the process, and hence reduce design costs. (3) The formal representation of the dependencies among different types of models of the same product enabled us to automatically check the consistency among different models, as well as to propagate changes from one model to others.

In our approach, we also encountered several challenges. For example, representing and checking the mathematical dependencies in the current ontology representation languages is difficult, and hardly any automated reasoner s deal with this aspect. In our approach, we were able to use the mathematical support available in $\mathcal{F}$LORA-2, but a complete solution might need to use a hybrid approach, in which different specialized reasoner s deal with different aspects of a model. For example, a DL–reasoner might handle the consistency check and classification tasks, while a constraint solver, or other mathematical algorithms deal with the numerical calculations in the model. Another challenge when modeling engineering systems that are closed in nature is the Open World and Unique Name Assumptions in OWL, which we discussed in Sect. 12.3.2.

While research is under way to address the current challenges, there are already numerous advantages of using ontologies in the engineering domain, as demonstrated by their wider adoption in industry.

References

Alani, L.I.: Template-basierte Erfassung von Produktanforderungen in einem CAD system. Ph.D. thesis, Technical University of Berlin, Germany (2007). http://goo.gl/pV4pSI

Biffl, S., Kovalenko, O., Lüder, A., Schmidt, N., Rosendahl, R.: Semantic mapping support in automationml. In: Proceedings of the 2014 IEEE Emerging Technology and Factory Automation, ETFA 2014, pp. 1–4. Barcelona, Spain, 16–19 Sept 2014. doi:10.1109/ETFA.2014.7005276

Borst, P.: Construction of engineering ontologies for knowledge sharing and reuse. Ph.D. thesis, Universiteit Twente, Sept 1997

Breuker, J., van de Velde, W. (eds.): Common KADS Library for Expertise Modelling. Reusable Problem Solving Components. Frontiers in Artificial Intelligence and Applications, vol. 21. IOS Press, Amsterdam, Oct 1994

Chaudhri, V.K., Farquhar, A., Fikes, R.E., Karp, P.D., Rice, J.P.: Open Knowledge Base Connectivity 2.0.3, 9th Apr 1998

Chen, W., Kifer, M., Warren, D.S.: HILOG: a foundation for higher-order logic programming. J. Logic Program. **15**(3), 187–230 (1993). http://citeseer.ist.psu.edu/chen89hilog.html

Dahchour, M., Pirotte, A.: The semantics of reifying N-ary relationships as classes. In: ICEIS, pp. 580–586 (2002)

Falkenhainer, B., Forbus, K.D.: Compositional modeling: finding the right model for the job. Artif. Intell. **51**(1–3), 95–143, Oct 1991. http://citeseer.ist.psu.edu/falkenhainer91compositional.html

Gómez-Pérez, A., Fernández-López, M., Corcho, O.: Ontological Engineering. Advanced Information and Knowledge Processing, 1st edn. Springer (2004)

Grabowski, H., Rude, S., Gebauer, M., Rzehorz, C.: Modelling of requirements: The key for cooperative product development. In: Flexible Automation and Intelligent Manufacturing 1996, (Proceedings of the 6th International FAIM Conference May 13–15, 1996 Atlanta, Georgia USA), pp. 382–389, (1996)

Gruber, T.R.: Towards principles for the design of ontologies used for knowledge sharing. In: Guarino, N., Poli, R. (eds.) Formal Ontology in Conceptual Analysis and Knowledge Representation. Kluwer Academic Publishers, Deventer (1993). http://citeseer.ist.psu.edu/gruber93toward.html

Gruber, T.R., Olsen, G.R.: An ontology for engineering mathematics. In: Doyle, J., Sandewall, E.J., Torasso, P. (eds.) Proceedings of the Fourth International Conference on Principles of Knowledge Representation (KR'94), pp. 258–269. Morgan Kaufmann Publishers (1994)

ISO/IEC 15288. ISO/IEC 15288:2002 Systems engineering—System life cycle processes. International Organization for Standardization (ISO)/International Electrotechnical Commission (IEC) (2002)

Karsai, G., Láng, A., Neema, S.K.: Tool integration patterns. In: Proceedings of the Workshop on Tool Integration in System Development (TIS 2003) at ESEC/FSE 2003, pp. 33–38. Helsinki, Finland, Sept 2003

Kim, H.M., Fox, M.S., Gruninger, M.: An ontology for quality management—enabling quality problem identification and tracing. BT Technol. J. **17**(4), 131–140 (1999)

Kovalenko, O., Debruyne, C., Serral, E., Biffl, S.: Evaluation of technologies for mapping representation in ontologies. In: On the Move to Meaningful Internet Systems: OTM 2013 Conferences—Proceedings of the Confederated International Conferences: CoopIS, DOA-Trusted Cloud, and ODBASE 2013, pp. 564–571. Graz, Austria, Sept 9–13, 2013. doi:10.1007/978-3-642-41030-7_41

Lin, J., Fox, M.S., Bilgic, T.: A requirement ontology for engineering design. Concurr. Eng.: Res. Appl. **4**(4), 279–291 (1996)

Madhavan, J., Bernstein, P.A., Domingos, P., Halevy, A.Y.: Representing and reasoning about mappings between domain models. In: Dechter, R., Kearns, M.J., Sutton, R.S. (eds.) Eighteenth National Conference on Artificial Intelligence, pp. 80–86, July 2002

Maedche, A., Motik, B., Silva, N., Volz, R.: MAFRA—a mapping framework for distributed ontologies. In: Gómez-Pérez, A., Benjamins, V.R. (eds.) Knowledge Engineering and Knowledge Management. Ontologies and the Semantic Web, Proceedings of 13th International Conference, EKAW 2002, Siguenza, Spain, October 1–4, 2002, Lecture Notes in Computer Science, vol. 2473, pp. 235–250. Springer (2002)

Mauss, J., Seelisch, F., Tatar, M.M.: A relational constraint solver for model-based engineering. In: CP, pp. 696–701 (2002)

Modelica Specification. Modelica—A Unified Object-Oriented Language for Physical Systems Modeling. Language Specification. Modelica Association, February 2005. http://www.modelica.org/documents/ModelicaSpec22.pdf

Moser, T., Mordinyi, R., Winkler, D., Melik-Merkumians, M., Biffl, S.: Efficient automation systems engineering process support based on semantic integration of engineering knowledge. In: IEEE 16th Conference on Emerging Technologies & Factory Automation, ETFA 2011, pp. 1–8. Toulouse, France, 5–9 Sept 2011. doi:10.1109/ETFA.2011.6059098

Motik, B., Rosati, R.: A faithful integration of description logics with logic programming. IJCAI **7**, 477–482 (2007)

Motik, B., Horrocks, I., Sattler, U.: Bridging the gap between OWL and relational databases. Web Semant.: Sci. Serv. Agents World Wide Web **7**(2), 74–89 (2009)

Motta, E.: Reusable components for knowledge models. Ph.D. thesis, Knowledge Media Institute, The Open University, Milton Keynes, UK (1998)

Motta, E., Fensel, D., Gaspari, M., Benjamins, V.R.: Specifications of knowledge components for reuse. In: 11th International Conference on Software Engineering and Knowledge Engineering (SEKE '99), June 1999

Noy, N., Rector, A.: Defining N-ary relations on the semantic web. In: W3C Working Group Note, World Wide Web Consortium (W3C). http://www.w3.org/TR/swbp-n-aryRelations/ (2006). Accessed Apr 2006

Noy, N.F., Fergerson, R.W., Musen, M.A.: The knowledge model of Protégé-2000: Combining interoperability and flexibility. In: Dieng, R., Corby, O. (eds.) Knowledge Acquisition, Modeling and Management, Proceedings of 12th International Conference, EKAW 2000, Juan-les-Pins, France, October 2–6, 2000, Lecture Notes in Computer Science, vol. 1937, pp. 17–32. Springer (2000). http://citeseer.ist.psu.edu/noy01knowledge.html

UML 2.0.: Unified Modeling Language: Superstructure—Specification v2.0. Object Management Group (OMG), Aug 2005. http://www.omg.org/docs/formal/05-07-04.pdf

SysML.: OMG Systems Modeling Language (OMG SysML™) Specification. Object Management Group (OMG), May 2006. http://www.sysml.org/docs/specs/OMGSysML-FAS-06-05-04.pdf

Pahl, G., Beitz, W.: Engineering Design—A Systematic Approach, 2nd edn. Springer, London (1996). Second Edition

Park, J.Y., Gennari, J.H., Musen, M.A.: Mappings for reuse in knowledge-based systems. In: Proceedings of the 11th Workshop on Knowledge Acquisition, Modeling, and Management (KAW '98). Banff, Canada, Apr 1998

Russell, S., Norvig, P.: Artificial Intelligence: A Modern Approach, 2nd edn. Prentice-Hall, Englewood Cliffs, NJ (2003)

Rzehorz, C.: Wissensbasierte Anforderungsentwicklung auf der Basis eines integrierten Produktmodells. Berichte aus dem Institut RPK, vol. 3. Shaker Verlag, June 1998

Sirin, E., Smith, M., Wallace, E.: Opening, closing worlds-on integrity constraints. In: OWLED (2008)

Sowa, J.F.: Knowledge Representation: Logical, Philosophical, and Computational Foundations. Brooks Cole Publishing Co., Pacific Grove, CA (1999)

Stevens, R., Brook, P., Jackson, K., Arnold, S.: Systems Engineering: Coping with Complexity. Prentice Hall PTR, June 1998

Top, J.L., Akkermans, H.: Tasks and ontologies in engineering modelling. Int. J. Hum.-Comput. Stud. **41**(4), 585–617 (1994)

Tudorache, T.: Engineering ontologies (2006a). http://protegewiki.stanford.edu/wiki/Engineering_ontologies

Tudorache, T.: Employing ontologies for an improved development process in collaborative engineering. Ph.D. thesis, Technical University of Berlin, Germany (2006b). http://goo.gl/8EeTm2

Tudorache, T.: Ontologies in Engineering: Modeling, Consistency and Use Cases. VDM Publishing (2008)

VDI 2206. Design methodology for mechatronic systems. Verein Deutscher Ingenieure (VDI), Berlin: VDI-Richtlinie 2206. Beuth Verlag, Berlin (2004)

Wache, H.: Semantische Mediation für heterogene Informationsquellen. Dissertationen zur Künstlichen Intelligenz, vol. 261. Akademische Verlagsgesellschaft, Berlin (2003). http://www.cs.vu.nl/~holger/Papers/phd-03.pdf

Wache, H., Vögele, T.J., Visser, U., Stuckenschmidt, H., Schuster, G., Neumann, H., Hübner, S.: Ontology-based integration of information—a survey of existing approaches. In: Gómez-Pérez, A., Gruninger, M., Stuckenschmidt, H., Uschold, M. (eds.) Proceedings of the IJCAI-01 Workshop on Ontologies and Information Sharing, pp. 108–117. Seattle, USA (2001)

Wielinga, B.J., Akkermans, H., Schreiber, G.: A formal analysis of parametric design. In: Gaines, B.R., Musen, M.A. (eds.) Proceedings of the 9th Banff Knowledge Acquisition for Knowledge-Based Systems Workshop. Alberta, Canada (1995)

World Wide Web Consortium.: OWL 2 Web Ontology Language, December 2012. http://www.w3.org/TR/owl2-syntax/ (2015). Accessed July 2015

World Wide Web Consortium.: Shapes Constraint Language (SHACL), W3C First Public Working Draft 08 October 2015, December 2015. http://www.w3.org/TR/2015/WD-shacl-20151008/ (2015). Accessed Nov 2015

Yang, G., Kifer, M., Zhao, C., Chowdhary, V.: Flora-2: User's Manual, 30th Apr 2005. http://citeseer.ist.psu.edu/yang01flora.html

Zimmermann, J.U.: Informational integration of product development software in the automotive industry. Ph.D. thesis, University of Twente (2005)

Chapter 13
Leveraging Semantic Web Technologies for Consistency Management in Multi-viewpoint Systems Engineering

Simon Steyskal and Manuel Wimmer

Abstract Systems modeling is an important ingredient for engineering complex systems in potentially heterogeneous environments. One way to deal with the increasing complexity of systems is to offer several dedicated viewpoints on the system model for different stakeholders, thus providing means for system engineers to focus on particular aspects of the environment. This allows them to solve engineering tasks more efficiently, although keeping those multiple viewpoints consistent with each other (e.g., in dynamic multiuser scenarios) is not trivial. In the present chapter, we elaborate how Semantic Web technologies (SWT) may be utilized to deal with such challenges when models are represented as RDF graphs. In particular, we discuss current developments regarding a W3C Recommendation for describing structural constraints over RDF graphs called Shapes Constraint Language (SHACL) which we subsequently exploit for defining intermodel constraints to ensure consistency between different viewpoints represented as RDF graphs. Based on a running example, we illustrate how SHACL is used to define correspondences (i.e., mappings) between different RDF graphs and subsequently how those correspondences can be validated during modeling time.

Keywords Consistency management · Multi-viewpoint systems engineering · Shapes Constraint Language · SHACL · Constraint checking · Ontology mapping · Ontology integration

---

S. Steyskal (✉)
Siemens AG Austria, 1210 Vienna, Austria
e-mail: simon.steyskal@wu.ac.at

S. Steyskal
Institute for Information Business, WU Vienna, 1020 Vienna, Austria

M. Wimmer
Institute of Software Technology and Interactive Systems,
TU Vienna, 1040 Vienna, Austria
e-mail: wimmer@big.tuwien.ac.at

S. Biffl and M. Sabou (eds.), *Semantic Web Technologies for Intelligent Engineering Applications*, DOI 10.1007/978-3-319-41490-4_13

13.1 Introduction

Large-scale heterogeneous systems are inherently much more complex to design, develop, and maintain than classical, homogeneous, centralized systems. Model-driven Engineering (MDE) is an emerging paradigm to cope with the complexity of such new systems (Brambilla et al. 2012; Bézivin et al. 2014). Especially the possibility to describe large and complex software systems at different levels of abstraction and from different, partly overlapping viewpoints (cf. *Multi-Viewpoint Modeling* (IEEE Architecture Working Group 2000)), is of utter importance for realizing the *Industry 4.0* vision (VDI 2014).[1] This approach allows stakeholders to focus on particular development aspects in more detail which in turn allows them to solve their engineering tasks more efficiently. However, due to the large variety and diversity of domain-specific tools used, one has to deal with a much higher heterogeneity between modeling languages, data formats, and tool technologies (Barth et al. 2012; Biffl et al. 2009; Drath et al. 2008).

Unfortunately, this separation of concerns comes with the price of having to keep those viewpoints consistent (Finkelstein et al. 1993). Since one or more engineers are working on the same system but may model it from different perspectives (i.e., viewpoints), changes to models may happen concurrently, thus have to be propagated to all other corresponding models for solving potential inconsistencies that might have arisen from those updates (Giese and Wagner 2009; Orejas et al. 2013). That is primarily due to the fact that viewpoint definitions are not completely independent from each other (i.e., elements of each viewpoint definition may correspond to elements of others). Such relationships are normally specified by means of correspondences, which are statements that define some concepts of a viewpoint to be related to ones in other viewpoints. Prominent examples that advocate such architectural decomposition are the Reference Model of Open Distributed Processing (RM-ODP) (ISO/IEC 2010), the Model-Driven Web Engineering (MDWE) initiative (Moreno et al. 2008), UML (OMG 2011), SysML (OMG 2012), or AutomationML (AutomationML Consortium 2014), which provide different diagram types or languages to represent all the diverse aspects of a system.

Challenges in Multi-Viewpoint Systems Engineering Software systems in complex technological environments, such as production/power plants, railway systems, or medical diagnosis devices, consist of various components and subsystems. Vendors of such systems face the challenge to design models that allow configuring individual components and incorporating them into a complete system configuration (cf. Fig. 13.1). Because components usually interact in diverse ways by exchanging data and using services from each other, designing and integrating individual models into a coherent and consistent system model are highly complex tasks that impose a couple of major challenges that have to be taken care of, namely:

[1]The interested reader may be refer to Chap. 2 for a detailed discussion on potential application scenarios of Semantic Web technologies in multidisciplinary engineering.

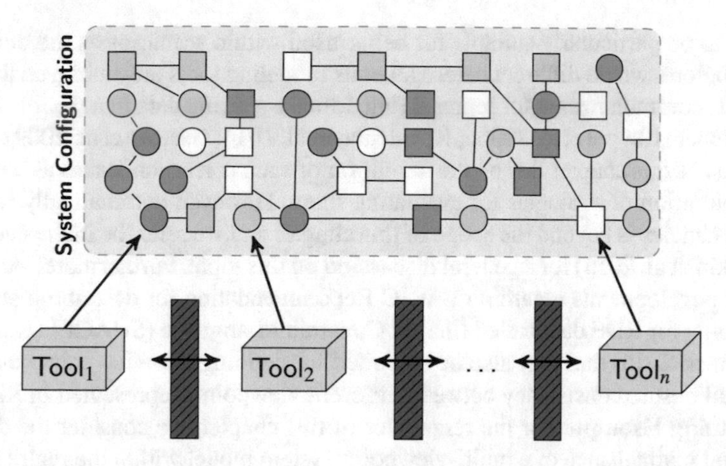

Fig. 13.1 Multi-viewpoint system modeling in heterogeneous settings

Babylonian language confusion While each tool uses its own (local) vocabulary for representing components, correspondences between different vocabularies' representations are not clear. Thus, it is a major challenge to develop a common vocabulary that is powerful enough to sufficiently describe all individual component models.

Tedious model exchange Different teams of engineers use the most adequate tools for developing and maintaining their components. However, these tools are usually neither developed for interoperability purposes nor for easily exchanging models among each other. Therefore, exchanging models and data between individual tools is a challenging task.

Incompatible components As components are used and maintained within different concrete products, multiple versions of components will coexist. If such a component is updated, combinations of components deployed within products may become incompatible. Detecting those incompatibilities among different components is complex and usually not adequately supported in settings where heterogeneous tools are being used.

Overlapping models Different component models usually have a significant knowledge overlap, which results in redundancies when those components are integrated with others for a specific product.

Inconsistent changes When teams of engineers develop and maintain their component models, concurrent changes may become inconsistent when integrated into a concrete product that is based on several interdependent components. If inconsistent changes are detected in a late phase of a project, costs of fixing them may increase significantly.

In the present chapter, we investigate strategies for dealing with such challenges by utilizing Semantic Web technologies for encoding viewpoints and their relationships (i.e., correspondences) among each other. Approaches based on SWT have proven to be particularly suitable for being used within scenarios as the ones mentioned before, where different heterogeneous modeling tools are employed that shall have one common format for representing domain-specific data in a flexible but precise manner (Kappel et al. 2006b; Kovalenko et al. 2014; Gasevic et al. 2009). Hence, the focus of this chapter lies on the definition of such correspondences as well as on the exploration of strategies for evaluating them. However, automatically repairing inconsistencies is beyond the scope of this chapter and we refer the interested reader to (Diskin et al. 2015) for a general discussion on this topic. Furthermore, we discuss current developments regarding a W3C Recommendation for describing structural constraints for RDF data called Shapes Constraint Language (SHACL) (Knublauch and Ryman 2015) that may also be exploited for defining interviewpoint constraints that shall ensure consistency between different viewpoints represented in RDF.

Running Example For the remainder of this chapter, we consider the development and maintenance of a multi-viewpoint system model within the realm of computer networks as our running example. This fictitious system is based on three individual system components/tools that all rely on partially different but overlapping conceptualizations of the domain of interest, which are expressed as ontologies as depicted in Fig. 13.2.

While ontologies O_{1-3} are all used for representing elements of a computer network and all model the same core concepts (albeit possibly named differently), they partially operate on different levels of abstraction (**Overlapping models**) which makes it difficult to properly incorporate them into a coherent system (**Tedious model exchange**). Without proper knowledge about the problem domain in hand, it is also not easy to spot that o_1:Computer and o_2:Connectable are actually conceptualizations of the same real-world resource, i.e., to devices that are related to some sort of cables/connections (**Babylonian language confusion**). As mentioned earlier, since one or more engineers are working on the same system but may model it from different viewpoints, changes to ontologies may be done concurrently which in turn leads to inconsistencies if those changes affected overlapping concepts (**Inconsistent changes**) or the changed ontologies were already deployed (**Incompatible components**).

The rest of the chapter is structured as follows. In Sect. 13.2, we discuss the task of validating integrated ontologies utilizing both reasoning under Open World Assumption (OWA) and integrity constraint checking under Closed World Assumption (CWA). In order to be able to perform the latter, recent developments regarding a W3C Recommendation for describing and checking structural constraints over RDF graphs are explored in Sect. 13.3 before we, based on a particular use case, elaborate how those techniques can be applied in practice (cf. Sect. 13.4). Finally, we discuss related work in Sect. 13.5 and conclude the chapter in Sect. 13.6 with an outlook on future work.

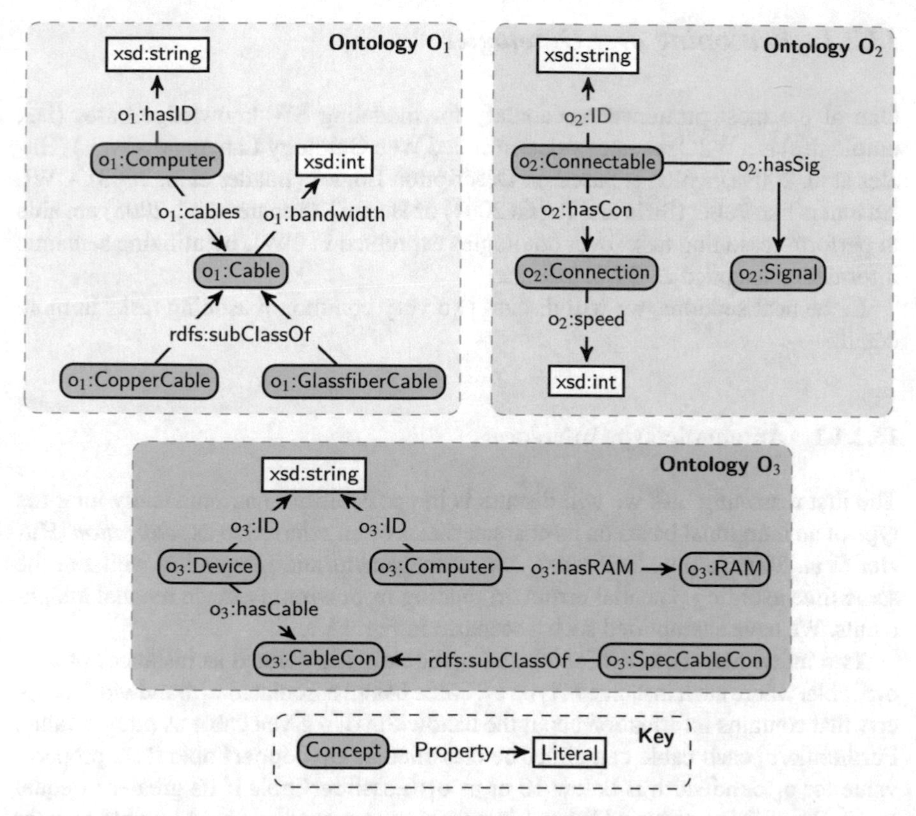

Fig. 13.2 Three ontologies, each representing a different viewpoint on a computer network

13.2 Utilizing Semantic Web Technologies for Validating Integrated System Components

The Semantic Web (SW) was initially conceptualized for enhancing data in the World Wide Web (WWW) with additional semantic information using knowledge representation and reasoning techniques, to allow machines to process this enhanced data more efficiently in an automated manner. But also more and more industrial applications, which might not be directly adhering to paradigms of the WWW, are adopting SWT for representing data too, thus making it accessible through an inference system e.g., for deducing implicit knowledge, which is often referred to as *Ontology-Based Data Access (OBDA)* (Wache et al. 2001).

In the following, we will first discuss two major aspects of OBDA, namely: reasoning within knowledge bases (cf. Sect. 13.2.1), and validation of RDF data (cf. Sect. 13.2.2) in more detail.

13.2.1 Reasoning over Ontologies

One of the most prominent vocabulary for modeling SW knowledge bases (i.e., ontologies) is a W3C recommendation called Web Ontology Language (OWL) (Hitzler et al. 2009a) which is based on Description Logics (Baader et al. 2003). OWL reasoners like Pellet (Sirin and Parsia 2004) or HermiT (Shearer et al. 2008) are able to perform reasoning tasks over ontologies expressed in OWL by utilizing semantic information encoded as OWL axioms.

In the next sections, we will discuss two very common reasoning tasks in more detail.

13.2.1.1 Automatic Type Inference

The first reasoning task we will discuss is the possibility to automatically infer the type of an individual based on its characteristics often referred to as *realization* (Hitzler et al. 2009b), thus facilitating data management and integration whilst at the same time avoiding potential errors originating from wrongly made manual assignments. We have exemplified such a scenario in Fig. 13.3.

Two individuals (i.e., o_1:Cable1 and o_1:Cable2) are defined as instances of type o_1:Cable, where each instance of type o_1:Cable has an associated o_1:bandwidth property that contains information about the bandwidth of a given cable as integer value. Furthermore, each cable can either be classified as o_1:CopperCable if its property value for o_1:bandwidth is below 10 or as o_1:GlassfiberCable if its greater or equal to 10. By utilizing these additional information, a reasoner would be able to infer that o_1:Cable1 is of type o_1:CopperCable (because its bandwidth is < 10) whereas o_1:Cable2 is of type o_1:GlassfiberCable (because its bandwidth is ≥ 10).

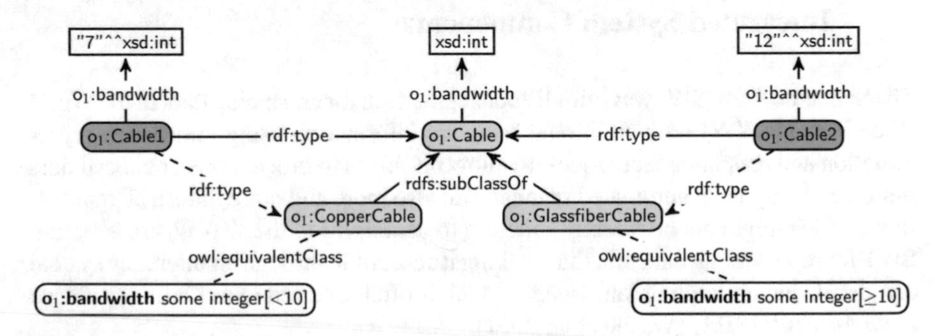

Fig. 13.3 Automatic type inference of individuals of type o_1:Cable

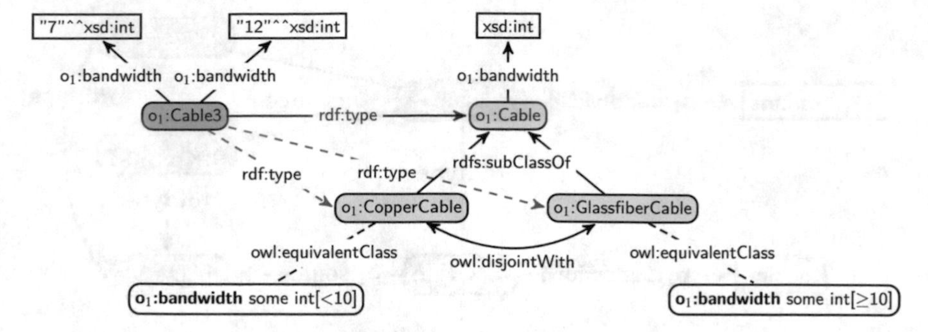

Fig. 13.4 Detection of inconsistent individual o_1:Cable3

13.2.1.2 Consistency and Satisfiability

Apart from the task of inferring new knowledge, reasoners can furthermore be used to detect inconsistencies and unsatisfiable individuals within the ontology as depicted in Fig. 13.4. In that example, two concepts o_1:CopperCable and o_1:GlassfiberCable are defined as being disjoint from each other (i.e., an individual cannot be of both types at the same time) and a new individual of type o_1:Cable called o_1:Cable1 is introduced having two values for o_1:bandwidth. Since each of those property values match a owl:equivalentClass axiom, a reasoner will declare o_1:Cable1 as being both, of type o_1:CopperCable and o_1:GlassfiberCable, which cannot be the case due to stated disjointness of those concepts. Thus, a reasoner will generate an error and define the ontology to be inconsistent based on the wrongly defined individual o_1:Cable1.

The example in Fig. 13.4 represents only one possible situation for which a reasoner could detect inconsistency of an ontology. Common reasoning tasks include the identification of *unsatisfiable concepts* (i.e., concepts which are defined in a way so that they cannot be instantiated), declaring the ontology to be *incoherent* (i.e., the ontology contains unsatisfiable concepts which are not instantiated but there are other instantiated concepts that are satisfiable) or (like in Fig. 13.4) declaring the ontology to be *inconsistent* if there exists no model of the ontology which makes all axioms hold.

13.2.2 Validation of RDF Data

In the Semantic Web, the tasks of (*i*) validating RDF data against a set of integrity constraints and (*ii*) performing RDFS/OWL reasoning, are grounded on different semantics. While the latter operates under the *Open World Assumption (OWA)* (i.e., a statement cannot be assumed to be false if it cannot be proven to be true) and the *Nonunique Name Assumption (nUNA)* (i.e., the same object/resource can have

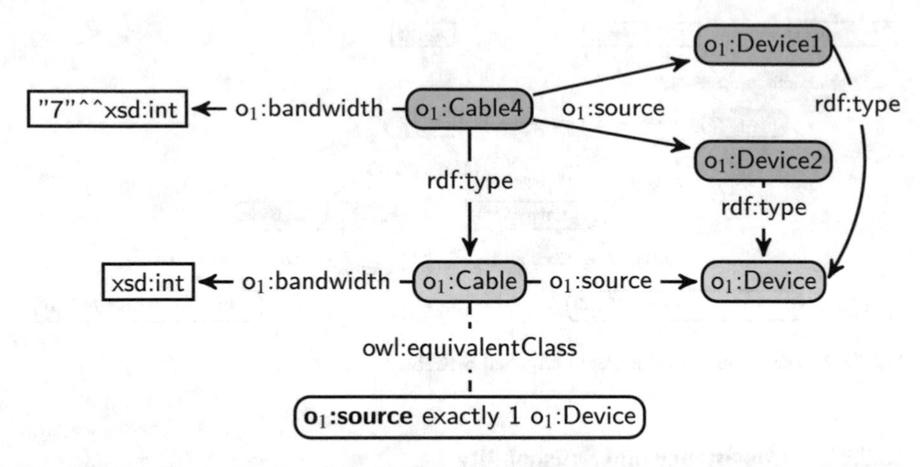

Fig. 13.5 Violated cardinality constraint of o_1:Cable4

multiple names), validation adheres to the *Closed World Assumption (CWA)* (i.e., a statement is assumed to be false if it cannot be proven to be true) and requires that different names identify different objects (= *Unique Name Assumption (UNA)*).

Since there does not exist any official standard for describing integrity constraints on RDF graphs, practitioners had to fall back on other formalisms for expressing and validating their constraints. Consider the example illustrated in Fig. 13.5 where OWL is used for specifying a cardinality constraint for class o_1:Cable.

Class o_1:Cable is defined to be equal to an anonymous class having exactly one o_1:Device associated via o_1:source, hence every individual of type o_1:Cable is only allowed to refer to one specific device. Even though o_1:Cable4 is violating its cardinality constraint (because it is associated to more than one device via o_1:source) an OWL reasoner would not be able to detect a violation, but instead infers that o_1:Device1 and o_1:Device2 are representing the same resource. While this inference is certainly reasonable for open environments like the WWW, there are also application scenarios where the CWA, or at least the partial closure of the knowledge base, is a more natural choice. For example, if ontology-based reasoning is done in conjunction with data stored in a database—or in general—if stored data is considered to be complete, thus missing statements should be taken as false (Knorr et al. 2011).

Another very common strategy is to use SPARQL for constraint checking by querying for nondesired behavior as exemplified in Listing 13.1. If such a query returns any results, a constraint violation can be assumed.

Listing 13.1 ASK Query which returns **true** if a o_1:Cable does not have exactly 1 property value of type o_1:Device associated via o_1:source.

```
ASK WHERE {
 {
   SELECT (count(?device) AS ?numbOfDevs) ?cable
   WHERE {
     ?cable a o1:Cable .
```

```
      ?cable o1:source ?device .
      ?device a o1:Device .
    } GROUP BY ?cable
  } FILTER(?numbOfDevs != 1)
}
```

However, also SPARQL is not completely suitable to serve as constraint checking formalism for the Semantic Web. For example, it is not possible to express and ship constraint checking queries together with the ontology they have to be queried against. This necessity has led to the development of a variety of approaches for closing the gap of missing validation standards for the SW, such as SPIN (Knublauch 2009), OSLC Resource Shapes (Ryman et al. 2013), Shape Expressions (Prud'hommeaux et al. 2014), or RDFUnit (Kontokostas et al. 2014). While experts in these individual tools are easily able to utilize them for validation tasks, others would benefit greatly from official validation standards and commodity tools which implement them.

13.3 Shapes Constraint Language (SHACL)

Originating from the RDF Validation Workshop[2] that was held by the W3C for identifying a first set of use cases and requirements a potential RDF constraint language must meet, the W3C RDF Data Shapes Working Group (WG) was launched in 2014 with the mission of producing a language for defining structural constraints on RDF graphs and enabling the definition of graph topologies for interface specification, code development, and data verification.[3] Eventually, the WG released a First Public Working Draft of such an RDF constraint language, named Shapes Constraint Language, in October 2015 (Knublauch and Ryman 2015).

Currently, there are two implementations of SHACL engines available that can be used for validating shape definitions. One is part of TopQuadrant's TopBraid Composer 5.0.1 Beta[4] and the other one is a Java open source API based on Jena[5] which is almost always up to date with recent WG resolutions.

Remark: We have to emphasize that the specification of SHACL is still work in progress, hence all SHACL constructs we discuss in this chapter may be subject to change at any time and should be treated with caution.

---

[2]https://www.w3.org/2012/12/rdf-val/.

[3]http://www.w3.org/2014/data-shapes/charter.

[4]http://www.topquadrant.com/downloads/topbraid-composer-install/.

[5]https://github.com/TopQuadrant/shacl/.

13.3.1 Preliminaries

In general, a validation process in SHACL differentiates between two types of input, namely:

Data Graph: A graph that contains the data that has to be validated.
Shapes Graph: A graph containing shape definitions and other information that is used to perform the validation of the data graph.

For example, SHACL can be used to check whether all nodes in ontology $\mathbf{O}_1$ (cf. Fig. 13.2) that are of type o1:Computer do only have one property value for o1:hasID, with this value being a string literal. SHACL can also be used to specify that a particular individual of type o1:Computer (e.g., o1:comp1) must have at least two associated cables, each of them having a bandwidth of less or equal 10. Listing 13.3 contains a potential data graph that could be validated against shapes illustrated in Listing 13.2 that encodes aforementioned constraints using SHACL. In the following, we will discuss the syntax of SHACL in more detail.

Listing 13.2 Sample shapes graph

```
o1:ComputerShape
        a sh:Shape ;
        sh:scopeClass o1:Computer ;
        sh:property [
                sh:predicate o1:hasID ;
                sh:datatype xsd:string ;
                sh:minCount 1 ;
                sh:maxCount 1 ;
    ] .

o1:Comp1Shape
        a sh:Shape ;
        sh:scopeNode o1:comp1 ;
        sh:property [
                sh:predicate o1:cables ;
                sh:valueShape [
                        sh:property [
                                sh:predicate o1:bandwidth ;
                                sh:maxInclusive 20 ;
                        ]
                ] ;
                sh:minCount 2 ;
        ] .

o1:GlassFiberCableShape
        a sh:Shape ;
        sh:scopeClass o1:Cable ;
        sh:filterShape [
                sh:property [
                        sh:predicate o1:bandwidth ;
                        sh:minInclusive 10 ;
                ]
```

```
        ] ;
        sh:property [
                sh:predicate o1:bandwidth ;
                sh:maxCount 1 ;
        ] .
```

Listing 13.3 Sample data graph

```
o1:cable1
        a o1:Cable ;
        o1:bandwidth 7 .

o1:cable2
        a o1:Cable ;
        o1:bandwidth 14 .

o1:comp1
        a o1:Computer ;
        o1:hasID "C1"^^xsd:string ;
        o1:cables o1:cable1, o1:cable2 .

o1:comp2
        a o1:Computer ;
        o1:hasID "C2"^^xsd:string .
```

13.3.2 Identifying Nodes for Validation

As mentioned before, a shape is a group of constraints that can be validated against RDF nodes. In order to decide which nodes (henceforth called *focus nodes*) shall be considered for validation, shapes can have different types of scopes that instruct a SHACL processor on how to select them, namely:

Node Scopes A shape can point to individual nodes that are supposed to be validated against it using the property sh:scopeNode. For example, shape o1:Comp1-Shape defined in Listing 13.2 has only one individual node (o1:comp1) in its scope, hence only that one node will be validated against o1:Comp1Shape.

Class-based Scopes The property sh:scopeClass can be used to link a shape with a class. The scope includes all instances of the sh:scopeClass and its subclasses. For example, shape o1:ComputerShape applies to all individuals of type o1:Computer (i.e., o1:comp1 and o1:comp2).

General Scopes Shapes can also point to one or more instances of sh:Scope. Those scopes can either be based on already predefined common scope patterns expressed as subclasses of sh:Scope or individually specified using sh:sparql and referred to via sh:scope.

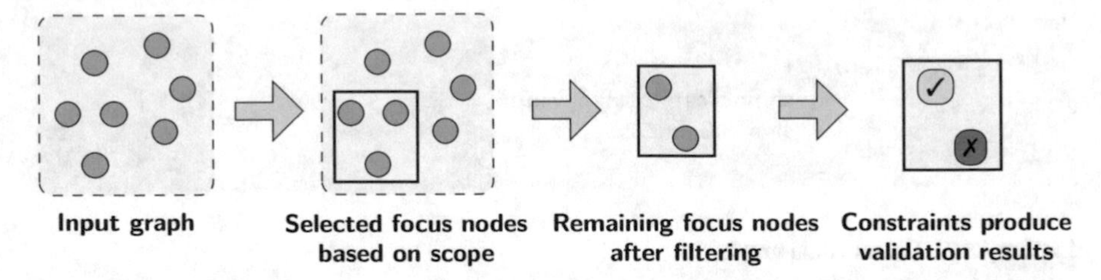

Fig. 13.6 Scoping and filtering process of shapes

If a specific constraint (or shape) should not be applied to all focus nodes produced by a scope, each constraint (or shape) may define specific preconditions a node must fulfill in order to be eligible for evaluation, by using filter shapes. Consider for example the shape o1:GlassFiberCableShape defined in Listing 13.2 which requires a node to have at most one associated o1:bandwidth. Although its scope includes all individuals of type o1:Cable, only o1:cable2 passes the filter shape, hence will be validated against o1:GlassFiberCableShape.

A schematic representation of such a scoping and filtering process is depicted in Fig. 13.6.

13.3.3 SHACL Constraint Types

SHACL differentiates between different types of constraints, namely:

(Inverse) Property Constraints: A (inverse) property constraint (sh:property, sh:inverseProperty) is a constraint that defines restrictions on the values of a given property in the context of the focus node. The focus node is the subject (or object for inverse property constraints) and the property represents the predicate of respective triples (sh:predicate). Each (inverse) property constraint must constrain exactly one property at a time. An example of such a property constraint is defined in sh:ComputerShape illustrated in Listing 13.2. It states that every individual of type o1:Computer must have exactly one (sh:minCount,sh:maxCount) property value for property o1:hasID and that this one property value must be of type xsd:string (sh:datatype).[6]

Node Constraints: A node constraint (sh:constraint) can be used to define constraints about the focus node itself rather than on values of its properties.

(SPARQL-based) Native Constraints: In some situations, concepts of SHACL's core vocabulary might not be sufficient enough to express a certain constraint. Therefore, constraints can also be expressed using native executable languages

[6]Note that SHACL (at the time of writing) considers multiple property constraints on the same predicate as being *conjunctive*, meaning that any triple with that predicate is expected to match all of property constraints.

such as SPARQL, e.g., a native constraint defined for a shape using sh:constraint can define a respective SPARQL query identified by sh:sparql that shall be used for validation. The query has to take care of producing the required bindings, where each result set row corresponds to a validation result.

Template Constraints: While constraints defined in a language like SPARQL provide a lot of flexibility and expressiveness, they may also be hard to understand for people not familiar with SPARQL. Furthermore, if a certain constraint pattern has to be checked for several different shapes, a shapes graph containing the same SPARQL query over and over again becomes very verbose. Thus, SHACL provides a means for encapsulating such native queries into constraint templates which can be parametrized. Constraint templates can be instantiated anywhere where a native constraint may appear (i.e., at sh:constraint).

13.3.4 SHACL Constraint Components

SHACL offers a variety of different constraint components (CC) for defining shapes. In the following, we will give short examples for each of the different types of CC[7]:

Value Type CC: e.g., for verifying that each value node is an instance of a given type (sh:class).

Cardinality CC: e.g., for restricting the minimum number of allowed property values for certain properties of a focus node (sh:minCount).

Value Range CC: for representing range restrictions on value nodes, e.g., for defining the minimum exclusive value of a property (sh:minExclusive).

String-based CC: e.g., for verifying whether value nodes match a certain regular expression (sh:pattern).

Logical CC: SHACL includes the notion of negating shapes (sh:not) and building a conjunction (sh:and) or disjunction (sh:or) of shapes.

Shape-based CC: for verifying that all value nodes are complying to a certain shape (sh:valueShape).

Property Pair CC: e.g., for validating whether each of the value sets of a given pair of properties at the same focus node are equal (sh:equal).

For a complete list of SHACL constraint components please consult the latest version of its specification.[8]

[7] Note, that most CC can be applied within multiple contexts, hence the concept of *value nodes* either refers to objects, subjects or the focus node itself depending on the context its respective CC is used in.

[8] http://w3c.github.io/data-shapes/shacl/.

Table 13.1 Available severity types in SHACL

Severity	Description
sh:Info	An informative message, not a violation
sh:Warning	A noncritical constraint violation indicating a warning
sh:Violation	A constraint violation that should be fixed

13.3.5 Reporting of Validation Results

A SHACL constraint validation process produces a set of validation results which can potentially be empty if no constraints were violated. A constraint violation consists of structural information providing insights on the RDF nodes that caused the violation, as well as human-readable messages. Each violation result must have exactly one severity level indicated via property sh:severity, such a level is based on the respective severity level of the constraint that has been violated (cf. Table 13.1 for a list of available severity types).

If the previously defined data graph (cf. Listing 13.3) is validated against its respective shape definitions (cf. Listing 13.2), following validation message shown in Listing 13.4 has to be produced by a SHACL engine.

Listing 13.4 Returned validation message

```
o1:ExampleConstraintViolation
        a sh:ValidationResult ;
        sh:severity sh:Violation ;
        sh:focusNode o1:comp1 ;
        sh:subject o1:comp1 ;
        sh:predicate o1:cables ;
        sh:object o1:cable2 ;
        sh:message "Value does not fulfill constraints of shape." .
```

13.4 Use Case: Integrating Heterogeneous Views on a Computer Network

As already mentioned in the introduction, a growing number of industrial companies is utilizing the benefits of MDE and SWT for developing their software artifacts and systems, e.g., by leveraging the possibility to describe large and complex software systems at different levels of abstraction and from different, partly overlapping viewpoints. Unfortunately, such a separation of concerns, i.e., the fact that one or more developers are working on the same software artifact but may model it from different viewpoints hence changes to models may happen concurrently, requires that all individual viewpoints must be kept consistent with each other (Giese and Wagner 2009).

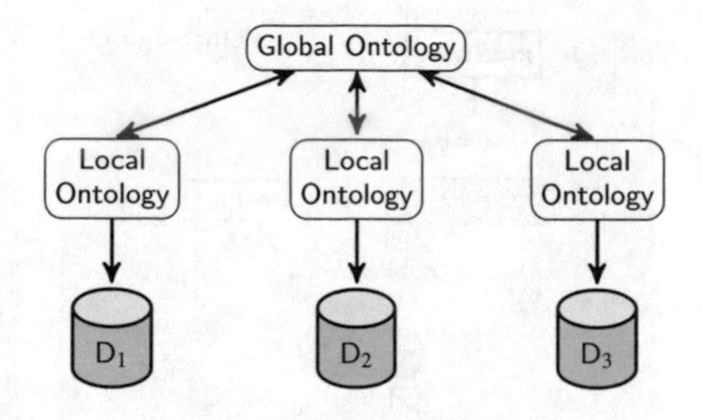

Fig. 13.7 (Adapted) Hybrid ontology integration approach (Wache et al. 2001)

In the following, we will describe first steps towards an approach for validating the consistency of multiple integrated views expressed as ontologies using SWT and most notably SHACL. Relating to the nomenclature defined in Chap. 6, we will use the term *mapping constraint* for referring to SHACL shapes that are used to express and validate mappings between concepts of an ontology.

13.4.1 Integration of Heterogeneous Viewpoints

As ontology integration architecture we have chosen the well known *hybrid ontology integration approach* illustrated in Fig. 13.7. Each local ontology, which represents a particular viewpoint of the system, is mapped to the same global ontology containing all overlapping concepts. Although we currently only focused on expressing and validating mappings between ontologies and have not implemented any repair mechanisms yet,[9] due to potential complexity of change propagation caused by such repairs (Eramo et al. 2012), we do not allow any direct mappings between individual views.

More precisely, consider the adapted version of our running example which is illustrated in Fig. 13.8. In addition to ontologies $\mathbf{O}_{1-3}$, a global ontology $\mathbf{O}_G$ was defined consisting of abstractions of overlapping concepts (i.e., oG:Device, oG:Connection, and their respective properties oG:ID and oG:hasBandw). Furthermore, we have introduced the notion of weaving ontologies $\mathbf{W}_{1-3G}$ which contain mappings between individual views and their global abstraction.

13.4.2 Defining Mappings Between Viewpoint Definitions using SHACL

It is difficult to keep a system as illustrated in Fig. 13.8 consisting of various local heterogeneous ontologies that are integrated with each other through a common global

[9]cf. Sect. 13.5 for an outlook on future work.

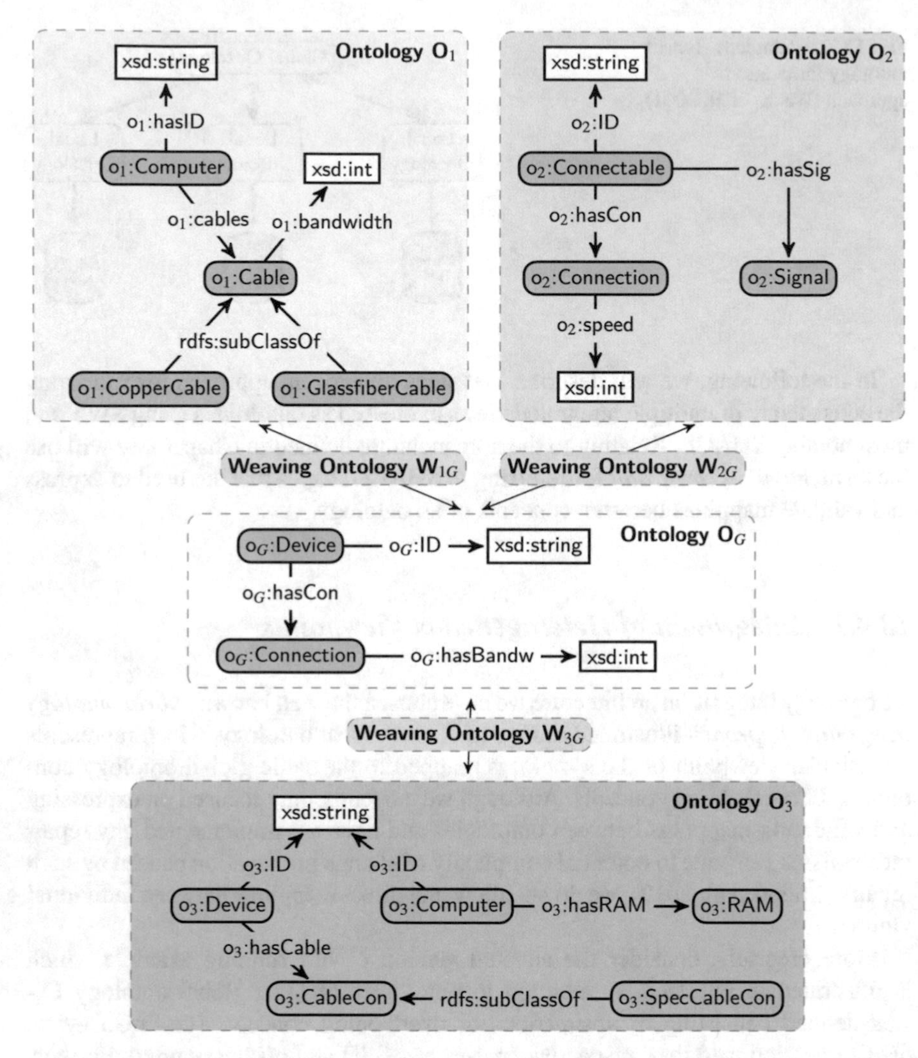

Fig. 13.8 Adapted version of the running example, now including a global ontology ($\mathbf{O}_G$) containing abstractions of overlapping concepts, and respective mappings between local ontologies and their global abstraction ($\mathbf{W}_{1-3G}$)

representation of their overlapping concepts consistent. Since different viewpoints may overlap, it is not sufficient enough to solely check consistency of each ontology alone, but to verify the validity of mappings that connect overlapping concepts too.

As mentioned above, we introduced the notion of weaving ontologies which contain the mappings between individual views and their global abstraction. Each weaving ontology consists of one or more mappings of type wm:WeavingClass that relate concepts of a view, i.e., local ontology, via wm:LHS to concepts of the global abstraction via wm:RHS. We have exemplified such mappings in Fig. 13.9, where weaving

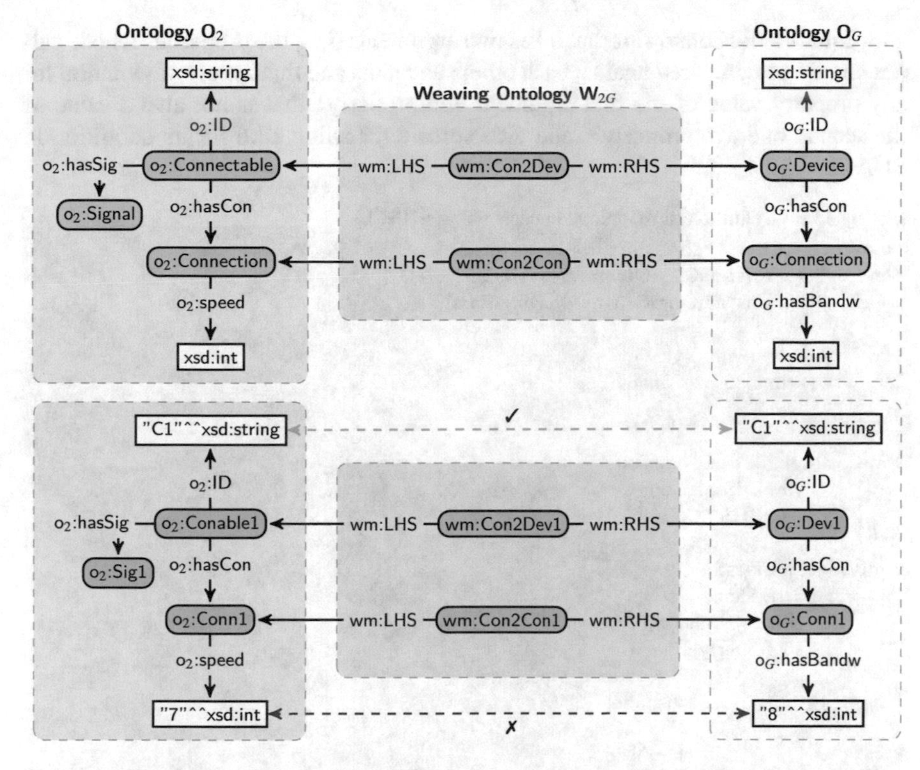

Fig. 13.9 Weaving ontologies utilizing SHACL for defining mapping constraints between ontologies (*top*) and their validation on concrete data (*bottom*)

ontology $\mathbf{W}_{2G}$ contains two concepts for defining mapping constraints between (i) o_2:Connectable and o_G:Device, and (ii) between o_2:Connection and o_G:Connection (cf. top part of Fig. 13.9).

Those exemplary mappings require that the ID of each mapped individual of type o_2:Connectable is equal to the one of the individual of type o_G:Device it is mapped to. Furthermore, all mapped individuals of type o_2:Connection o_G:Connection must have the same values for their properties o_2:speed and o_G:hasBandw, respectively. In order to express those constraints in RDF, we define a SHACL shape for each individual mapping in the weaving ontology (cf. Listing 13.5). While **wm:WeavingClass Shape** expresses the general requirement that each mapping must map exactly one concept of the LHS to exactly one concept of the RHS,[10] **wm:Con2DevShape** states that each mapping of type wm:Con2Dev must relate individuals of type o_2:Connectable to ones of type o_G:Device and that the IDs of mapped individuals must be equal (**wm:Con2ConShape** is defined respectively). To validate the equality of property values, we have defined a new constraint template called

[10]As of now, we are only considering one-to-one mappings, but plan to support one-to-many and also many-to-many mappings in the near future.

wm:EqualValuesConstraint that takes two arguments (i.e., the properties whose values should be compared against each other) as inputs and that reports a violation for any property value of the first argument wm:property1 that is not also a value of the second one wm:property2, and vice versa (cf. Listing 13.6 for its definition in SHACL).

Listing 13.5 Definition of mapping concepts using SHACL

```
wm:WeavingClass a rdfs:Class .
wm:Con2Dev rdfs:subClassOf wm:WeavingClass .
wm:Con2Con rdfs:subClassOf wm:WeavingClass .

wm:Con2Dev1 a wm:Con2Dev ;
        wm:LHS o2:Conable1 ;
        wm:RHS oG:Dev1 .

wm:Con2Con1 a wm:Con2Con ;
        wm:LHS o2:Conn1 ;
        wm:RHS oG:Conn1 .

wm:WeavingClassShape
        a sh:Shape ;
        sh:scopeClass wm:WeavingClass;
        sh:property [
                sh:predicate wm:LHS ;
                rdfs:label "LHS"^^xsd:string ;
                sh:maxCount 1 ;
                sh:minCount 1 ;
        ] ;
        sh:property [
                sh:predicate wm:RHS ;
                rdfs:label "RHS"^^xsd:string ;
                sh:maxCount 1 ;
                sh:minCount 1 ;
        ] .

wm:Con2DevShape
        a sh:Shape ;
        sh:scopeClass wm:Con2Dev;
        sh:property [
                sh:predicate wm:LHS ;
                sh:class o2:Connectable ;
        ] ;
        sh:property [
                sh:predicate wm:RHS ;
                sh:class oG:Device ;
        ] ;
        sh:constraint [
                a wm:EqualValuesConstraint ;
                wm:property1 o2:ID ;
                wm:property2 oG:ID ;
        ] .
```

```
wm:Con2ConShape
        a sh:Shape ;
        sh:scopeClass wm:Con2Con;
        sh:property [
                sh:predicate wm:LHS ;
                sh:class o2:Connection ;
        ] ;
        sh:property [
                sh:predicate wm:RHS ;
                sh:class oG:Connection ;
        ] ;
        sh:constraint [
                a wm:EqualValuesConstraint ;
                wm:property1 o2:speed ;
                wm:property2 oG:hasBandw ;
        ] .
```

Listing 13.6 Definition of custom constraint template **wm:EqualValuesConstraint**

```
wm:EqualValuesConstraint
        a sh:ConstraintTemplate ;
        rdfs:subClassOf sh:TemplateConstraint ;
        rdfs:label "Equal values constraint" ;
        rdfs:comment "Reports a violation for any value of property1 of the LHS
        that is not also a value of property2 of the RHS and vice-versa, for
        the same focus node." ;
        sh:argument [
                sh:predicate wm:property1 ;
                sh:class rdf:Property ;
                rdfs:label "property 1" ;
                rdfs:comment "Represents a property of the concept pointed at by the
                LHS of the respective mapping." ;
        ] ;
        sh:argument [
                sh:predicate wm:property2 ;
                sh:class rdf:Property ;
                rdfs:label "property 2" ;
                rdfs:comment "Represents a property of the concept pointed at by the
                RHS of the respective mapping." ;
        ] ;
        sh:message "Value sets of wm:LHS/{?property1} and
        wm:RHS/{?property2} must be equal";
        sh:sparql """
        SELECT $this ($this AS ?subject) ?predicate ?object
        WHERE {
        {
                $this wm:LHS ?tmp .
                ?tmp $property1 ?object .
                FILTER NOT EXISTS {
                        $this wm:RHS ?tmp2 .
                        ?tmp2 $property2 ?object .
                }
                BIND ($property1 AS ?predicate) .
        }
```

```
        UNION
        {
                $this wm:RHS ?tmp .
                ?tmp $property2 ?object .
                FILTER NOT EXISTS {
                        $this wm:LHS ?tmp .
                        ?tmp $property1 ?object .
                }
                BIND ($property2 AS ?predicate) .
        }
    }""" .
```

If SHACL's predefined constraint types are not sufficient enough to express certain types of constraints, one can always define its own constraint template which can then be used like any other SHACL constraint. In Listing 13.6, we defined such a custom constraint template (**wm:EqualValuesConstraint**). The constraint takes two properties as arguments, i.e., values of wm:property1 and wm:property2, and injects those properties values together with the respective focus node for which the constraint is evaluated for into its predefined SPARQL query.[11] Note that equivalence of property values represents only one possible dependency between concepts. Due to the expressiveness of SPARQL, hence SHACL, any other kind of mapping relationship could of course be expressed too.

If the example of Fig. 13.9 would have been validated against the shapes defined in Listing 13.5, a SHACL validator would be able to detect the inconsistency of mapping wm:Con2Con1 where the property value of o2:speed of o2:Conn1 is not equal to the one of oG:hasBandw of oG:Conn1. Subsequently, validation results like the one depicted in 13.4 would be reported by the validation engine and could guide system engineers by resolving those inconsistencies.

Clearly, an approach based on Semantic Web technologies, such as the one we discussed within this chapter, is capable of dealing with the previous identified challenges of *Multi-Viewpoint Systems Engineering* (cf. Sect. 13.1). While thoroughly defined mappings between individual views and their global abstraction help to integrate overlapping models and as a result reduce ambiguities between concepts and foster data exchange among models, constraints expressed as SHACL shapes provide means to define undesired behavior, i.e., allow a SHACL engine to detect inconsistencies caused by changes to individual components of the integrated system.

13.5 Related Work

With respect to the contribution of this chapter, we will discuss three lines of work more thoroughly: (i) approaches for handling inconsistencies in software engineering, (ii) approaches for bridging models and ontologies, and (iii) approaches for specifying SW-based management strategies for modeling artifacts.

[11] To distinguish variables that are bound by SPARQL from those that are bound by the SHACL engine itself, the latter are prefixed by $ (e.g., $this, $property1, ...).

Handling Inconsistencies in Software Engineering. A large number of approaches address the problem of integrating and synchronizing multiple viewpoints (Diskin et al. 2010). For example, works on synchronizing artifacts in software engineering which are highly influenced by approaches on multi-view consistency (Finkelstein et al. 1993; Grundy et al. 1998; Feldmann et al. 2014) using a generic representation of modifications and relying on users to write code to handle various types of modifications in all of the views. This idea influenced later efforts on model synchronization frameworks (Ivkovic and Kontogiannis 2004; Johann and Egyed 2004), in particular bidirectional model transformations (Song et al. 2011; Xiong et al. 2007). Other approaches use so-called correspondence rules for synchronizing models in the contexts of RM-ODP and MDWE (Cicchetti and Ruscio 2008; Eramo et al. 2008; Ruiz-Gonzalez et al. 2009). More theoretical approaches make use of algebraic structures called *lenses* for defining bidirectional transformations between models (Diskin et al. 2011a, b; Foster et al. 2008; Hofmann et al. 2011) or based on Triple Graph Grammars (TGGs) (Schürr and Klar 2008). A more detailed discussion on inconsistency management approaches exemplified by a mechatronic manufacturing system design case study is provided in (Feldmann et al. 2015).

Bridging Models and Ontologies. Strategies to combine modeling approaches stemming from MDE with ones related to the Semantic Web have been extensively studied over the last years (Gasevic et al. 2009). There are several approaches for transforming Ecore-based models to OWL and vice versa (Walter et al. 2010; Kappel et al. 2006a) and there exist approaches that allow defining ontologies in software modeling languages such as UML by using dedicated profiles (Milanovic et al. 2006). Moreover, some works have focused on identifying and combining benefits of (conventional) models with those of ontologies (Parreiras et al. 2007; Parreiras and Staab 2010). However, not only the purely structural part of UML is considered, some works also target the translation of constraints between these two technical spaces by using an intermediate format (Djuric et al. 2004). We build on these mentioned approaches, e.g., for generating RDF graphs from models, but focus on correspondence definitions and their respective validation.

Managing Models utilizing Semantic Web Technologies. The authors of (Parreiras et al. 2008) propose to use an ATL-inspired language for defining mappings between ontologies. Thus, unidirectional transformations are implementable for ontologies as it is known for model transformations. In (Kappel et al. 2007), ontology matching tools are utilized to search for correspondences between metamodels and for deriving relevant correspondences based on model transformations. Another approach relates to the one presented in (Wagelaar 2010) which translates parts of ATL transformations to ontologies for checking consistency of transformation rules, e.g., overlaps between rules in terms of overlapping matches. In our previous work (Bill et al. 2014) we follow this line of research, but considered bidirectional transformations specified in TGGs. Thus, in our translations to ontologies we did not solely consider source to target transformations, but used SPARQL to encode comparison and synchronization transformations. In this chapter, we illustrated a similar approach for specifying correspondences, by leveraging SHACL, W3C's recent advancement in the area of constraint checking of RDF data. Since transformation and synchronization transformation support will not be part of the main SHACL specification, we will investigate means for extending SHACL's functionality in this regards, e.g., by utilizing the theoretical foundation of TGGs in the future. Finally, there are approaches shifting the task of defining and performing consistency checks to SWTs. For instance, the authors of (Rieckhof et al. 2010) propose to use the Semantic Web Rule Language (SWRL)[12] for defining correspondences between models, hence facilitate basic model synchronization. SPARQL and OWL have been applied in (Feldmann et al. 2014; Kovalenko et al. 2014) to define and check consistency between different models used in engineering processes. Also more and more companies (e.g., large infrastructure providers) adopted SWTs in the recent past to deal with their highly complex use cases, such as knowledge management and enterprise data integration (Breslin et al. 2010; Schenner et al. 2014), especially due to the increasing amount of tool support available (cf. Grünwald et al. 2014). To sum up, to the best of our knowledge, we are not aware of any other approach utilizing SHACL to define and check intermodel/viewpoint consistency constraints as of now.

13.6 Conclusion

In the present chapter, we have presented an approach for defining correspondences between RDF graphs presenting different viewpoints on a system based on an emerging W3C standard called Shapes Constraint Language (SHACL). The correspondences are encoded as structural constraints which have to be satisfied by the referenced RDF data graphs in order to build up a coherent and complete system model. We also discussed how SHACL may be used in hybrid integration approaches which base on a global schema in order to reduce the overall complexity of bilateral viewpoint integrations. The declarative nature of SHACL allows to express intermodel/viewpoint correspondences without any further need of boilerplate code for validation execution and reporting.

As a next step, we plan to explore various possible means of extending SHACL with rule support. Whereby one could e.g., define simple constraint repair patterns that should be executed whenever their surrounding shape/constraint fails, or create mappings between ontologies based on rules that should be executed before the actual validation process starts. In this context we also consider to apply the theory of Triple Graph Grammars for developing transformation and synchronization support for SHACL. Finally, we plan to extend our mapping approach for supporting one-to-many and many-to-many mappings as well as to explore the capability of SHACL to deal with structural heterogeneities between modeling languages and viewpoints (Wimmer et al. 2010).

[12]http://www.w3.org/Submission/SWRL.

Acknowledgments This work has been partially funded by the Vienna Business Agency (Austria) within the COSIMO project (grant number 967327), the Christian Doppler Forschungsgesellschaft, and the BMWFW, Austria.

References

AutomationML Consortium: AutomationML Whitepaper Part 1—Architecture and general requirements (2014). http://www.automationml.org

Baader, F., Calvanese, D., McGuinness, D., Nardi, D., Patel-Schneider, P.: The Description Logic Handbook: Theory, Implementation and Applications. Cambridge University Press (2003)

Barth, M., Drath, R., Fay, A., Zimmer, F., Eckert, K.: Evaluation of the openness of automation tools for interoperability in engineering tool chains. In: Proceedings of the Conference on Emerging Technologies Factory Automation (ETFA) (2012)

Bézivin, J., Paige, R.F., Aßmann, U., Rumpe, B., Schmidt, D.C.: Manifesto—model engineering for complex systems (2014). CoRR abs/1409.6591

Biffl, S., Schatten, A., Zoitl, A.: Integration of heterogeneous engineering environments for the automation systems lifecycle. In: Proceedings of the International Conference on Industrial Informatics (INDIN) (2009)

Bill, R., Steyskal, S., Wimmer, M., Kappel, G.: On synergies between model transformations and semantic web technologies. In: Proceedings of the 8th Workshop on Multi-Paradigm Modeling co-located with the 17th International Conference on Model Driven Engineering Languages and Systems, MPMMODELS 2014, pp. 31–40. Valencia, Spain, 30 Sept 2014. http://ceur-ws.org/Vol-1237/paper4.pdf

Brambilla, M., Cabot, J., Wimmer, M.: Model-Driven Software Engineering in Practice. Morgan & Claypool (2012)

Breslin, J.G., O'Sullivan, D., Passant, A., Vasiliu, L.: Semantic web computing in industry. Comput. Ind. **61**(8), 729–741 (2010)

Cicchetti, A., Ruscio, D.D.: Decoupling web application concerns through weaving operations. Sci. Comput. Progr. **70**(1), 62–86 (2008)

Diskin, Z., Xiong, Y., Czarnecki, K.: Specifying overlaps of heterogeneous models for global consistency checking. In: MoDELS Workshops. LNCS, vol. 6627, pp. 165–179. Springer (2010)

Diskin, Z., Xiong, Y., Czarnecki, K.: From state- to delta-based bidirectional model transformations: the asymmetric case. J. Object Technol. **10**(6), 1–25 (2011a). doi:10.5381/jot.2011.10.1.a6

Diskin, Z., Xiong, Y., Czarnecki, K., Ehrig, H., Hermann, F., Orejas, F.: From state- to delta-based bidirectional model transformations: The symmetric case. In: Model Driven Engineering Languages and Systems, 14th International Conference, MODELS 2011, Proceedings, pp. 304–318. Wellington, New Zealand, 16–21 Oct 2011 (2011b). doi:10.1007/978-3-642-24485-8_22

Diskin, Z., Gholizadeh, H., Wider, A., Czarnecki, K.: A three-dimensional taxonomy for bidirectional model synchronization. J. Syst. Softw. **111**, 298–322 (2015)

Djuric, D., Gasevic, D., Devedzic, V., Damjanovic, V.: A UML Profile for OWL Ontologies. In: Proceedings of MDAFA. LNCS, vol. 3599, pp. 204–219. Springer (2004)

Drath, R., Lüder, A., Peschke, J., Hundt, L.: AutomationML—the glue for seamless automation engineering. In: Proceedings of on the Conference on Emerging Technologies and Factory Automation (ETFA) (2008)

Eramo, R., Pierantonio, A., Romero, J.R., Vallecillo, A.: Change management in multi-viewpoint system using ASP. In: Workshops Proceedings of the International IEEE Enterprise Distributed Object Computing Conference (EDOCW), pp. 433–440. IEEE (2008)

Eramo, R., Malavolta, I., Muccini, H., Pelliccione, P., Pierantonio, A.: A model-driven approach to automate the propagation of changes among architecture description languages. Softw. Syst. Model. **11**(1), 29–53 (2012)

Feldmann, S., Kernschmidt, K., Vogel-Heuser, B.: Combining a SysML-based modeling approach and semantic technologies for analyzing change influences in manufacturing plant models. In: Proceedings of 47th CIRP Conference on Manufacturing Systems (CMS) (2014)

Feldmann, S., Herzig, S.J.I., Kernschmidt, K., Wolfenstetter, T., Kammerl, D., Qamar, A., Lindemann, U., Krcmar, H., Paredis, C.J.J., Vogel-Heuser, B.: A comparison of inconsistency management approaches using a mechatronic manufacturing system design case study. In: Proceedings of CASE, pp. 158–165. IEEE (2015)

Finkelstein, A., Gabbay, D.M., Hunter, A., Kramer, J., Nuseibeh, B.: Inconsistency handling in multi-perspective specifications. In: Proceedings of ESEC. LNCS, vol. 717, pp. 84–99. Springer (1993)

Foster, J.N., Pilkiewicz, A., Pierce, B.C.: Quotient lenses. In: Proceedings of ICFP, pp 383–396. ACM (2008)

Gasevic, D., Djuric, D., Devedzic, V.: Model Driven Engineering and Ontology Development, 2nd edn. Springer (2009)

Giese, H., Wagner, R.: From model transformation to incremental bidirectional model synchronization. Softw. Syst. Model. **8**(1), 21–43 (2009)

Grundy, J.C., Hosking, J.G., Mugridge, W.B.: Inconsistency management for multiple-view software development environments. IEEE Trans. Softw. Eng. **24**(11), 960–981 (1998)

Grünwald, A., Winkler, D., Sabou, M., Biffl, S.: The semantic model editor: efficient data modeling and integration based on OWL ontologies. In: Proceedings of the 10th International Conference on Semantic Systems (SEMANTICS), pp. 116–123 (2014)

Hitzler, P., Krötzsch, M., Parsia, B., Patel-Schneider, P.F., Rudolph, S.: OWL 2 Web Ontology Language Primer. W3C Recommendation (2009a). http://www.w3.org/TR/2009/REC-owl2-primer-20091027/

Hitzler, P., Sebastian, R., Krtzsch, M.: Foundations of Semantic Web Technologies. Chapman & Hall/CRC, London (2009b). http://www.worldcat.org/isbn/142009050X

Hofmann, M., Pierce, B.C., Wagner, D.: Symmetric lenses. In: Proceedings of POPL, pp. 371–384. ACM (2011)

IEEE Architecture Working Group: IEEE Std 1471–2000. Recommended practice for architectural description of software-intensive systems. Technical Report, IEEE (2000)

ISO/IEC: Reference Model for Open Distributed Processing (RM-ODP). iSO/IEC 10746-1 to 10746-4, ITU-T Recs. X.901 to X.904 (2010)

Ivkovic, I., Kontogiannis, K.: Tracing evolution changes of software artifacts through model synchronization. In: Proceedings of ICSM, pp 252–261. IEEE (2004)

Johann, S., Egyed, A.: Instant and incremental transformation of models. In: Proceedings of ASE, pp. 362–365. IEEE (2004)

Kappel, G., Kapsammer, E., Kargl, H., Kramler, G., Reiter, T., Retschitzegger, W., Schwinger, W., Wimmer, M.: Lifting metamodels to ontologies: a step to the semantic integration of modeling languages. In: Proceedings of MODELS, LNCS, vol. 4199, pp. 528–542. Springer (2006a)

Kappel, G., Kargl, H., Wimmer, M., Kapsammer, E., Kramler, G., Reiter, T., Retschitzegger, W., Schwinger, W.: On models and ontologies—a semantic infrastructure supporting model integration. In: Modellierung, GI, pp. 11–27 (2006b)

Kappel, G., Kargl, H., Kramler, G., Schauerhuber, A., Seidl, M., Strommer, M., Wimmer, M.: Matching metamodels with semantic systems—an experience report. In: Proceedings of BTW Workshops, pp. 38–52 (2007)

Knorr, M., Alferes, J.J., Hitzler, P.: Local closed world reasoning with description logics under the well-founded semantics. Artif. Intell. **175**(9–10), 1528–1554 (2011)

Knublauch, H.: SPIN—SPARQL Inferencing Notation (2009). http://spinrdf.org/, http://spinrdf.org/, visited 10-10-2015

Knublauch, H., Ryman, A.: Shapes Constraint Language (SHACL). W3C First Public Working Draft (2015). http://www.w3.org/TR/2015/WD-shacl-20151008/

Kontokostas, D., Westphal, P., Auer, S., Hellmann, S., Lehmann, J., Cornelissen, R., Zaveri, A.: Test-driven evaluation of linked data quality. In: Proceedings of the 23rd International World Wide Web Conference (WWW), pp. 747–758 (2014)

Kovalenko, O., Serral, E., Sabou, M., Ekaputra, F.J., Winkler, D., Biffl, S.: Automating cross-disciplinary defect detection in multi-disciplinary engineering environments. In: Proceedings of EKAW, LNCS, vol. 8876, pp. 238–249. Springer (2014)

Milanovic, M., Gasevic, D., Giurca, A., Wagner, G., Devedzic, V.: Towards Sharing Rules Between OWL/SWRL and UML/OCL. ECEASST 5 (2006)

Moreno, N., Romero, J.R., Vallecillo, A.: An overview of model-driven web engineering and the MDA. In: Web Engineering: Modelling and Implementing Web Applications, pp. 353–382. Springer (2008)

OMG: Unified Modeling Language (UML). Version 2.4.1 (2011). http://www.omg.org/spec/UML/2.4.1

OMG: Systems Modeling Language (SysML). Version 1.3 (2012). http://www.omg.org/spec/SysML/1.3

Orejas, F., Boronat, A., Ehrig, H., Hermann, F., Schölzel, H.: On propagation-based concurrent model synchronization. ECEASST 57 (2013)

Parreiras, F.S., Staab, S.: Using ontologies with UML class-based modeling: the TwoUse approach. Data Knowl. Eng. **69**(11), 1194–1207 (2010)

Parreiras, F.S., Staab, S., Winter, A.: On marrying ontological and metamodeling technical spaces. In: Proceedings of FSE, pp. 439–448 (2007)

Parreiras, F.S., Staab, S., Schenk, S., Winter, A.: Model driven specification of ontology translations. In: Proceedings of ER, LNCS, vol. 5231, pp. 484–497. Springer (2008)

Prud'hommeaux, E., Gayo, J.E.L., Solbrig, H.R.: Shape expressions: an RDF validation and transformation language. In: Proceedings of the 10th International Conference on Semantic Systems (SEMANTICS), pp. 32–40 (2014)

Rieckhof, F., Seifert, M., Aszmann, U.: Ontology-based model synchronisation. In: Proceedings of TWOMDE Workshop (2010)

Ruiz-Gonzalez, D., Koch, N., Kroiss, C., Romero, J.R., Vallecillo, A.: Viewpoint synchronization of UWE models. In: Proceedings of MDWE Workshop (2009)

Ryman, A.G., Hors, A.L., Speicher, S.: OSLC Resource Shape: A language for defining constraints on Linked Data. In: Proceedings of LDOW, CEUR Workshop Proceedings, vol. 996 (2013)

Schenner, G., Bischof, S., Polleres, A., Steyskal, S.: Integrating distributed configurations with RDFS and SPARQL. In: Proceedings of the 16th International Configuration Workshop, pp. 9–15 (2014)

Schürr, A., Klar, F.: 15 years of triple graph grammars. In: Proceedings of the 4th International Conference Graph Transformations (ICGT), pp. 411–425 (2008)

Shearer, R., Motik, B., Horrocks, I.: Hermit: A highly-efficient OWL reasoner. In: Proceedings of the OWLED Workshop on OWL: Experiences and Directions (2008)

Sirin, E., Parsia, B.: Pellet: an OWL DL reasoner. In: Proceedings of the International Workshop on Description Logics (DL) (2004)

Song, H., Gang, H., Chauvel, F., Wei, Z., Sun, Y., Shao, W., Mei, H.: Instant and incremental QVT transformation for runtime models. In: Proceedings of MoDELS, LNCS, vol. 6981, pp. 273–288. Springer (2011)

VDI: Statusreport Referenzarchitekturmodell Industrie 4.0 (RAMI4.0) (2014). http://www.vdi.de/industrie40

Wache, H, Voegele, T., Visser, U., Stuckenschmidt, H., Schuster, G., Neumann, H., Hübner, S.: Ontology-based integration of information—a survey of existing approaches. In: Proceedings of IJCAI Workshop: Ontologies and Information, pp. 108–117 (2001)

Wagelaar, D.: Towards using OWL DL as a metamodelling framework for ATL. In: Proceedings of MtATL Workshop, pp. 79–85 (2010)

Walter, T., Parreiras, F.S., Gröner, G., Wende, C.: OWLizing: transforming software models to ontologies. In: Proceedings of ODiSE, pp. 7:1–7:6 (2010)
Wimmer, M., Kappel, G., Kusel, A., Retschitzegger, W., Schoenboeck, J., Schwinger, W.: From the heterogeneity jungle to systematic benchmarking. In: Proceedings of Workshops and Symposia at MODELS, pp. 150–164 (2010)
Xiong, Y., Liu, D., Hu, Z., Zhao, H., Takeichi, M., Mei, H.: Towards automatic model synchronization from model transformations. In: Proceedings of ASE, pp. 164–173. ACM (2007)

Chapter 14
Applications of Semantic Web Technologies for the Engineering of Automated Production Systems—Three Use Cases

Stefan Feldmann, Konstantin Kernschmidt and Birgit Vogel-Heuser

Abstract The increasing necessity to adapt automated production systems rapidly to changing requirements requires a better support for planning, developing and operating automated production systems. One means to improve the engineering of these complex systems is the use of models, thereby abstracting the view on the system and providing a common base to improve understanding and communication between engineers. However, in order for any engineering project to be successful, it is essential to keep the created engineering models consistent. We envision the use of Semantic Web Technologies for such consistency checks in the domain of Model-Based Engineering. In this chapter, we show how Semantic Web Technologies can support consistency checking for the engineering process in the automated production systems domain through three distinct use cases: In a first use case, we illustrate the combination of a Systems Modeling Language-based notation with Web Ontology Language (OWL) to ensure compatibility between mechatronic modules after a module change. A second use case demonstrates the application of OWL with the SPARQL Query Language to ensure consistency during model-based requirements and test case design for automated production systems. In a third use case, it is shown how the combination of the Resource Description Framework (RDF) and the SPARQL Query Language can be used to identify inconsistencies between interdisciplinary engineering models of automated production systems. We conclude with opportunities of applying Semantic Web Technologies to support the engineering of automated production systems and derive the research questions that need to be answered in future work.

Keywords Automated production systems · Semantic Web Technologies · Model-Based Engineering · Knowledge-based systems · Inconsistency management

---

S. Feldmann (✉) · K. Kernschmidt · B. Vogel-Heuser
Institute of Automation and Information Systems, Technische Universität München, Boltzmannstraße 15, 85748 Garching near Munich, Germany
e-mail: feldmann@ais.mw.tum.de

K. Kernschmidt
e-mail: kernschmidt@ais.mw.tum.de

B. Vogel-Heuser
e-mail: vogel-heuser@ais.mw.tum.de

S. Biffl and M. Sabou (eds.), *Semantic Web Technologies for Intelligent Engineering Applications*, DOI 10.1007/978-3-319-41490-4_14

14.1 Introduction

Rapidly evolving system and product requirements impose an increasing need regarding the efficiency and effectivity in engineering of automated production systems. Whereas such systems were mainly dominated through mechanical and electrical/electronic parts within the last decades, the significance of software is and will be increasing (Strasser et al. 2009). As a consequence, automated production systems need to fulfill ever increasing functionality and, thus, are becoming more and more complex. It is obvious that the engineering of automated production systems must follow this trend: methods and tools to better support engineers in developing and operating automated production systems need to be developed and further improved.

One means to improve the engineering of automated production systems is the use of Model-Based Engineering, thereby abstracting the view on the interdisciplinary system and providing a common base to improve understanding and communication. However, the multitude of disciplines and persons involved in the engineering process of automated production systems requires the use of a variety of different modeling languages, formalisms, and levels of abstraction and, hence, a number of disparate, but mostly overlapping models is created during engineering. Therefore, there is a need for tool support for, e.g., finding model elements within the models and keeping the engineering models consistent among each other.

In order to overcome this challenge, this Chapter illustrates how applying Semantic Web Technologies can support Model-Based Engineering activities in the automated production systems domain. Based on an application example (Sect. 14.2), the challenges that arise during engineering of automated production systems are presented in Sect. 14.3, followed by a short introduction of related works in the field of inconsistency management in Sect. 14.4. Subsequently, the technologies being used throughout this Chapter, namely, Resource Description Framework (RDF), Web Ontology Language (OWL), and SPARQL Query Language, are briefly introduced in Sect. 14.5. Three distinct use cases that illustrate how Semantic Web Technologies can support the engineering are introduced in Sect. 14.6:

- *Use case 1* (Sect. 14.6.1) describes the combination of a Systems Modeling Language (SysML)-based notation with OWL to ensure compatibility between mechatronic modules after a change of modules.
- *Use case 2* (Sect. 14.6.2) demonstrates how consistency can be ensured during model-based requirements and test case design by means of OWL and the SPARQL Query Language.
- In *use case 3* (Sect. 14.6.3) we show how the combination of RDF and the SPARQL Query Language can be used to identify inconsistencies between interdisciplinary engineering models of automated production systems.

The paper closes with a summary of the opportunities that can be gained for engineering of automated production systems by means of Semantic Web Technologies and with directions for future research.

14.2 Application Example: The Pick and Place Unit

In the following, as a basis to illustrate our research results, we introduce an application example, namely, the *Pick and Place Unit (PPU)*. The PPU is a bench-scale academic demonstration case derived from industrial use cases to evaluate research approaches and results at different stages of the life cycle in the automated manufacturing systems domain. Although the PPU is a simple demonstration case, it is complex enough to demonstrate an excerpt of the challenges that arise during engineering and operation of automated production system. To provide an evaluation environment as close to reality as possible, the PPU solely consists of industrial components. Furthermore, all scenarios and models that have been developed for the PPU were derived from real industrial use cases (Vogel-Heuser et al. 2014).

We assume that, in its initial configuration (Fig. 14.1), the PPU consists of four modules: a *stack*, a *crane*, a *stamp*, and a *ramp* module. The source of work pieces is represented through the stack module. Work pieces are pushed from the stack into a handover position, at which sensors are installed to identify the type of work pieces provided by the stack. Subsequently, depending on the type of work piece—either plastic or metallic—work pieces are transported to the stamp or ramp modules. In the stamp module, work pieces are detected at the handover position, subsequently positioned and clamped below a bistable cylinder, and finally stamped by applying pressure to the work piece. The final work piece depot is represented through the ramp module.

The transport between the different modules is realized by the crane module. Therein, a vacuum gripper, which is mounted on a boom, grips, or releases work

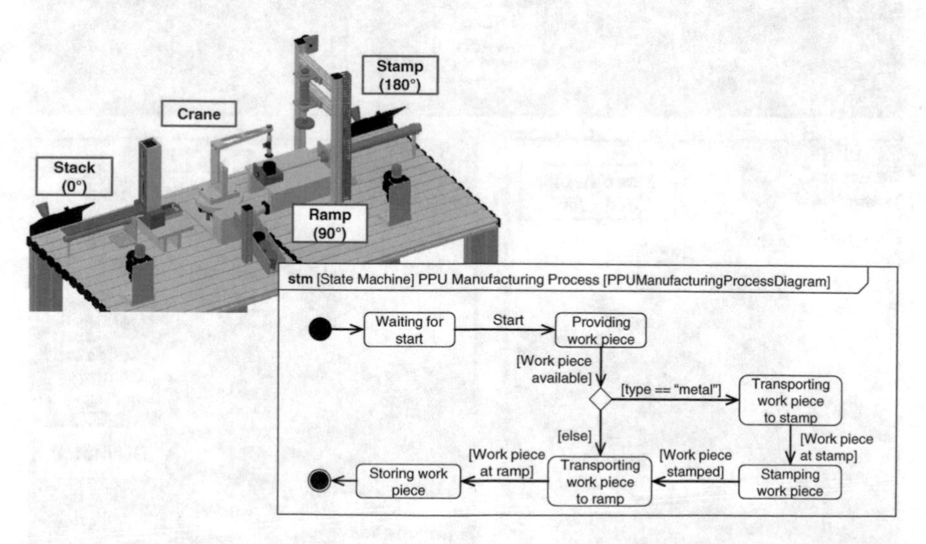

Fig. 14.1 Overview on the Pick and Place Unit (PPU) and its manufacturing process, extended from Feldmann et al. (2015a)

pieces. Using a bistable cylinder, the boom and, hence, the work pieces held by the vacuum gripper are lifted or lowered. The entire assembly is mounted on a turning table. A DC motor attached to the turning table allows for moving the crane to the respective modules. In order to detect the current angular position of the crane, three micro switches are attached to the bottom of the turning table: one at 0° (stack), one at 90° (ramp), and one at 180° (stamp). Figure 14.2 provides a detailed overview of the modules of the PPU.

During the engineering of the PPU, a multitude of different engineering models is created. For instance, disparate discipline-specific models are used by different engineers to cover aspects from the mechanical, electrical/electronic, or software engineering discipline (cf. use case 1 in Sect. 14.6.1) as well as to capture requirements and generate test cases from these (cf. use case 2 in Sect. 14.6.2). All these disciplines require distinct modeling approaches, formalisms as well as tools in order to cover the aspects of interest for the respective discipline. Although one common

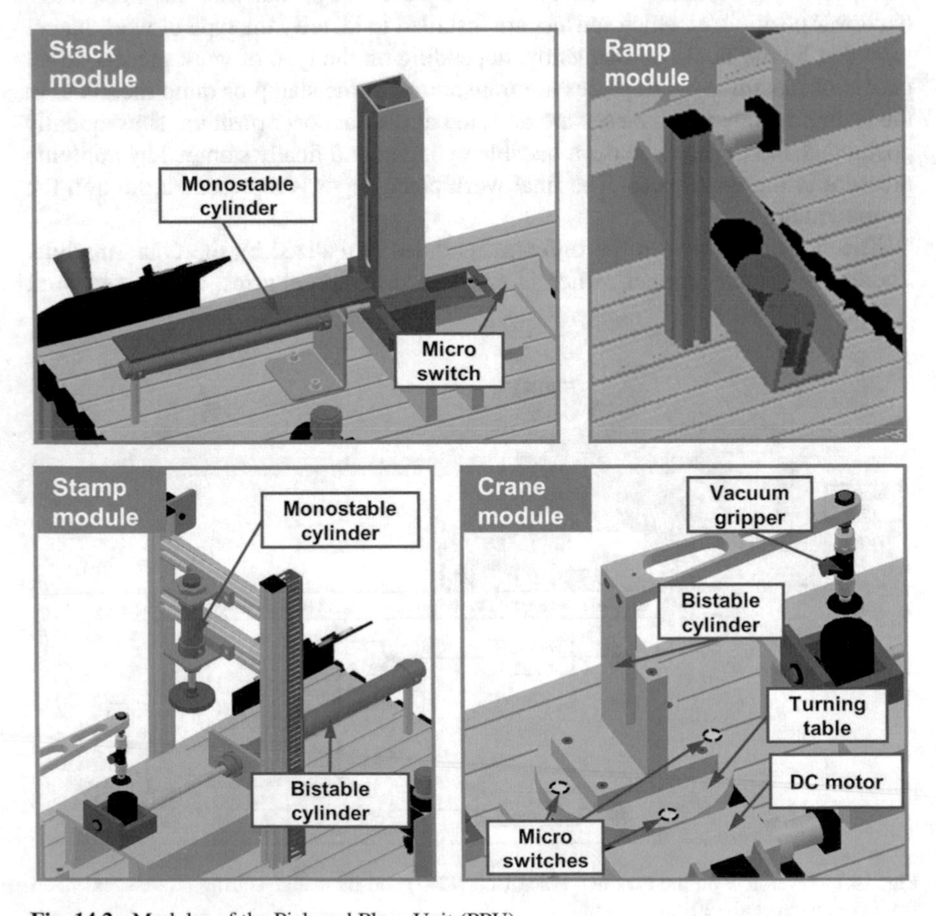

Fig. 14.2 Modules of the Pick and Place Unit (PPU)

principle in Model-Based Engineering is to separate the resulting views on the system as much as possible, complete separation cannot be achieved, leading to overlaps in the models. One example for such an information overlap is the representation of components in the different disciplines: For instance, a sensor is represented as a geometrical block in a CAD system, as a digital input in an electrical circuit diagram, and as a respective software variable in the control software. As a consequence of creating these interdisciplinary engineering models, the risk of introducing inconsistencies arises, which makes appropriate strategies for identifying and resolving inconsistencies necessary (cf. use case 3 in Sect. 14.6.3).

14.3 Challenges in the Automated Production Systems Domain

Requirements on the system and product are changing rapidly in the automated production systems domain, especially in the context of Industrie 4.0 (cf. Chap. 2). As a consequence, a number of challenges arise that need to be addressed in the engineering phase of automated production systems. An excerpt of these challenges has been introduced in (Feldmann et al. 2015b) and is briefly described in the following:

Challenge 1: **Heterogeneity of Models**. In order to address the concerns of the various stakeholders involved in the design and development of automated production systems adequately, a multitude of different formalisms, modeling languages, and tools is required. As an example for the PPU demonstrator, a requirements model could be used for early phases of the design, whereas a detailed CAD model could be applied for the detailed planning of the mechanical plant layout. Moreover, while some models are needed to specify particular aspects of a system (e.g., the mechatronic structure of the PPU), others are applied for the purpose of analysis (e.g., to predict the work piece throughput of the PPU in a simulation model). As a consequence, a set of disparate models is created that make use of fundamentally different formalisms, different abstraction levels, and whose expressiveness is restricted to concepts that are relevant to a specific application domain. This heterogeneity of models poses a major challenge (Gausemeier et al. 2009), as the composition of these models is hard, if not impossible to define a priory. Consequently, mechanisms for symbolic manipulation across these heterogeneous models are required.

Challenge 2: **Semantic Overlaps Between Models**. Although heterogeneous models are created during the life cycle of automated production systems, these models are overlapping, as different stakeholders have overlapping views on the system under study. These overlaps result from the presence of either duplicate or related information. For instance, a requirement on the minimum work piece throughput could be imposed in an early design stage of the PPU. In later verification phases, a simulation model could be used to analyze the work piece throughput that can be predicted under the given circumstances and system configuration. Clearly, there is a relation between the specified minimum work piece throughput and the predicted work piece throughout. Therefore, we say that statements are made about semantically related entities. We refer to such overlaps as *semantic overlaps*. As a consequence, such semantic overlaps between models need to be identified and specified as a basis for inconsistency management.

Challenge 3: **Automated Inconsistency Management within and Across Models**. Especially because of the heterogeneity of engineering models that are created for automated production systems (challenge 1) and the resulting overlap between these models (challenge 2), the risk of inconsistencies appearing in and across engineering models is high. Two strategies are commonly employed in order to manage these inconsistencies: (1) avoiding inconsistencies and (2) frequently searching for and handling inconsistencies. The first strategy requires to provide modelers with necessary and sufficient information about the different decisions, which are made by other stakeholders and impact their models. The second strategy is often employed in process models that include review (and testing) activities.

As models often consist of thousands of entities, at least some degree of automation and, hence, mechanisms for automated inconsistency management are required. Consequently, we argue that if a mechanism for symbolic manipulation across heterogeneous models is used and an appropriate method for identifying and defining semantic overlaps can be identified (i.e., if both challenges 1 and 2 can be addressed), both strategies become feasible. However, due to the lesser amount of knowledge that needs to be encoded and processed, we argue that the second strategy is likely to be more effective and less costly.

Challenge 4: **Integrated Tool Support**. In practice, especially in industrial environments, it is essential to keep the number of tools and methods that are applied as low as possible. In particular, for technicians and non-trained personnel, who are often involved during maintenance, start-up and operation of automated production systems, trainings are often costly. As a consequence, it is inevitable to integrate such support systems into existing tools instead of providing additional tools for the special purpose of managing inconsistencies. Thereby, stakeholders can work with the models and tools they are familiar with, without having to deal with additional models and tools focusing on inconsistency management. One essential challenge therein is to provide the mappings between the discipline-specific models and the respective symbolic manipulation mechanisms (cf. challenge 1) as a basis for inconsistency management.

14.4 Related Works in the Field of Inconsistency Management

In practice, inconsistency management is often included in complex review and verification tasks. However, the main challenge in identifying inconsistencies among the multitude of heterogeneous engineering models is that tools and models are often loosely coupled (see, e.g., (Kovalenko et al. 2014) as well as Chaps. 8 and 13). Consequently, even if modeling tools can import and export models serialized in standardized formats such as XML (World Wide Web Consortium 2008) or AutomationML (International Electrotechnical Commission 2014), such tool-specific implementations may differ and, thus, lead to compatibility issues. In some cases, point-to-point integrations (e.g., using model transformations) and tool-specific wrappers are used to alleviate this challenge but, nevertheless, are fragile and costly to maintain (Feldmann et al. 2015b).

A comparison of inconsistency management approaches in the related literature (Feldmann et al. 2015a) revealed that these approaches can be broadly classified into *proof-theory-based* (i.e., deductive), *rule-based* (e.g., inductive and abductive) as well as model *synchronization-based* (i.e., model transformation-based) approaches (Feldmann et al. 2015a). In proof-theoretic approaches, consistency to an underlying formal system can be shown and, thus, proof-theoretic approaches provide logically correct results. However, they have several practical limitations. For instance, proofs of engineering models require a complete and consistent definition of the underlying formal system, which in most cases is labor-intensive (if even possible). Model synchronization-based approaches require the transformation between different formal systems, which is not always possible without encoding large amounts of additional knowledge and information in transformations. Due to the effort in creating these transformations, this is unlikely to be feasible (practically and economically). While being less formal than proof-theoretic approaches, we conclude that rule-based approaches are a more flexible alternative as rules can be added without complete knowledge of an underlying formal system. Nevertheless, the rules used in rule-based approaches need to be maintained (revised and grown), resulting in possibly time-consuming and costly work. However, the completeness of such sets of rules can be varied, allowing for an economic trade-off. Therefore, for the purpose of inconsistency management in heterogeneous engineering models, we envision the use of a rule-based approach.

14.5 Semantic Web Technologies in a Nutshell

In order to address the aforementioned challenges in an appropriate manner, a highly flexible and maintainable (software) system is required. As a basis to achieve flexibility and maintainability, we envision the application of Semantic Web Technologies. In the following, the core technologies to providing effective and efficient engineering support in the automated production systems domain are presented. For a more detailed overview and introduction to basics of Semantic Web Technologies, please refer to Chap. 3.

From a Procedural Software System to a Knowledge-Based System

An exclusively procedural software system requires the explicit inclusion of knowledge about the structure and semantics of the various models involved during engineering within the code (Feldmann et al. 2015b). As a consequence, especially when a variety of disparate models is created (cf. challenge 1), maintaining and evolving such a software system is costly: in the worst case, the management of n models requires $n \cdot (n-1)/2$ bi-directional model integrations. A practical realization of a framework for supporting the engineering of automated production systems therefore requires a high degree of flexibility and extensibility. One means to achieve such a flexible and extensible framework is to represent models in a common representational formalism (Estévez and Marcos 2012) and to put appropriate mechanisms for, e.g., identifying and resolving inconsistencies in place. Consequently, we envision a *knowledge-based system* to be used to support the engineering in the automated production systems domain (Feldmann et al. 2015b). Among others, knowledge-based systems typically consist of two essential parts: a *knowledge base* that is used to represent the facts about the world (i.e., the knowledge modeled in the different models) and an *inference mechanism* that provides a set of logical assertions and conditions to process the knowledge base (i.e., to identify and resolve inconsistencies). Further typical parts of knowledge-based systems are the *explanation* and *acquisition* components as well as the *user interface*, which are, however, not in focus of this Chapter.

Representing Knowledge in Knowledge Bases: RDF(S) and OWL

One formal language used to describe structured information and, hence, to represent knowledge in the context of Semantic Web Technologies is the *Resource Description Framework (RDF)* (World Wide Web Consortium 2014). Originally, the goal of RDF is to allow applications to "exchange data on the Web while preserving their original meaning" (Hitzler et al. 2010), thereby allowing for further processing knowledge. Hence, the original intention of RDF is close to the challenge of heterogeneous models—to describe heterogeneous knowledge in a common representational formalism. Therein, RDF is similar to conceptual modeling approaches such as class diagrams in that it allows for statements to be made about entities, e.g., *stack is a module and consists of a monostable cylinder and a micro switch*. Such statements about entities are formulated by means of subject–predicate–object triples (e.g., *stack – is a – module*, *stack – consists of – micro switch*), thereby forming a directed graph. An exemplary RDF graph is visualized in Fig. 14.3. To leave no room for ambiguities, RDF makes use of so-called *Unified Resource Identifies* as unique names for entities (e.g., *ex:stack* and *ex:Module*) and properties (e.g., *ex:consistsOf*) being used. By means of so-called RDF vocabularies, collections of identifiers with a clearly defined meaning can be described. For such a meaning to be described in a machine-interpretable manner, besides specifying knowledge on instances (i.e., *assertional knowledge*), the RDF recommendation allows for specifying background information (i.e., *terminological knowledge*) by means of *RDF*

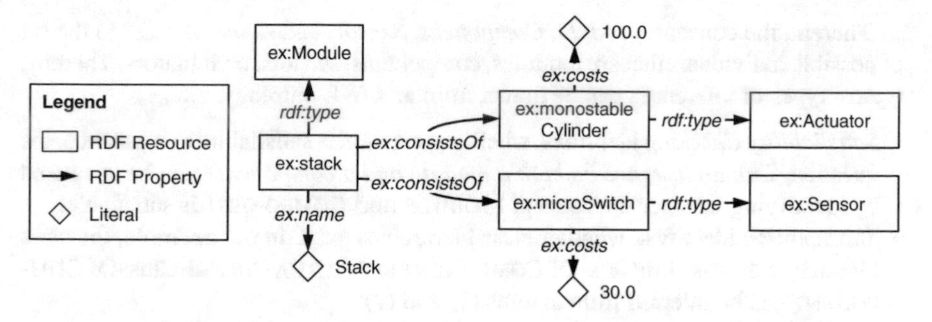

Fig. 14.3 Exemplary RDF graph

Schema (RDFS). RDF(S) provides language constructs to formulate simple graphs containing class and property hierarchies as well as property restrictions. With its formal semantics, RDF(S) leaves no room for interpretation of what conclusions can be drawn from a given graph, thereby providing standard inference mechanisms for any RDF(S)-compliant graph. RDF(S) can, hence, be used as a language to model simple ontologies, but provides limited expressive means and is not suitable to formulate more complex knowledge (Hitzler et al. 2010). Examples for knowledge that cannot be formulated in RDF(S) are the phrases *Each Module consists of at least one component* and *Components are either actuators or sensors*.

One mean to formulate more complex knowledge are rules that can be used to draw *conclusions* from a *premise* statement, i.e., by applying rules in the form of *IF premise THEN conclusion*. Another mean to formulate complex knowledge is the use of the *Web Ontology Language (OWL)* (World Wide Web Consortium 2009), which provides further language constructs defined with description logics based semantics. OWL moreover contains two sub-languages[1] to provide a choice between different degrees of expressivity, scalability, and decidability, namely, OWL Full and OWL DL. Therein, for the purposes of this Chapter, OWL DL[2] enables maximum expressivity while maintaining decidability (Hitzler et al. 2010). The OWL DL formal (description logics based) semantics allow to define what conclusions can be drawn from an OWL DL ontology. For instance, the aforementioned phrases *Each Module consists of at least one component* and *Components are either sensor or actuators* can be formulated as specified in OWL[3] axioms (1) and (2).

MODULE **EquivalentTo** (*consistsOf* **some** COMPONENT) (1)

COMPONENT **EquivalentTo** (SENSOR **or** ACTUATOR) (2)

---

[1]Note that, in addition to OWL DL and OWL Full, there are three profiles for a variety of applications, namely, OWL EL, QL, and RL which, however, are out of the scope of this Chapter.

[2]For reasons of simplicity, we use the term *OWL* when referring to *OWL DL*.

[3]To enhance readability, OWL Manchester Syntax is used throughout the paper.

Therein, the concepts *Module*, *Component*, *Sensor*, and *Actuator* refer to the set of possible individuals that are modules, components, sensors, or actuators. Thereby, typical types of inferences can be drawn from an OWL ontology, e.g.,

- *Satisfiability Checking* identifies, whether a concept is satisfiable. For instance, the question, *Can an instance be both a module and a component?* can be answered by identifying whether the concept MODULE **and** COMPONENT is satisfiable.
- *Subsumption* identifies, whether class hierarchies exist. In our example, the class hierarchy SENSOR **SubClassOf** COMPONENT and ACTUATOR **SubClassOf** COMPONENT can be inferred from axioms (1) and (2).
- *Consistency Checking* identifies whether inconsistencies in the model exist.

Accessing Knowledge in Knowledge Bases: SPARQL Query Language

A set of specifications providing the means to retrieve and manipulate information represented in RDF(S) (or OWL, respectively) is the *SPARQL Protocol and RDF Query Language* (World Wide Web Consortium 2013). The primary component of the standard is the *SPARQL Query Language*.[4] SPARQL is in many regards similar to the well-known *Structured Query Language (SQL)*, which is supported by most relational database systems.

A query consists of three major parts: *namespace* definitions being used within the query, a *clause* identifying the type of the query and a *pattern* to be matched against the RDF data. SPARQL is highly expressive and allows for the formulation of required and optional patterns, negative matches, basic inference (e.g., property paths to enable transitive relations), conjunctions, and disjunctions of result sets as well as aggregates, i.e., expressions over groups of query results. Four disparate query types can be used in SPARQL

- *SELECT queries* return values for variable identifiers, which are retrieved by matches to a particular pattern against the RDF graph,
- *ASK queries* return a Boolean variable that indicates whether or not some result matches the pattern,
- *CONSTRUCT queries* allow for substituting the query results by a predefined template for the RDF graph to be created, and
- *DESCRIBE queries* return a single RDF graph containing the relevant data about the result set. As the "relevance" of data is strongly depending on the specific application context, SPARQL does not provide normative specification of the output being generated by DESCRIBE queries.

An example for a SPARQL SELECT query is shown in Fig. 14.4. Using this query, the RDF graph in Fig. 14.3 is queried for entities x that consist of an entity y, which, in turn, is described by the cost value c. By using the BIND form, the Boolean result of the formula *?c > 50.0* can be assigned to variable *moreThan50*, which denotes, whether the cost value c is greater than *50* or not.

---

[4]For reasons of simplicity, we use the term *SPARQL* when referring to *SPARQL Query Language*.

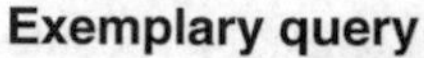

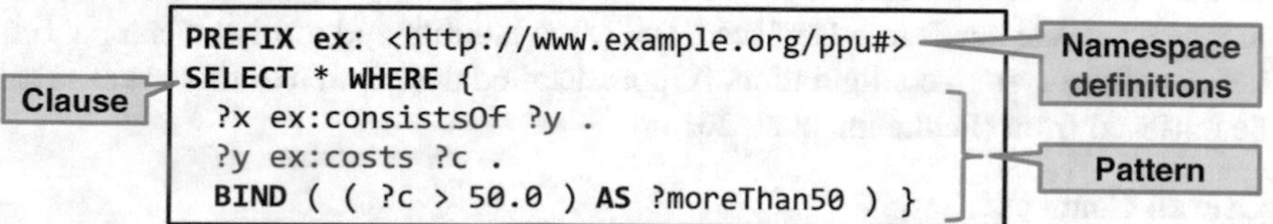

Results of query execution

?x	?y	?c	?moreThan50
ex:stack	ex:monostableCylinder	100.0	true
ex:stack	ex:microSwitch	30.0	false

Fig. 14.4 Exemplary SPARQL query (*top*) and results when executing the query

14.6 Use Cases for Applying Semantic Web Technologies in the Automated Production Systems Domain

Within the following, three distinct use cases are presented, which aim at supporting the engineering in the automated production systems domain by means of Semantic Web Technologies.

14.6.1 Use Case 1: Ensuring the Compatibility Between Mechatronic Modules

Automated production systems are characterized by a multitude of mechanical, electrical/electronic and software components with tight interrelations between them. In order to facilitate the development of automated production systems and reduce costs of the engineering process, companies usually define mechatronic modules including components from different disciplines, which can be reused in various systems.

During the life cycle of such a system, frequent changes have to be carried out to different system components or modules, e.g., if new customer requirements have to be fulfilled or if specific components/modules have to be replaced but are not available on the market any more. A challenge in carrying out changes during the life cycle of an automated production system is to ensure the compatibility of the exchanged component/module with the existing system (e.g., regarding data ranges of specified properties, type compatibility, etc., (Feldmann et al. 2014a)). Lacking consideration of change influences can lead to further necessary changes in the system, which are costly and can prolong the downtime unnecessarily.

Therefore, this use case describes how a model-based approach can be used to analyze changes before they are implemented in the real system (Feldmann et al. 2014a). Consequently, we aim at combining such a model-based approach with

Semantic Web Technologies to provide the means (1) to identify compatible modules in case a module needs to be replaced, and (2) to identify and resolve incompatibilities in a given system configuration. A more detailed description of this use case can be retrieved from (Feldmann et al. 2014a).

Overall Concept

In order to specify the relevant aspects for checking the system for incompatibilities, an information model is defined that contains the information necessary for identifying whether two modules are compatible or not. By that, any Model-Based Engineering (MBE) approach can be combined with our compatibility information model, which can directly be used to compute the formal knowledge base. Figure 14.5 shows the relation between the MBE approach, the formal knowledge base and the information model with its elements and relations. As shown in the information model, structural aspects of mechanical, electrical/electronic, and software components, which can be combined to mechatronic modules, and the respective interfaces in the different disciplines are considered for the compatibility check. Furthermore, the functionalities, which are fulfilled by a component/module, are considered.

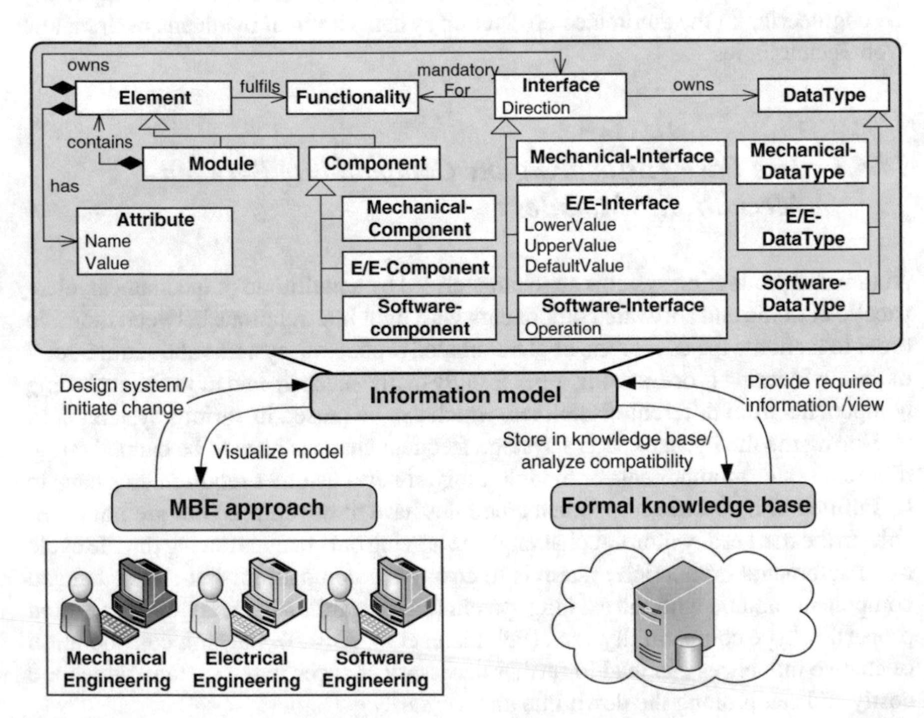

Fig. 14.5 Combination of MBE approach and a formal knowledge base to analyze changes, extended from Feldmann et al. (2014a)

Visual Model for Modeling Systems Comprehensibly

Regarding the visual model, Model-Based Engineering (MBE) approaches gained more and more influence over the past years. Especially for systems engineering, the *Systems Modeling Language (SysML)* (Object Management Group 2012) was developed as a graphical modeling language to represent structural and behavioral aspects during development. Through specific modeling approaches, based on SysML, the relevant aspects for analyzing changes can be integrated into the model. In this use case, SysML4*Mechatronics* (Kernschmidt and Vogel-Heuser 2013) is used as modeling language, as it was developed specifically for the application of mechatronic production systems. An exemplary SysML4*Mechatronics* model for the PPU application example is shown in Fig. 14.6. Next to the mechatronic modules (*stack*, *crane*, *stamp*, and *ramp*), the *bus coupler*, the *PLC* and the *mounting plate* are depicted as separate blocks, which are required by the modules, e.g., all modules are connected to the mounting plate of the system mechanically and, if required, communicate through a Profibus DP interface.

OWL and SPARQL for Compatibility Checking

For identifying incompatibilities between mechatronic modules, we argue that two disparate types of compatibility rules exists

- *Inherent compatibility rules* apply for arbitrary types of automated production systems and must not be violated by any system under study. Examples for such inherent compatibility rules are type and direction compatibility.
- *Application-specific compatibility rules* apply within a given context (e.g., for specific types of systems or for a concrete project). For instance, project-specific naming conventions are often applied for specific projects or applications.

In order to allow for flexibly maintaining and extending a software system for checking compatibility of mechatronic modules, we envision the application of

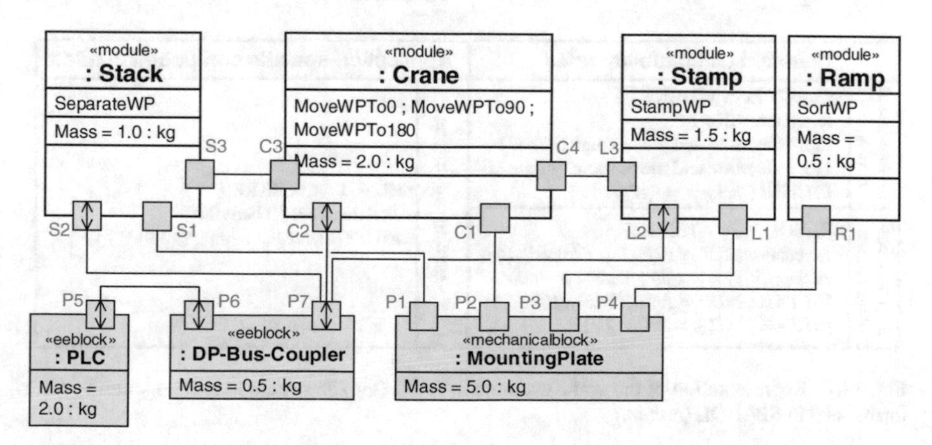

Fig. 14.6 Exemplary SysML4*Mechatronics*model for the PPU application example

OWL, which provides the means to formulate the knowledge in a compatibility information model shown in Fig. 14.5, and SPARQL, which allows identifying whether certain compatibility rules are violated or not. Within OWL, a domain ontology is used, which defined the concepts and relations necessary to represent the knowledge in our compatibility information model. Using this domain ontology, knowledge on available modules, interface types, etc., can be formulated in the ontology's terminological knowledge. Accordingly, knowledge on the instances available for the system under study is represented in the ontology's assertional knowledge. As a consequence, SPARQL queries can easily be formulated using the terms defined in the compatibility domain ontology. We argue that SPARQL queries can, hence, be defined, maintained, and managed more efficiently using such a domain ontology.

The representation of the crane application example in an OWL ontology as well as some exemplary compatibility rules are illustrated in Fig. 14.7. Using SPARQL SELECT queries, incompatibility patterns are described; any result returned by querying the ontology stands for incompatible elements within the model. Through

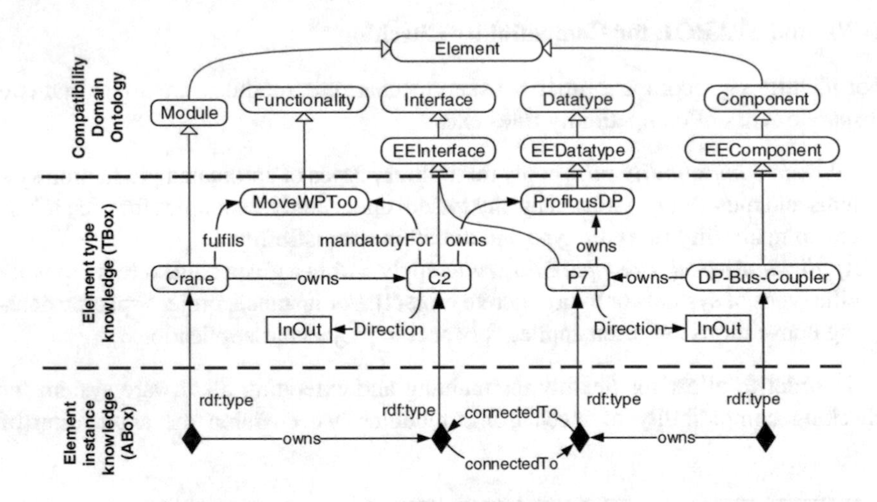

	Inherent compatibility rules		Application-specific compatibility rules
1 – Interfaces' data types	**SELECT** *?x ?y* **WHERE** { *?x* :connectedTo *?y* . *?x* **a** (:Interface **and** :owns **some** *?xType*) . *?y* **a** (:Interface **and** :owns **some** *?yType*) . **FILTER** (*?xType* != *?yType*) . }	3 – Naming conventions	**SELECT** *?x* **WHERE** { *?x* **a** :Element ; :Name ?n . **FILTER** (! regex (?n , „^[^0-9].*")) . }
2 – Interfaces' data ranges	**SELECT** *?x ?y* **WHERE** { *?x* :connectedTo *?y* . *?y* :UpperValue *?yUpp* . *?x* :UpperValue *?xUpp* ; :Direction *?xDir* . **FILTER** (*?xDir* = „In"||*?xDir* = „InOut") . **FILTER** (*?yUpp* > *?xUpp*) . }		

Fig. 14.7 Representation of the crane example in OWL (*top*) and exemplary compatibility rules formulated in SPARQL (*bottom*)

queries 1 and 2, it can be identified whether data types and ranges of two ports are compatible or not. Query 3 defines an application-specific compatibility rule that requires entities' names not to start with a number.

Proof-of-Concept Implementation

The conceptual architecture introduced previously was realized in terms of a proof-of-concept implementation, see Fig. 14.8. The models are defined in the Eclipse Modeling Framework (EMF).[5] The necessary model transformations, i.e., between SysML4*Mechatronics*, the compatibility information language and OWL, were realized by means of the Query/View/Transformation Operational (QVTo) standard (Object Management Group 2011) and executed using the Eclipse QVTo implementation.[6] For OWL, a meta model was developed in the well-known Ecore format based on the respective W3C standard (World Wide Web Consortium 2009). Subsequent XSL Transformations allow for transforming between the EMF-specific XML format and the respective tool-specific XML formats. Pellet[7] was used as the reasoner and querying engine for executing the queries.

Validation

In the current state (shown in Fig. 14.6), the crane has four distinct interfaces for the connection to the rest of the system. During the life cycle of the system, the crane has to be replaced. Due to the aspired shift to a Profinet bus system, a crane module with Profinet interface shall be used for the exchange. Except for the change of the bus system, the crane module shall fulfill the same functionalities. Figure 14.9 shows the SysML4*Mechatronics* model of the system with the replaced crane module.

Having transformed the modeled information into the OWL DL ontology of the formal knowledge base, the respective SPARQL queries can be executed. Using the

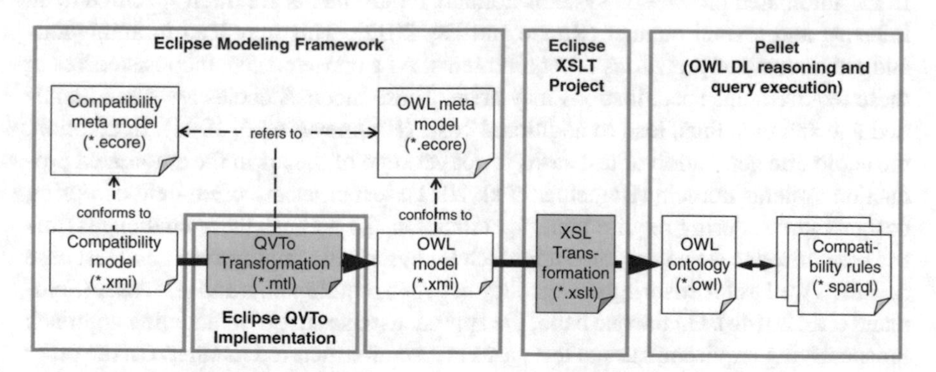

Fig. 14.8 Overview on the architecture for compatibility checking

[5] https://eclipse.org/modeling/emf/, retrieved on 12/11/2015.

[6] https://wiki.eclipse.org/QVTo, retrieved on 12/11/2015.

[7] https://github.com/complexible/pellet, retrieved on 12/11/2015.

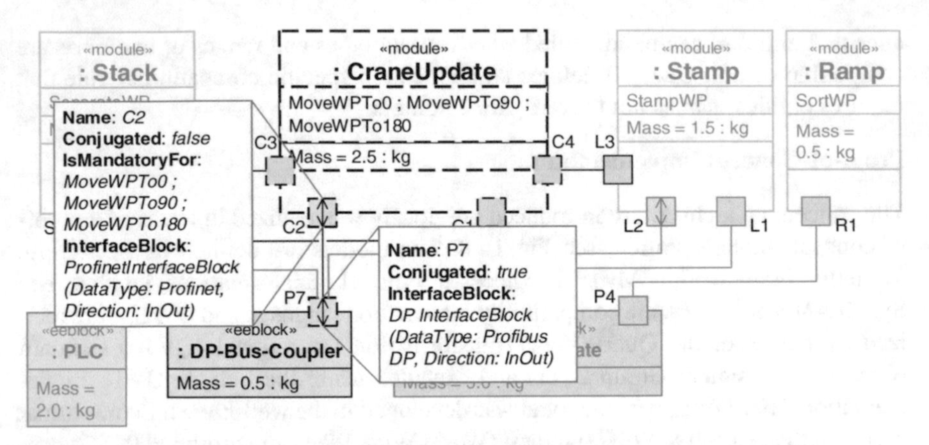

Fig. 14.9 Exchanged crane module (represented as SysML4*Mechatronics* module)

example queries as described above, an incompatibility between the Profinet and the Profibus DP interface can be identified (cf. query 1), as the respective datatypes are not compatible. Queries 2 and 3 do not identify incompatibilities, as data ranges are not applicable for the illustrated example (cf. query 2) and no violations of the naming conventions for entities can be identified (cf. query 3).

14.6.2 Use Case 2: Keeping Requirements and Test Cases Consistent

In the automated production systems domain, requirements are often specified in an informal and textual manner (Runde and Fay 2011). This may lead to ambiguous and erroneous interpretations of requirements. As a consequence, inconsistencies in these requirements specifications may arise. These inconsistencies are often identified late and can, thus, lead to additional costs (Heitmeyer et al. 1996). In addition, the automatic generation of test cases is not yet state of the art in the automated production systems domain (Hametner et al. 2011): test cases are often defined during design and not during requirements specification. To address these challenges, this use case introduces an integrated approach for systematic requirements and test case design as well as for ensuring consistency between requirements and test cases (Feldmann et al. 2014b). On the one hand, the approach makes use of a modeling approach for specifying requirements and test cases in a semi-structured manner. On the other hand, Semantic Web Technologies are applied to allow for consistency checking of the models and, thus, of the modeled requirements and test cases. For a detailed description of the models being used in this use case, please refer to (Feldmann et al. 2014b).

Modeling Requirements and Test Cases

The approach for modeling requirements and test cases systematically and graphically consists of three primary modeling elements, namely, *feature*, *requirement*, and *test case* (cf. Fig. 14.10).

Features. Features refer to a plant component or functionality, e.g., the capability of transporting or detecting a work piece. In the PPU application example, the feature *PPU* as well as its sub-features *Stack*, *Stamp*, *Ramp*, and *Crane* are defined. In addition, features are characterized by parameters that either represent sensors or actuators of the system under study (namely, in- and out-parameters) or parameters that describe the product and environment. By means of these parameters, requirements on features can be further specified. For instance, the crane feature is further defined by the in-parameters *source*, which defines the source of the transport function (either "ramp", "stack," or "stamp"), *velocity*, which characterizes the crane module's velocity, as well as the work piece's *Mass* and *type*. Out-parameters

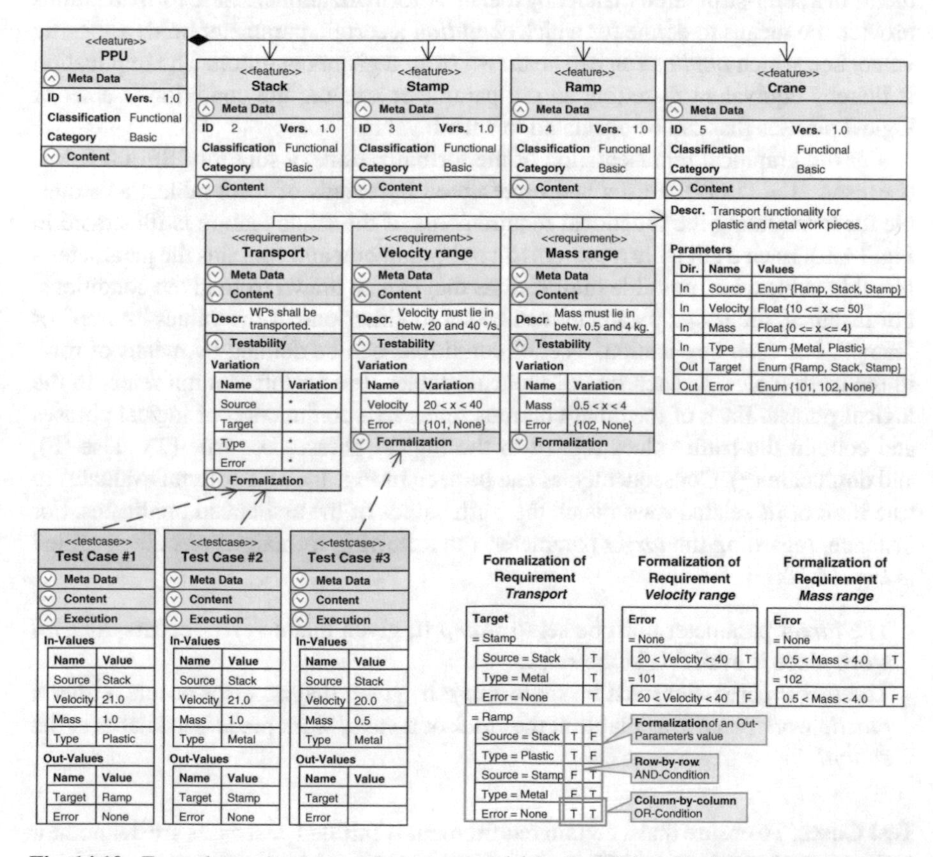

Fig. 14.10 Example requirements and test case model for the PPU application example, extended from Feldmann et al. (2014b)

that are available for the crane feature are *target*, which specifies the target of the transport function, and *error*, which defines an error code (either "101", "102," or "None").

Requirements. A requirement is associated to a set of features, thereby characterizing their intended behavior by means of parameter restrictions. In our PPU application example, three distinct requirements are defined: a *transport* requirement defines how work pieces are transported, and the *velocity* and *mass range* requirements specify applicable ranges for the crane's velocity and a work piece's mass. Parameter variations in requirement specifications define in which range a parameter needs to be tested for a given requirement. For instance, the crane must be able to operate for work pieces that weigh between *0.5* and *4* kg. These variations can furthermore be used for generating test cases, e.g., by generating all possible combinations of in-parameter values with respect to their variations. As requirements can either be functional or nonfunctional, the modeling approach allows modelers to either textually specify a (nonfunctional) requirement or to specify their (functional) requirements in a semi-structured manner by means of formalizations. These formalizations provide the means to define for which *condition* a certain parameter holds a specific value, i.e., which *implication* can be drawn from a given condition. The implication is thereby equivalent to setting an out-parameter's value; the condition is a set of logical phrases that can be evaluated to a truth value.

For the graphical representation of the formalizations, a subset of SPECTR-ML (Leveson et al. 1999) is used and represented by means of truth tables; an example for formulating the functional requirements of the crane feature is illustrated in Fig. 14.10. Each truth table represents an out-parameter and contains the parameter's possible values, i.e., possible implications that can be drawn from given conditions. For instance, the *target* parameter can be set to either one of the values "stamp" or "ramp". For each implication, a set of conditions can be defined by means of rows in the truth tables. In each row of the truth tables, the far left column refers to the logical phrase. Each of the other columns refers to a conjunction of logical phrases and contain the truth values related to the logical phrase, i.e., true (T), false (F), and don't care (*). Consequently, as can be seen in Fig. 14.10, a column evaluates to true if all of its related rows match the truth values of the associated predicates. For instance, regarding the *target* parameter, the following formalizations are specified in Fig. 14.10:

- The *target* parameter must be set to *stamp* if, given that no error occurs, a *metal* work piece is available at the *stack*.
- The *target* parameter must be set to *ramp* if, given that no error occurs, either a *plastic* work piece is available at the *stack* or a *metal* work piece is available at the *stamp*.

Test Cases. To ensure that a certain requirement is fulfilled, test cases are defined for a given set of requirements. These test cases can either be specified manually (e.g., for non-functional requirements) or generated automatically (e.g., from formalizations of functional requirements). A test case therefore consists of a set of parameter values to be considered for a test: based on given in-parameter values, expected out-parameter values are defined. Three exemplary test cases for the PPU application example are shown in Fig. 14.10.

Ensuring Consistency Between Requirements and Test Cases

Although the previously introduced modeling concept provides a starting point for discussing and communicating the system under study, potential inconsistencies in the requirement and test case specification may arise. Among others, the following types of inconsistencies are likely to occur:

- *Inconsistencies between features and (functional) requirements*: Parameters are specified by valid parameter ranges in the feature. These parameters are used and further detailed in requirements associated to the feature by means of parameter variations that define in which range a parameter must be tested. In order for a model to be valid, the parameter ranges defined in the feature must be consistent with the parameter variations specified for a requirement.
- *Inconsistencies in between (functional) requirements*: The set of requirements defined for a feature is formalized by means of logical conditions and implications. Consequently, the entire set of requirements defined for a feature describes the system behavior as desired by the modeler. These requirements may consist of contradicting formalizations, leading to potential inconsistencies between requirements. Consistency must therefore be ensured between the formalizations of a feature's requirements.
- *Inconsistencies between (functional) requirements and test cases*: Based on the parameter variations defined in a feature's requirements, test cases can be generated, e.g., by identifying all permutations of available in-parameters. Nevertheless, for each given permutation of in-parameters, a unique combination of out-parameters must be determinable in order for a model to be consistent.

It is obvious that all of these types of inconsistencies can be identified as a logical consequence of contradiction statements. Hence, the use of OWL, its description logics based semantics and the inference types that can be drawn from an OWL model (e.g., whether a concept is satisfiable or whether inconsistencies in the model exist) is an appropriate solution approach. As a consequence, based on the parameter ranges in a feature we formulate a set of possible feature states by means of an OWL concept. Each implication defined in the requirements' truth tables can then be specified as a sub-concept of this feature state—an OWL reasoner can consequently determine whether or not this concept is satisfiable, i.e., whether the requirement is consistent to the feature. If implications are over-specified (see, for instance, the specification of the implication *Error* = *None* in Fig. 14.10), the respective implication is defined as the intersection of the existing set of implications. If the set of implications is inconsistent, an OWL reasoner identifies the model to be inconsistent, i.e., that there is an inconsistency between the requirements. Finally, test cases are regarded as states of the feature concept and, hence, if contradictions between a test case and the requirement occur, an OWL reasoner identifies the model to be inconsistent. Consequently, using such an OWL model, all aforementioned types of inconsistencies can be identified.

In the resulting OWL model, parameters defined in respective features are represented by OWL data properties. As we assume that for each parameter only one value exists in a feature's state, the corresponding data properties are defined to be functional. Consequently, we formulate a feature's state as an OWL concept, which is equivalent to the intersection of the respective parameter ranges defined in the feature. For instance, the crane feature's state can be formulated within a respective concept CraneState (cf. axiom (3)).

CraneState **EquivalentTo** (*Error* **some** {"101", "102", "None"})
and (*Source* **some** {"Stack", "Stamp", "Ramp"})
and (*Target* **some** {"Stack", "Stamp", "Ramp"})
and (*Type* **some** {"Metal", "Plastic"})
and (*Velocity* **some float**[≥10, ≤50]) **and** (*Mass* **some float**[≥0, ≤4]) (3)

Respectively, formalizations in truth tables of requirements are defined as OWL concepts being sub-concepts of the feature's state. For instance, the formalization of requirements *velocity* and *mass range* are specified in axioms (4)–(6):

Error101 **EquivalentTo** CraneState **and** (**not** *Velocity* **some float**[>20, <40]) (4)

Error102 **EquivalentTo** CraneState **and** (**not** *Mass* **some float**[>0.5, <4]) (5)

ErrorNone **EquivalentTo** CraneState **and** (*Velocity* **some float**[>20, <40])
and (*Mass* **some float**[>0.5, <4]) (6)

Accordingly, the implications being defined in the requirements' truth tables can be formulated as the necessary conditions for the respective concepts, see axioms (7)–(9).

Error101 **SubClassOf** (*Error* **value** "101") (7)

Error102 **SubClassOf** (*Error* **value** "102") (8)

ErrorNone **SubClassOf** (*Error* **value** "None") (9)

The formalizations of the requirement *transport* are specified accordingly. Using the model, it can be identified whether contradictions exist between requirements and features as well as in between requirements. In the present PPU application example, it is obvious that no inconsistency exists and, hence, an OWL reasoner infers the model to be consistent. However, if we formulate the test cases as depicted in Fig. 14.10 to be instances of the feature state's concept CraneState, an inconsistency can be detected for test case 3: as both the *velocity* and *mass* parameters violate the formalizations imposed in the *velocity* and *mass range* requirements, an OWL reasoner infers the *error* parameter to hold both the values "101" and "102". Nevertheless, as all properties were defined to be functional, the OWL reasoner identifies an inconsistency. This ambiguity is identified and the engineer is notified that the requirements need to be specified further; it is identified that the test case 3 cannot be consistent to the requirements.

Proof-of-Concept Implementation

In order to evaluate the concept for modeling requirements and test cases and for ensuring consistency, a proof-of-concept implementation was developed. An overview on the architecture of the prototypical tool is given in Fig. 14.11. Model instances are created by the modeler using the Eclipse Modeling Framework (EMF). Therefore, a metamodel was defined in the well-known Ecore format, which allows for providing a simple tree editor in EMF. The mapping to the respective OWL ontology was defined by means of the MOF Model to Text Standard (MOFM2T) (Object Management Group 2008), which can, e.g., be executed in the Acceleo[8] MOFM2T implementation. It has to be noted that the domain ontology (that, e.g., defines the base properties and classes to be used) is independent from the created models and, hence, is imported by the respective model-specific application ontologies. The resulting OWL ontology can then be processed by available OWL reasoners—in our implementation, Pellet was used.

14.6.3 *Use Case 3: Identifying Inconsistencies in and Among Heterogeneous Engineering Models*

During the development and operation of automated production systems, a multitude of stakeholders from different disciplines is involved. In order for the specific concerns of these stakeholders to be addressed, views on the system under study are

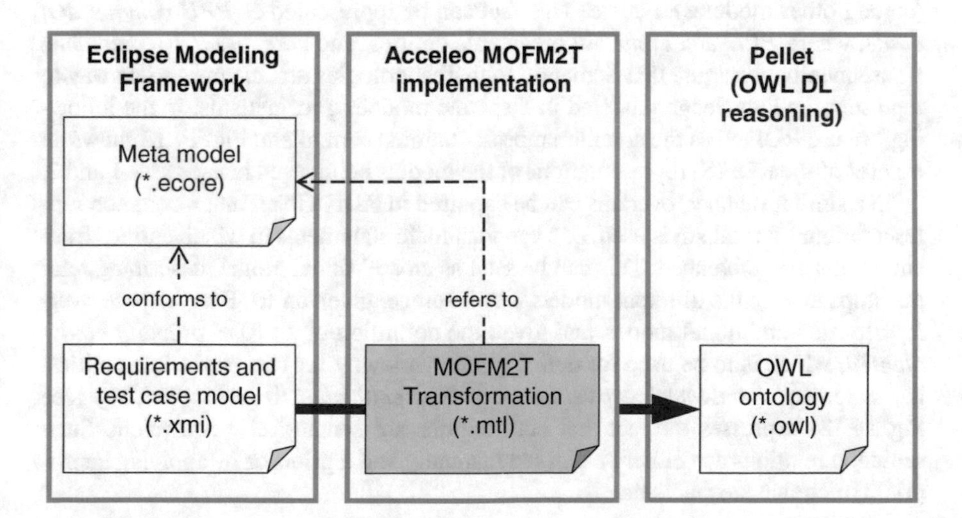

Fig. 14.11 Overview on the architecture for consistency checking among features, requirements and test cases

[8]http://www.eclipse.org/acceleo/, retrieved on 12/11/2015.

formed. To adequately address these concerns, a variety of different formalisms, modeling languages and tools is necessary (Broy et al. 2010). Two distinct formalisms that address different stakeholders' concerns were shown in the previous sections: one for the detailed design of automated production systems (cf. use case 1) and one for requirements and test case management (cf. use case 2). Nevertheless, although it is considered good practice to separate concerns as much as possible, completely separating concerns is impossible. As a consequence, some concerns are addressed by the various stakeholders and, hence, lead to overlaps in the models. An example of such a model overlap can be seen in use cases 1 and 2: both models refer to the different modules of the PPU, therefore specifying overlapping information on the system under study and, hence, leading to the potential for inconsistencies to occur. Consequently, it is inevitable to ensure that the set of models is free of inconsistencies, which is being addressed in this use case. A detailed description of the use case can be found in (Feldmann et al. 2015b).

Using RDF to Represent Heterogeneous Models in a Common Formalism

As argued beforehand, effectively and efficiently handling inconsistencies necessitates a high degree of flexibility. Consequently, we argue that representing the heterogeneous models in a common representational formalism is inevitable for any inconsistency management framework to be economical.

Information in knowledge bases is typically represented using predicated statements about entities (Giarratano and Riley 1994). This is similar to how information and knowledge are represented in the abstract syntax of models (Herzig et al. 2011): For instance, one fact encoded in use case 1 asserts that the PPU is composed of (among other modules) a crane. This fact can be represented as *PPU composedOf crane*, where *PPU* and *crane* are predicable entities, and *composedOf* a predicate. Consequently, we argue that Semantic Web Technologies are an appropriate way to represent the knowledge modeled in disparate modeling formalisms. In the following, we use RDF(S) as the common representational formalism; Fig. 14.12 shows an excerpt of the RDF(S) representations of the models being used in use cases 1 and 2.

In a similar manner, overlaps can be captured in RDF. Given that a common representational formalism is used, one can formulate statements in which entities from any model are referenced. This can be seen as an additional model, describing relationships among the different models. For instance, given an RDF namespace *overlaps* for this additional model, and given the definition of an RDF property *equivalentTo*, which is to be used for defining the synonymy for two predicable entities, the statement *sysml4mech:expVelocity overlaps:equivalentTo req:reqVelocity* (see Fig. 14.12) expresses the fact that both entities are semantically equivalent. Such semantic relations can either be defined manually and a priori or by applying appropriate inference mechanisms.

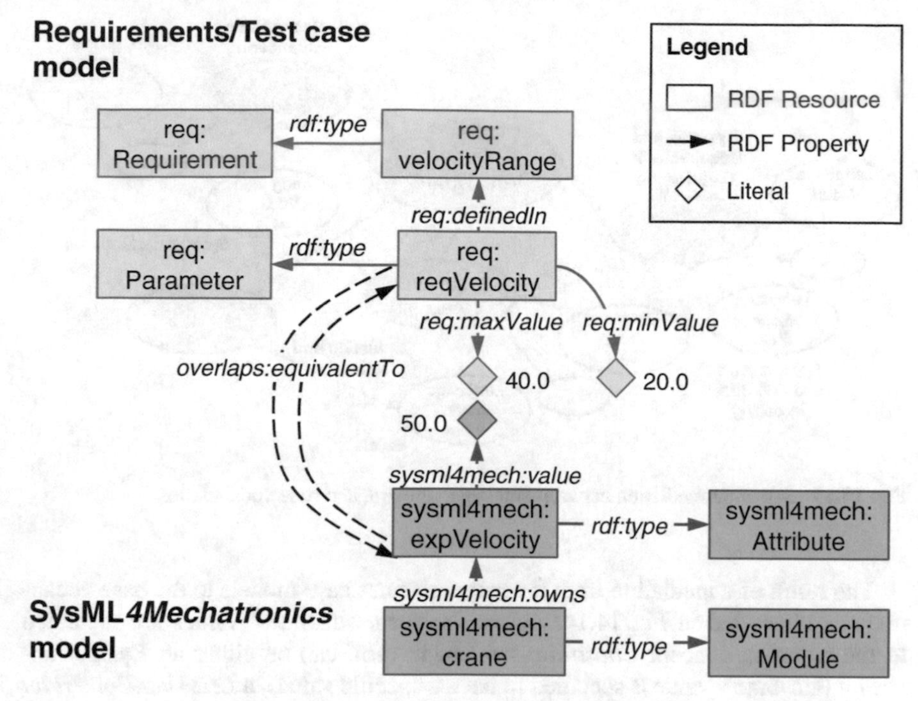

Fig. 14.12 Excerpt of the RDF representation for the use case models

Vocabularies for Identifying Cross-Model Inconsistencies

Given that RDF(S) as the common representational formalism is used, inconsistencies among the heterogeneous models can be identified. However, with an increasing number of models involved, the number of necessary integration between models arises, making mechanisms for efficient and effective handling of inconsistencies necessary. Therefore, we argue that a common terminology is required that bridges the gap between different domain models and, thus, (1) provides a common syntax, and (2) allows for defining the semantic relations between the modeled information. We hypothesize that—at least at some level of (semantic) abstraction—there exist concepts common to specific domains, which can be represented using different languages, and that concepts exist that are common to all domains, e.g., the term *value* that is used in both use cases 1 and 2. Using such a *base vocabulary* (cf. Fig. 14.13), some common inconsistencies among and between multiple domain models can be managed. Clearly, the definition of *semantic overlaps* may not always require the full expressiveness of domain models; thus, the base vocabulary is semantically much weaker than *domain vocabularies* or *language vocabularies* and represents a "common denominator" to all other vocabularies. The base vocabulary ideally remains unchanged, while integrating a further type of domain model necessitates only a respective novel RDF representation and the definition of a *semantic mediation* between the novel domain vocabulary and the base vocabulary.

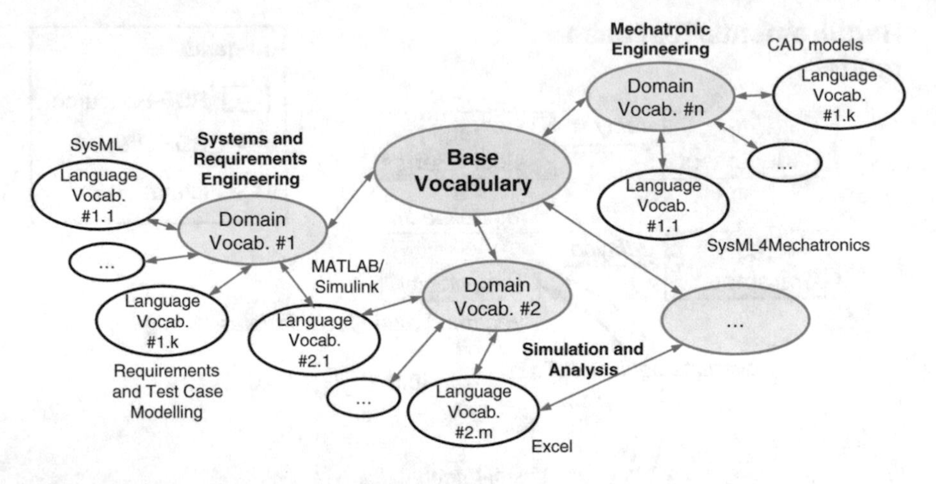

Fig. 14.13 Semantic mediation between language, domain, and base vocabularies

The result of a mediation from the respective use case models to the base vocabulary is illustrated in Fig. 14.14. As can be seen, attributes' values are mediated to the common concept *Constraint*, which, in turn, can be either an *EqualsConstraint* (attribute's value is specified to have a specific value), a *LessThanConstraint* (attribute's value is specified to be less than a specific value) and a *GreaterThanConstraint* (attribute's value is specified to be greater than a specific value). By that, inconsistency rules can be formulated by referring to the concepts defined in the base vocabulary.

SPARQL Queries for Identifying Inconsistencies

For engineering of physical systems such as automated production systems, it is impossible to say whether or not such systems are fully consistent (Herzig et al. 2011). The main reason is the lack of perfect knowledge about the processes and the

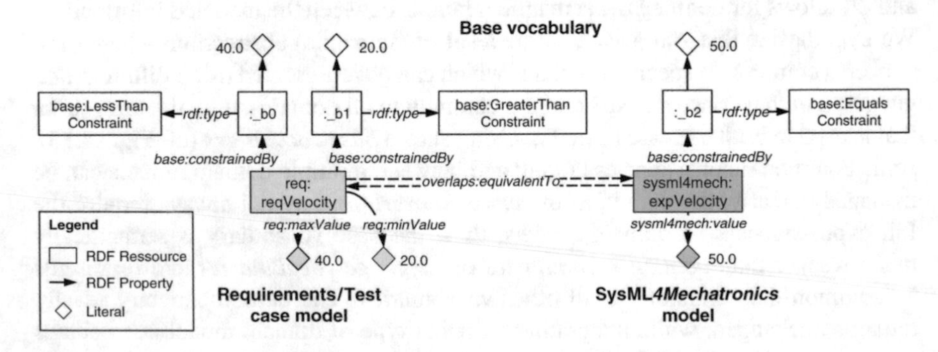

Fig. 14.14 Result of mediating the use case models to the base vocabulary

phenomena in nature (e.g., regarding precision of manufacturing processes). The best one can do is to identify specific types of inconsistencies defined a priori by an expert. Such inconsistencies can, for instance, result from logical contradictions (e.g., over- or under-specifications of attributes' values) or from violations of agreed heuristics and guidelines (Feldmann et al. 2015b). Thinking of types of inconsistencies in this manner makes the representation of inconsistencies as rules, i.e., in the form *IF condition THEN action*, feasible. Consequently, to identify specific instances of inconsistencies according to the known types of inconsistencies, we use SPARQL SELECT queries to (1) formulate the context of, and conditions for inconsistencies using a graph pattern (the *condition*) and (2) retrieve a list of those elements which were checked for a type of inconsistency and either met, or did not meet the condition for inconsistency (the *action*). In this manner, inconsistency checks are similar to unit tests.

Two exemplary inconsistency rules that serve the purpose of identifying, whether the constraints imposed in Fig. 14.14 contradict each other or not, are illustrated in Fig. 14.15. Therein, query 1 matches any two entities x and y identified as semantically equivalent, that are constraint by an *EqualsConstraint* (entity x) and a *LowerThanConstraint* (entity y), respectively. The comparison of the entities' values, namely, *?xVal* < *?yVal*, is bound to the variable *isInconsistent* and denotes whether the constraints are inconsistent or not. Query 2 is formulated accordingly to allow for comparing *EqualsConstraints* with *GreaterThanConstraints*.

In the presented use cases 1 and 2, the value of the SysML4*Mechatronics* property *expVelocity* is defined to be 50.0, whereas the value of *reqVelocity* in the requirements and test case model is defined to be greater than or equal to 20.0 as well as lower than or equal to 40.0. Consequently, using the queries 1 and 2, it can be identified that the respective *LowerThanConstraint* is violated (cf. query 1), whereas the *GreaterThanConstraint* is fulfilled.

Query #1

```
PREFIX : <http://www.example.org/base/ns#>
PREFIX overlaps: <http://www.example.org/overlaps/ns#>
PREFIX rdf: <http://www.w3.org/1999/02/22-rdf-syntax-ns#>
SELECT * WHERE {
  ?x overlaps:equivalentTo ?y .
  ?x :constrainedBy [ rdf:type :EqualsConstraint ; :value ?xVal ] .
  ?y :constrainedBy [ rdf:type :LowerThanConstraint; :value ?yVal ] .
  BIND ( ( ?xVal > ?yVal) AS ?isInconsistent ) }
```

Query #2

```
PREFIX : <http://www.example.org/base/ns#>
PREFIX overlaps: <http://www.example.org/overlaps/ns#>
PREFIX rdf: <http://www.w3.org/1999/02/22-rdf-syntax-ns#>
SELECT * WHERE {
  ?x overlaps:equivalentTo ?y .
  ?x :constrainedBy [ rdf:type :EqualsConstraint ; :value ?xVal ] .
  ?y :constrainedBy [ rdf:type :GreaterThanConstraint; :value ?yVal ] .
  BIND ( ( ?xVal < ?yVal) AS ?isInconsistent ) }
```

Fig. 14.15 Exemplary inconsistency queries for the use case models

Proof-of-Concept Implementation

As a proof-of-concept, a technology demonstrator was developed to demonstrate and evaluate the technical feasibility and viability of the conceptual approach. An overview on the basic architecture of the demonstrator is illustrated in Fig. 14.16. For the knowledge base, the RDF triple store Apache Fuseki[9] was used. For the purpose of demonstrating the technology, we assume that a mechanism exists for automatically transforming the models into an RDF representation (cf. use cases 1 and 2). The mediation between the different vocabularies was formulated and performed using the Apache Jena Rule Reasoning Framework.[10] The mechanism for identifying inconsistencies is composed of a set of SPARQL queries, as well as a SPARQL-compliant query engine. For the latter we have used ARQ,[11] which is also a part of the Apache Jena framework. Within the control and representation layer of our technology demonstrator, the queries can be managed, handed to the query engine and the query results can be interpreted and visualized to the user.

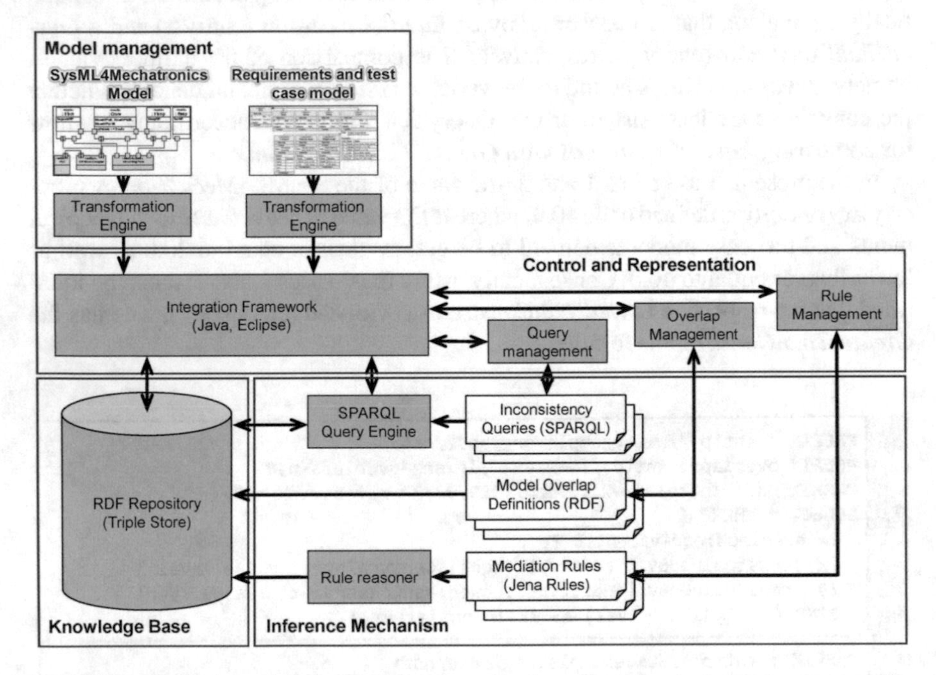

Fig. 14.16 Technology demonstrator for managing inconsistencies in heterogeneous models, extended from Feldmann et al. (2015b)

[9]http://jena.apache.org/documentation/serving_data/, retrieved on 12/11/2015.

[10]http://jena.apache.org/documentation/inference/, retrieved on 12/11/2015.

[11]http://jena.apache.org/documentation/query/, retrieved on 12/11/2015.

14.7 Conclusion and Directions for Future Research

This Chapter is motivated by the challenge of developing modern automated production systems that meet the requirements to manufacture smaller lot sizes up to customer specific products. Such systems have to enable changes in all phases of the system life cycle, to react on new customer or system requirements. Consequently, it is inevitable to support the engineering of such automated production systems by means of, e.g., ensuring consistency. As a basis to overcome this challenge, the Chapter presented three distinct use cases that make use of Semantic Web Technologies as one means to apply logical inference and, by that, to identify inconsistencies in and between the models. In particular, compatibility checking between mechatronic modules, consistency checking between requirements and test cases, as well as, inconsistency management between heterogeneous models were introduced in this Chapter. Consequently, we argue that Semantic Web Technologies can be used to support the engineering of automated production systems. Especially the application of a knowledge-based system, in which Model-Based Engineering is combined with Semantic Web Technologies, provides an integrated approach in which engineers can apply the notations and modeling approaches that are common to them.

Although some of the challenges introduced at the beginning of this Chapter can be addressed by the presented concepts, clearly much research is left to be done.

Research Direction 1: **Identification of semantic overlaps.** In our use cases, we assumed that semantic overlaps between model entities were defined a priori by a human. However, for more complex models and systems, we expect that manually defining and managing overlaps is too costly. However, explicit knowledge of semantic overlaps is indispensable for finding most types of inconsistencies. One means to identify semantic overlaps was proposed in (Herzig and Paredis 2014) using a probabilistic reasoning approach. We moreover argue that background knowledge of the domain (e.g., of the automated production systems domain) can be used to efficiently identifying and defining semantic overlaps.

Research Direction 2: **Flexible definition and execution of inconsistency rules.** As shown in the use cases, RDF(S), OWL and SPARQL provide the means to formulate and check for different types of inconsistencies. However, we expect that further types of inconsistencies exist, making the investigation of the appropriateness of the mechanism for identifying inconsistencies necessary. Furthermore, we argue that dependencies between different types of inconsistencies occur. For instance, in some cases, the detection of one inconsistency can make the execution of a second inconsistency rule obsolete. Hence, preconditions for executing inconsistency rules needs to be incorporated in future research.

Research Direction 3: **Support in resolving inconsistencies.** The ultimate goal of inconsistency management is to support stakeholders in resolving inconsistencies. Consequently, besides the visualization of detected inconsistencies, support for tracing and deciding on how an inconsistency should be resolved is indispensable. In this way, the development of automated production systems is facilitated through the use of Semantic Web Technologies and thus, enables engineers to create less failure-prone systems.

Research Direction 4: Estimating the suitability of available techniques for inconsistency management. In this Chapter, Semantic Web Technologies were used as the core technology for inconsistency management in the automated production systems domain. Further approaches that make use of Semantic Web Technologies are, e.g., the works of Kovalenko et al. (2014) that make use of OWL and SPARQL as well as Steyskal and Wimmer (cf. Chap. 13) that make use of the Shapes Constraint Language (SHACL) (World Wide Web Consortium 2015), which is a current working draft of the W3C. Further technologies that are being used for inconsistency management are, e.g., the Object Constraint Language (OCL) (Object Management Group 2014) as well as XML-based methodologies. Surely, each of these technologies has its advantages as well as disadvantages regarding, e.g., usability, comprehensibility as well as scalability and performance for different types and sizes of models. One aspect to be addressed in future work is, hence, to identify, which technology is most suitable for which use case.

Acknowledgments This work was supported in part by the German Research Foundation (DFG) Collaborative Research Centre 'Sonderforschungsbereich SFB 768—Managing cycles in innovation processes—Integrated development of product-service systems based on technical products'. Moreover, parts of this work were developed as part of the IGF-project 17259 N/1 of the Deutsche Forschungsgesellschaft für Automatisierung und Mikroelektronik (DFAM) e.V., funded by the AiF as part of the program to support cooperative industrial research (IGF) with funds from the Federal Ministry of Economics and Technology (BMWi) following an Order by the German Federal Parliament. We moreover thank Christiaan J.J. Paredis, Sebastian J.I. Herzig and Ahsan Qamar (Georgia Institute of Technology) for their support and fruitful discussions.

References

Broy, M., Feilkas, M., Herrmannsdoerfer, M., Merenda, S., Ratiu, D.: Seamless model-based development: from isolated tools to integrated model engineering environments. Proc. IEEE **98**(4), 526–545 (2010). doi:10.1109/JPROC.2009.2037771

Estévez, E., Marcos, M.: Model-based validation of industrial control systems. IEEE Trans. Ind. Inf. **8**(2), 302–310 (2012). doi:10.1109/TII.2011.2174248

Feldmann, S., Kernschmidt, K., Vogel-Heuser, B.: Combining a SysML-based modeling approach and semantic technologies for analyzing change influences in manufacturing plant models. In: CIRP Conference on Manufacturing Systems (2014a). doi:10.1016/j.procir.2014.01.140

Feldmann, S., Rösch, S., Legat, C., Vogel-Heuser, B.: Keeping requirements and test cases consistent: towards an ontology-based approach. In: IEEE International Conference on Industrial Informatics (2014b). doi:10.1109/INDIN.2014.6945603

Feldmann, S., Herzig, S.J.I., Kernschmidt, K., Wolfenstetter, T., Kammerl, D., Qamar, A., Lindemann, U., Krcmar, H., Paredis, C.J.J., Vogel-Heuser, B.: A comparison of inconsistency management approaches using a mechatronic manufacturing system design case study. In: IEEE International Conference on Automation Science and Engineering (2015a). doi:10.1109/CoASE.2015.7294055

Feldmann, S., Herzig, S.J.I., Kernschmidt, K., Wolfenstetter, T., Kammerl, D., Qamar, A., Lindemann, U., Krcmar, H., Paredis, C.J.J., Vogel-Heuser, B.: Towards effective management of inconsistencies in model-based engineering of automated production systems. In: IFAC Symposium on Information Control in Manufacturing (2015b). doi:10.1016/j.ifacol.2015.06.200

Gausemeier, J., Schäfer, W., Greenyer, J., Kahl, S., Pook, S., Rieke, J.: Management of cross-domain model consistency during the development of advanced mechatronic systems. In: International Conference on Engineering Design (2009)

Giarratano, J.C., Riley, G.: Expert Systems: Principles and Programming, 2nd edn. PWS Publishing Co., Boston (1994)

Hametner, R., Kormann, B., Vogel-Heuser, B., Winkler, D., Zoitl, A. Test case generation approach for industrial automation systems. In: IEEE International Conference on Automation, Robotics and Applications (2011). doi:10.1109/ICARA.2011.6144856

Heitmeyer, C.L., Jeffords, R.D., Labaw, B.G.: Automated consistency checking of requirements specifications. ACM Trans. Softw. Eng. Methodol. **5**(3), 231–261 (1996). doi:10.1145/234426.234431

Herzig, S.J.I., Paredis, C.J.J.: Bayesian reasoning over models. In: Workshop on Model-Driven Engineering, Verification, and Validation (2014). http://ceur-ws.org/Vol-1235/paper-09.pdf

Herzig, S.J.I., Qamar, A., Reichwein, A., Paredis, C.J.J.: A conceptual framework for consistency management in model-based systems engineering. In: ASME International Design Engineering Technical Conference & Computers and Information in Engineering Conference (2011)

Hitzler, P., Krötzsch, M., Rudolph, S.: Foundations of Semantic Web Technologies. CRC Press, Boca Raton, FL, USA (2010)

International Electrotechnical Commission: Engineering Data Exchange Format for Use in Industrial Automation Systems Engineering—Automation Markup Language (2014)

Kernschmidt, K., Vogel-Heuser, B.: An interdisciplinary SysML based modeling approach for analyzing change influences in production plants to support the engineering. In: IEEE International Conference on Automation Science and Engineering (2013). doi:10.1109/CoASE.2013.6654030

Kovalenko, O., Serral, E., Sabou, M., Ekaputra, F., Winkler, D., Biffl, S.: Automating cross-disciplinary defect detection in multi-disciplinary engineering environments. In: Janowicz, K., Schlobach, S., Lambrix, P., Hyvnen, E. (eds.) Knowledge Engineering and Knowledge Management, Lecture Notes in Computer Science, vol. 8876, pp. 238–249. Springer International Publishing (2014). doi:10.1007/978-3-319-13704-9_19

Leveson, N.G., Heimdahl, M.P.E., Reese, J.D.: Designing specification languages for process control systems: lessons learned and steps to the future? In: Software Engineering, Lecture Notes in Computer Science, vol 1687, pp. 127–146. Springer, Berlin (1999). doi:10.1007/3-540-48166-4_9

Object Management Group: MOF Model To Text Transformation Language (2008). http://www.omg.org/spec/MOFM2T/1.0/

Object Management Group: Meta Object Facility (MOF) 2.0 Query/View/Transformation (2011). http://www.omg.org/spec/QVT/1.1/

Object Management Group: Systems Modeling Language (SysML) (2012). http://www.omg.org/spec/SysML/1.3/

Object Management Group: Constraint Language (OCL), Version 2.4 (2014). http://www.omg.org/spec/OCL/2.4/

Runde, S., Fay, A.: Software support for building automation requirements engineering—an application of semantic web technologies in automation. IEEE Trans. Ind. Inf. **7**(4), 723–730 (2011). doi:10.1109/TII.2011.2166784

Strasser, T., Rooker, M., Hegny, I., Wenger, M., Zoitl, A., Ferrarini, L., Dede, A., Colla, M.: A research roadmap for model-driven design of embedded systems for automation components. In: IEEE International Conference on Industrial Informatics (2009). doi:10.1109/INDIN.2009.5195865

Vogel-Heuser, B., Legat, C., Folmer, J., Feldmann, S.: Researching Evolution in Industrial Plant Automation: Scenarios and Documentation of the Pick and Place Unit. Technical Report TUM-AIS-TR-01-14-02, Technische Universität München (2014). https://mediatum.ub.tum.de/node?id=1208973
World Wide Web Consortium: Extensible Markup Language (XML) 1.0 (2008). http://www.w3.org/TR/xml
World Wide Web Consortium: OWL 2 Web Ontology Language Document Overview (2009). http://www.w3.org/TR/owl2-overview/
World Wide Web Consortium: SPARQL Protocol and RDF Query Language 1.1 Overview (2013). http://www.w3.org/TR/sparql11-overview/
World Wide Web Consortium: Resource Description Framework (RDF) (2014). http://www.w3.org/RDF/
World Wide Web Consortium: Shapes Constraint Language (SHACL) (2015). http://www.w3.org/TR/shacl/

Chapter 15
Conclusions and Outlook

Marta Sabou and Stefan Biffl

Abstract This chapter summarizes and reflects on the material presented in this book. In particular, the chapter aims to conclude on the relation between *Industrie 4.0* needs and Semantic Web technologies (SWTs) based on the various intelligent engineering applications (IEAs) reported in Parts III and IV of the book. Concretely, this chapter seeks answers to the following questions: which *Industrie 4.0* scenarios are addressed by engineering applications described in this book? Which are the most and least used capabilities of SWTs? Which limitations of SWTs seem important and which alternative technologies can be used to compensate for those limitations? What is the technological blueprint of an IEA and what SWTs are typically needed to instantiate this blueprint? This analysis and reflections lead to an outlook on the future application of SWTs for building IEAs to address *Industrie 4.0* tasks.

15.1 Introduction

The aim of this book is to explore how *Semantic Web technologies* (SWTs) can be used for creating *Intelligent Engineering Applications* (IEAs), in particular in the context of the emerging *Industrie 4.0* vision. As detailed in Chap. 1, this aim is achieved by investigating five research questions. The first question (Q1) refers to identifying concrete needs in *Industrie 4.0* settings where SWTs can be used. To that end, Chap. 2 describes four typical *Industrie 4.0* scenarios where IEAs are particularly needed. From these scenarios, seven concrete needs for SWTs are distilled and contrasted against typical capabilities of SWTs which are described in Chap. 3 as a response to question Q2. Question Q3 about typical SWTs is covered in Part II of the book, while Parts III and IV answer question Q4 by describing

M. Sabou (✉) · S. Biffl
Institute of Software Technology and Interactive Systems, CDL-Flex,
Vienna University of Technology, Vienna, Austria
e-mail: Marta.Sabou@ifs.tuwien.ac.at

S. Biffl
e-mail: Stefan.Biffl@tuwien.ac.at

S. Biffl and M. Sabou (eds.), *Semantic Web Technologies for Intelligent Engineering Applications*, DOI 10.1007/978-3-319-41490-4_15

concrete examples of IEAs that aim to solve a wide range of engineering tasks, in diverse settings, by relying on SWTs.

Reflecting on all this material, in this last chapter of the book our aim is twofold. First, we draw conclusions about current practice in using SWTs for IEAs and as such we provide answers to question Q5 regarding SWT strengths and weaknesses as well as compatibilities with other technologies. Second, we provide an outlook on future research challenges for the use of SWTs in engineering settings. Accordingly, in Sect. 15.2 we summarize how the IEAs described in this book make use of SWTs, which *Industrie 4.0* scenarios they solve, which SWT capabilities are used, which SWT limitations are observed, and what are the most often employed technologies for overcoming SWT limitations. In Sect. 15.3, we distil common characteristics of the presented IEAs and describe an emerging technology blueprint of IEAs. In Sect. 15.4, we conclude with an outlook on future research.

15.2 Semantic Web Technologies for Building Intelligent Engineering Applications: Capabilities and Limitations

In this section, we analyze the IEAs described by this book to conclude on the scenarios they target (Sect. 15.2.1), the most and least used SWT capabilities (Sects. 15.2.2 and 15.2.3), as well as the observed SWT limitations (Sect. 15.2.4). Table 15.1 shows an overview of the analysis of the IEAs described in this book.

15.2.1 *Industrie 4.0 Scenarios and Tasks Solved with SWTs*

An analysis of the chapters describing IEAs shows a clear popularity of *Scenario 1, "Discipline-Crossing Engineering Tool Networks"*—which was introduced and described in Chapter 2. A few chapters also investigate tasks specific to *Scenario 2, "Use of existing artifacts for plant engineering."*

This book illustrates a wide range of tasks and settings pertaining to Scenario 1, where SWTs have been employed successfully. Chapters 12, 13, and 14 focus on the task of maintaining consistency among partially overlapping engineering models involved in the design of a complex system. Concretely, the second use case of Chap. 12 (use case *b*) looks at supporting the collaborative development process of an automatic transmission gearbox by distributed engineering groups from different engineering disciplines needed to realize this complex mechatronic object, namely: mechanical, hydraulic, electrical, and software engineering. This use case illustrates the same problem as described by Scenario 1, since several overlapping models of the same system must be kept consistent. An additional need was to enable change propagation among the interrelated models, meaning that a change in

Table 15.1 Overview of the main characteristics of IEAs described in this book

Ch.	Solved task	Scenario	SWTs	SWT limitations	Alternative technologies
9	Creation of control system setup to enable product ramp-up	Sc. 2	OWL, SWRL, SPARQL	Dealing with mathematical calculations	Statistical analysis, data mining
10	Simulation design and integration	Sc. 3	OWL, SPARQL		
12	(a) Design optimization; (b) Model consistency management and change propagation across engineering models	Sc. 1 (b)	F-Logic, Flora2	Lack of support for checking mathematical dependencies; OWA not suitable for engineering	Relational Constraint Solver (RCS)
13	Model consistency management	Sc. 1	RDF, SHACL	Lack of user-friendly interfaces	Model-Driven Engineering
14	(a) Module compatibility check; (b) compatibility check between requirements and test cases; (c) identifying inconsistencies in and among engineering models.	Sc.1 (c); Sc.2 (a)	RDF, OWL, SPARQL	Lack of user-friendly interfaces	Model-Driven Engineering

Letters in parenthesis after a chapter number refer to use cases in that chapter

one model could lead to changes in all related models whenever necessary (i.e., according to the interdependencies specified between models).

Chapter 13 describes how the emerging *Shapes Constraint Language* (SHACL) SWT standard can be used to solve the typical consistency management issues among different, overlapping views (or models) of the same complex systems. Chapter 13 is not linked to a specific industry, but rather focuses on the generic consistency management problem among models when using a *Multi-Viewpoint Modeling* approach, which is unavoidable when the design of a complex system is shared among different engineering teams.

The third use case (*c*) of Chap. 14 aims to identify consistencies between different engineering models of the same system and therefore it illustrates Scenario 1. Because the engineering models describe the same system, they overlap to some extent and, therefore, these overlapping parts should be kept consistent. To achieve this task, there is also a need to define which parts of the models correspond to each other as a basis for enabling compatibility checks.

Scenario 2, "Use of existing artifacts for plant engineering," is discussed in two chapters. Chapter 9 investigates the task of recommending a control system setup to enable efficient product ramp-up processes and illustrates this task in the context of semiconductor manufacturing. More concretely, a production plan needs to be created for a target production system based on the knowledge available about a source production system, so that the target system enables creating a product with the same quality as in the source system. A production plan describes what equipment and raw materials should be used as well as how these should be used and assessed to ensure a high quality product. This task is relevant for Scenario 2, as the task requires assembling a production plan for the target production system by modifying and adjusting the production system of the source system in a way that it reuses artifacts (e.g., devices, configurations, raw materials) from the target site. Then, the first use case *(a)* of Chap. 14 focuses on ensuring compatibility between mechatronic modules that need to be replaced in a given system configuration by (1) identifying modules compatible with a module that needs to be replaced and (2) identifying and resolving conflicts in a given system configuration as a follow-up of a module change.

Scenario 3 "Flexible Production System Organization" on run-time flexibility of production systems is addressed in Chap. 10 with an ontology-based system that supports simulation design and integration.

Various book chapters also describe solutions to tasks that do not fit any of the scenarios discussed in Chap. 2. These tasks give further insights into the rich variety of IEAs that could be built. These tasks are

- *Design optimization*. Chapter 12's use case *(a)* aims to support engineers in deriving an optimized design layout starting from a system specification, which is refined in iterative design steps. To enable this task, ontologies were used to represent requirements, to allocate them to parts of the systems, to attach constraints to requirements and parts of the system, and to keep track of different versions of requirements during the subsequent processing (i.e., design phases).
- *Compatibility checking between requirements and their defined test cases*. Use case *(b)* in Chap. 14 focuses on the task of ensuring consistency between requirements specified for a production system and test cases corresponding to checking those requirements. The authors use semantic modeling capabilities to formally represent domain knowledge (denoted with C1 in Chap. 3) and reasoning services offered by Semantic Web reasoners (C4) to check whether test cases are compatible with requirements.

None of the discussed IEAa address *Scenario 4 "Maintenance and Replacement Engineering"* about improved maintenance capabilities of production system components during the runtime of the system.

Based on the identified scenarios, seven common needs for solving these scenarios are identified in Chap. 2, including

- *N1-Explicit engineering knowledge representation*: the need for knowledge representation support, which allows analyzing the requirements for the knowledge to represent and provides domain experts with appropriate tools for knowledge representation and design.
- *N2-Engineering data integration*: the need for an engineering data integration approach, which provides an integrated data model of the common concepts of stakeholders across disciplines to enable the linking of engineering knowledge across disciplines.
- *N3-Engineering knowledge access and analytics*: effective and efficient mechanisms are needed (a) for querying engineering models, including versions and changes; and (b) for defining and evaluating engineering model constraints and rules across several views.
- *N4-Efficient access to semistructured data in the organization and on the Web* to go beyond current solutions that mostly focus on accessing structured data.
- *N5-Flexible and intelligent engineering applications* to support tasks such as defect detection and constraint checking on integrated engineering data.
- *N6-Support for multidisciplinary engineering process knowledge*: the need to represent engineering process responsibilities and process states linked to the production system plant model.
- *N7-Provisioning of integrated engineering knowledge at system runtime*: the need for providing integrated engineering knowledge at system runtime beyond simple PDF printouts of engineering plans. The knowledge has to be available in a sufficiently timely manner to support applications that depend on reacting in time to real-time processes.

Based on these needs of engineering scenarios, an important point to investigate is the extent to which they are addressed by SWTs, as we do in Sect. 15.2.2

15.2.2 Most Used Semantic Web Capabilities

To answer question Q2 announced in Chaps. 1, 3 identified five Semantic Web capabilities, including: (C1) Formal and flexible semantic modeling; (C2) Intelligent, Web-scale knowledge integration; (C3) Browsing and exploration of distributed data set; (C4) Quality assurance of knowledge with reasoning; and (C5) knowledge reuse. Table 2.2 from Chap. 2 depicts the *estimated* support of engineering scenario needs by SW capabilities and shows a potential good coverage of these needs by SWTs. In this section, we deepen the initial analysis from Chap. 2 by instantiating the same table with information derived from concrete IEA's presented in this book. Table 15.2 reproduces the content of Table 2.2 and extends it with observations about IEA's reported in this book.

We observe that none of the reported IEAs has addressed needs N4, N6, and N7. Need N1 for explicit engineering knowledge representation was present in all IEAs and was addressed by exploring the Semantic Web capability of formal and flexible

Table 15.2 Overview of engineering scenario needs and the SWT capabilities used to address those needs

Semantic Web Capabilities & Needs	C1. Semantic modelling	C2. Knowledge integration	C3. Browsing & exploration of distributed data	C4. Quality assurance of knowledge with reasoning	C5. Knowledge reuse
N1. Explicit engineering knowledge representation	++ Ch 9, 10, 12(a,b), 13, 14(a, b, c)	+			+
N2. Engineering data integration	+	++ Ch 12(b), 13, 14(c)			+
N3. Engineering knowledge access and analytics	++	++ Ch 12(b), 13, 14(c)	+	++ Ch 9, 10, 12 (a,b), 13,14 (a, b, c)	++ Ch 12(a)
N4. Efficient access to semistructured data in the organization and on the Web	+	++	++		
N5. Flexible and intelligent engineering applications	+	++ Ch 12(b), 13, 14(c)	+	++ Ch 9, 10, 12 (b), 13, 14 (a, b, c)	
N6. Support for multidisciplinary engineering process knowledge	+	++	+	++	++
N7. Provision of integrated engineering knowledge at production system runtime	+		+	++	+

The estimated level of support of a need by a capability is shown as in Table 2.2. strong (++); moderate (+). Letters in parenthesis after a chapter number refer to use cases in that chapter

semantic modeling (C1). Similarly, the need for data integration (N2) played a key role and was addressed with the Semantic Web capability focused on this need (C2). Need N3 for engineering knowledge access and analytics was the core focus of all the IEAs and has been addressed with capabilities for intelligent, Web-scale knowledge integration (C2), capabilities for reasoning (C4), or a combination thereof. All works that addressed need N3, did so as a prerequisite for building flexible and intelligent engineering applications (N5), therefore we see a similar use of C2 and C4 for addressing N5 as for addressing N3.

The Semantic Web solutions to the described applications mostly make use of (a combination of) the following Semantic Web capabilities C1, C2, and C4:

C1 "Formal and flexible semantic modeling"—is a basic prerequisite for applying SWTs. The applications in each chapter rely on the semantic representation of a variety of information relevant for each particular use case. The K-RAMP framework from Chap. 9 makes use of OWL ontologies to represent product and process knowledge. In Chap. 10, ontologies are used to enable simulation design and integration. Use case (a) of Chap. 12 takes advantage of ontologies for explicitly representing requirements in a machine-processable way. The second use case (b) makes use of classical knowledge representation techniques such as Frame-logic to represent engineering models. In Chap. 13, the authors propose a solution where RDF is used to encode different views on the system in a uniform manner.

In Chap. 14, the first use case (a) on ensuring compatibility between mechatronic modules creates ontologies for representing compatibility information. Domain knowledge about requirements and test cases is explicitly represented in the second use case (b) to allow consistency checking between these artifacts. Use case (c) in Chap. 14 aims to manage consistencies between different engineering models of the system by (1) defining a base vocabulary that contains the common concepts used by the various models considered and (2) using a common data representation language (RDF) to encode the various system models in a uniform way.

C2 "Intelligent, Web-scale knowledge integration"—is a capability used to integrate engineering knowledge. Chaps. 12, 13, and 14 propose a data integration solution that follows the same technical approach: local, domain-specific knowledge is integrated by reference to "global knowledge ontologies" which capture common concepts shared across engineering disciplines. In Chap. 13, SHACL is employed to define the inter-model dependencies.

C4 "Quality assurance of knowledge with reasoning"—querying and reasoning mechanisms enable making sense of and providing the intelligent services needed by each engineering application. Chapter 9 makes use of matchmaking functionalities to identify which target elements can be reused when recommending a production plan for the target manufacturing system. Matchmaking is achieved through SPARQL and SWRL rules. In Chap. 12, the Flora2 reasoning framework detects inconsistencies among engineering models. In Chap. 13, model dependencies described with SHACL can be automatically checked during modeling time to uncover potential inconsistencies between the various models. Reasoning capabilities of SWTs support this task. In Chap. 14, the first use case on ensuring compatibility between mechatronic modules encodes and checks compatibility rules through SPARQL queries. The second use case uses Semantic Web reasoners to check whether test cases are compatible with requirements.

15.2.3 Least Used Semantic Web Capabilities

From the discussed Semantic Web capabilities in Chap. 2, two were marginally used or not used at all. These are C3 and C5

- *C3 "Browsing and exploration of distributed data set"* is not currently used by any of the applications discussed in this book, with the exception of the AutomationML Analyzer tool (Sabou et al. 2016) briefly mentioned in Chap. 11.
- *C5 "Knowledge reuse."* None of the chapters in this book reports on creating IEAs by reusing already existing formally represented knowledge. Chapter 12's first use case considers that the knowledge created will be possibly reused, e.g., requirement templates and system structure information could be reused across design problems.

15.2.4 Semantic Web Limitations and Challenges

The match between production system engineering needs to Semantic Web capabilities in Table 15.2 shows an overall positive result. Semantic Web technologies seem to be a very good match for addressing the aspects of heterogeneity in engineering due to their capability to integrate data intelligently and flexibly on a large scale. However, some requirements and assumptions of production system engineering are fundamentally different from typical Semantic Web application areas, which focus on the better integration of Web-scale data and knowledge. The chapters in this book also highlighted some missing features of SWTs and propose alternative technologies for addressing these missing features. The most often highlighted aspects are.

Lack of knowledge acquisition interfaces that are easy to use by engineers. As a solution, Chap. 13 and 14 explore the use of *Model-Driven Engineering* techniques, in particular *SysML* and *SysML4Mechatronics*, as a "front-end" to acquire engineering models. These models are then translated into ontologies, on which reasoning is performed.

Lack of support for mathematical calculations. Engineering-specific settings often require the processing of numeric data. For example, constraints can be represented as numeric equations or numeric measurements about the quality of a process are recorded. SWTs focus mostly on logics-based knowledge representation and do not have a strength in processing numeric data. This book illustrates that hybrid solutions can be provided that combine SWTs with techniques more suitable to mathematical data processing, such as *data mining, statistical analysis* (Chap. 9) and *Relational Constraint Solvers* (RCS) for solving cardinality problems (Chap. 12).

The OWA underlying SWTs is not a natural fit to the engineering domain. Traditional engineering approaches, e.g., databases, planning methods, and quality assurance methods, rely, in general, on a *Closed-World Assumption* (CWA): if a fact is not in the knowledge base and cannot be deduced from the knowledge base, the answer is negative. In contrast, Semantic Web technologies rely on *Open-World Assumption* (OWA). This implies that facts that cannot be deduced from the knowledge base are unknown, not necessarily wrong. Therefore, IEAs using SWTs have to carefully explain their result in a way that makes sense to an engineer. The application has to either operate in the user interface on a CWA or explain the rationale for a result that has been obtained following an OWA, e.g., for an unknown result it has to be explained whether this is likely to be a negative result or missing data in order to support a useful interpretation for the engineer. Several mechanisms are currently being investigated for combining open- and closed-world reasoning (Pan 2012), including: expressing negations in SPARQL 1.1 queries; or modified reasoning mechanisms that rely on notions such as *DBox* (Seylan et al. 2009) or *NBox* (Ren et al. 2010).

Dealing with dynamic engineering data. During the production system engineering process, engineering plan data frequently change and even small changes can make a big difference in the overall meaning of a plan. The Semantic Web community has considered motivating use cases and research challenges for dealing with dynamic (i.e., streaming) data for several years (Della Valle et al. 2009) and has been engaged in a rich set of research activities on the topic of stream reasoning (Margara et al. 2014). Large-scale use cases where SWTs were applied to deal with dynamic, streaming data were reported from areas as diverse as observing large-scale city-level events through social media monitoring (Balduini et al. 2013), manufacturing (Wenzel et al. 2011), or the detection of malfunctioning turbines at Siemens (Kharlamov et al. 2014). Some of the flagship technologies in this area include ontologies for describing fast changing streaming data such as that harvested from sensors (Compton et al. 2012) and streaming solutions such as C-SPARQL (Barbieri et al. 2010), CQELS (Le-Phuoc et al. 2011) or STARQL, a SPARQL-like language for querying streaming data (Kharlamov et al. 2014). The efficient use of these technologies depends on a set of factors that should be clarified at the beginning of each project, including the size of the streaming knowledge (e.g., how many new triples arrive per second to the system), the type of background (i.e., slow-changing) knowledge, and tolerable delay in results delivery. Although several technology solutions and concrete applications that efficiently process streaming data have been showcased, Margara et al. (2014) identify the need to perform research on "theoretical foundations, algorithms, techniques, and implementation of stream reasoning that could enable building efficient and scalable tools." Several open research questions still have to be addressed both at the level of system models (for modelling data and operations on data) and at system implementations as explained in detail in (Margara et al. 2014).

There are two major practical challenges that come from the requirement to use any Semantic Web application in the context of existing IT landscapes and personnel (Oberle 2014), which we discuss next.

Technical integration of Semantic Web technologies with existing enterprise systems. In a typical business application at BBC, ontology mappings as well as an ontology store had to be integrated with the existing landscape of BBC's enterprise system (Oberle 2014). Paulheim et al. (2011) found several differences between the traditional object-oriented methods in the business environment and SWTs, which hamper an easy integration of SWTs within a host enterprise system.

- *Conceptual model versus task-specific model.* Ontologies often play the role of a reference model, i.e., a generic, commonly agreed upon conceptual model of a domain, while object-oriented class models are task-specific, with the focus on an efficient implementation of an application. Therefore, reference ontologies and class models may be incompatible in the sense that a 1:1 mapping between them does not exist, e.g., *Ecore* relies on the unique naming assumption unlike OWL (see Chap. 3 for a detailed explanation of the Nonunique Naming Assumption). As a potential integration solution between ontologies and class models, Oberle (2014) proposed an approach for mapping pragmatic class models and ontologies with declarative mapping instructions that can be interpreted by mapping execution engines as a way to bridge conceptual and task-specific models.
- *Modeling may depend on use cases.* Different use cases typically have different representation needs. Therefore, different use cases may require different mappings, stores, and reasoners, needing *n*-time technical integration effort.

Training engineers and developers. In many engineering environments, there is a lack of SWT experts. The training of existing employees to familiarize them with the new technology or the acquisition of SWT experts is difficult. Oberle (2014) proposed a partial solution approach for enabling software engineers to develop enterprise systems on the basis of an ontology in their familiar environment (Rahmani et al. 2010) with an adjustable transformation from OWL to Ecore, which allows authoring of and programmatic access to a reference ontology, e.g., the BBC *programme* ontology, by his or her familiar development environment (e.g., Eclipse). However, the introduction of SWTs to stakeholders in a traditional systems engineering environment will need special care to minimize the risk of insufficient understanding and support for realizing a successful application.

15.2.5 Alternative Technologies

To overcome some of the limitations of SWTs, alternative technologies can be used. For example, SWT-based approaches might be combined with techniques more suitable to mathematical data processing, such as *data mining, statistical analysis* (Chap. 9), and *Relational Constraint Solvers* (Chap. 12). Additionally, engineers can use alternative solution approaches, which we hereby briefly compare to SWTs.

General-purpose end-user approaches. In engineering environments, there is a wide variety of tools and data formats used, and these tool networks often use general-purpose end-user approaches, such as databases, spreadsheets, and scripting, to integrate, transform, and reuse data. While these general-purpose end-user approaches are widely used, they suffer, in general, from low formality and flexibility and, therefore, rely heavily on domain experts to apply and maintain the code and to interpret the results (Biffl et al. 2012) (Biffl et al. 2015) (Fay et al. 2013) (Winkler and Biffl 2012). Therefore, general-purpose end-user approaches fall short in addressing the needs identified in Chap. 2, Sect. 2.5. However, when introducing SWTs into an engineering environment, the careful analysis of general-purpose end-user approaches in use is a prerequisite for understanding the existing expertise and for minimizing the risk of failing to provide the benefits expected from applying SWTs.

Model-driven engineering (MDE). MDE and Semantic Web are different approaches to creating IEAs that provide some similar capabilities, but also capabilities that differ in important ways. Similar capabilities include

- *Making knowledge explicit with conceptual modelling*. In MDE, this is achieved with metamodels, models, and transformations (Weilkiens and Lamm 2015); in Semantic Web with ontologies, ontology instances, and reasoning.
- *Direct interaction with knowledge bases*. Both MDE and Semantic Web communities provide user-friendly generic tools for creating, changing, and populating knowledge bases: in MDE with *Eclipse*-based tools and plug-ins (Brambilla et al. 2012); in Semantic Web based on an open development environment for semantic web applications (Knublauch et al. 2004).
- *Data integration*. Both MDE and Semantic Web communities provide mechanisms for creating and maintaining mappings to integrate heterogeneous data sources, in particular, links between schemas and between instances (Oberle 2014).

Semantic Web approaches have some advantages over MDE regarding agile schema development, i.e., schema evolution at runtime, reasoning-based checks (Kernschmidt and Vogel-Heuser 2013), knowledge reuse. Semantic Web approaches have strong advantages over MDE in the Semantic Web home grounds of linked data, e.g., with the *unique resource identifier* (URI) identification, and linking URIs with the *sameAs* mechanism, and of browsing and exploring distributed data sets, e.g., engineering models and external data sources (Gabrilovich and Markovitch 2007). Oberle (2014) discusses why and when to apply SWTs in enterprise systems, which are in some ways similar to engineering project support systems, and characterizes the state of Semantic Web usage as considerable academic interest and early industrial products.

MDE has advantages over Semantic Web with a strong open source community in business and industry, and a skill set that is better compatible with existing expertise in typical software engineering projects.

15.3 A Technology Blueprint for IEAa

The IEAs described in this book rely on a set of technical solutions, which can be synthesized into a technology blueprint. A first common characteristic is that data integration is a prerequisite for realizing most of the reported IEAs and that most authors chose an ontology-based data integration approach in line with Wache et al.'s *hybrid ontology model* (2001). In all these cases, an ontology is built that captures the common concepts among engineering disciplines. This ontology plays the role of a semantic bridge to integrate data described in terms of discipline-specific local ontologies. Chapter 4 describes the Engineering Knowledge Base Approach that is based on the same data integration paradigm as the other IEAs.

To realize IEAs based on this paradigm, a broad repertoire of technology problems needs to be addressed. Figure 15.1 depicts the main phases that need to be considered both in terms of technology-agnostic phases (left side) and in terms of concrete SWTs that can be used to implement these technology-agnostic phases. These are:

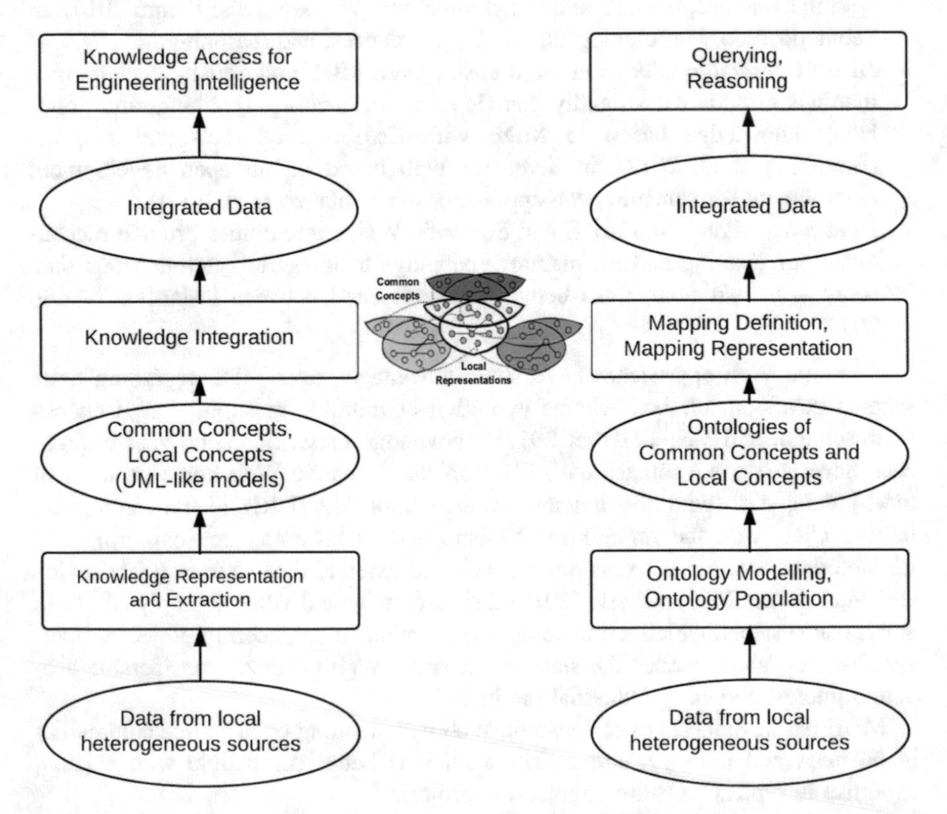

Fig. 15.1 Technology blueprint for realizing IEAs. Technology-agnostic (*left*) and Semantic Web-specific (*right*) versions

Knowledge Extraction denotes the phase in which, given use-case-specific engineering data, the relevant local and global ontologies are created. *Local ontologies* semantically describe the content of one data source (for example, the data within a single engineering model) while *global ontologies* capture common concepts, that is, terms that are shared across the boundaries of individual engineering disciplines. In most of this book's chapters, Semantic Web experts manually create the local and global ontologies, presumably in cooperation with domain experts, i.e., engineers. In Chap. 14, on the other hand, the authors translate *SysML4Mechatronics* models created by engineers into ontologies, thus shortening the knowledge extraction phase.

SWTs that are typically used in this phase are *Ontology Modeling* and *Ontology Population*. Ontology modeling is discussed in detail in the first part of Chap. 5 of this book, which provides an overview of how to semantically model engineering models. Chapter 5 describes the main ontology engineering methodologies, provides an overview of several ontologies built to support IEAs, and concludes on a set of frequently used knowledge design patterns when building these ontologies. The second part of Chap. 5 describes methods for the acquisition of semantic knowledge from engineering artifacts. This chapter part surveys a set of methods that allow extracting draft ontologies from engineering data sources, or that enable populating already created ontologies with instance data from engineering data saved as Excel files. These methods are often used to reduce the length and cost of the knowledge extraction phase.

Knowledge Integration focuses on creating mappings between common concepts and relevant local concepts. Mappings consist of explicitly specified relations between elements of two ontologies, stating, for example, that a concept in one ontology is the same as (or broader, or narrower than) a concept in the other ontology. As stated in Sect. 15.2.2, the IEAs described in this book use several diverse methods to encode mappings between local and global ontologies. For example, in Chap. 12 local ontologies extend concepts from so-called "Engineering ontologies" that capture common concepts. Chapter 13 describes how SHACL can be used to capture verifiable correspondences between models.

Chapter 6 investigates in detail the topic of establishing mappings between engineering data structures as an important SWT for data integration. The chapter provides a catalogue of frequent mapping types that typically occur when mapping engineering data models and concludes that the mapping needs are much more complex than what is enabled by standard SWTs where the focus is primarily on establishing equivalence mappings. Chapter 6 also provides an overview of the SWTs suitable for representing mappings and compares these in terms of their capabilities to represent the mappings in the proposed catalogue. Concluding that the existing technologies have complementary strengths, the chapter also introduces the *Expressive and Declarative Ontology Alignment Language (EDOAL),* an emerging approach to describe complex mappings on the Semantic Web.

Knowledge Access for Engineering Intelligence enables access mechanisms to the integrated data as an interface for supporting engineering intelligence, for example, enabling project-level technical and managerial coordination, such as technical constraint checking and defect detection or project scheduling. In a Semantic Web centric solution, this phase would involve managing and exploring ontology-based knowledge (i.e., knowledge evolution, querying, and inconsistency detection). The various IEAs described in this book rely on a combination of reasoning support and SPARQL queries to generate engineering intelligence.

Chapter 7 focuses specifically on needs for knowledge change management in settings where the hybrid ontology model is used for data integration. The main conclusion is that current approaches to change management in the Semantic Web support such settings only weakly, being traditionally geared toward change management within a single-ontology setting. A more comprehensive approach to knowledge change management is therefore needed, and Chap. 7 proposes a solution concept for achieving this task.

Chapter 8 presents and evaluates different technical architectures for realizing Semantic Web-based IEA's. Four different software architectures are proposed to achieve semantic integration: (A) using a single-ontology component, (B) using an ontology component and a relational database with a RDF2RDB[1] mapper that transforms relational database data to Semantic Web formats, (C) an ontology component and a graph database, and D) an ontology component and a versioning system for managing individuals. The four architectures have been evaluated and compared based on an industrial use case along dimensions such as usability, maintenance, and performance. The results of this evaluation lead to identify areas of future work as discussed in Sect. 15.4.

15.4 Outlook

Based on the content of this book and the conclusions discussed in Sects. 15.2 and 15.3, we see several trends and future research avenues.

We conclude that there are **ample opportunities for using SWTs in Industrie 4.0 settings**. Indeed, Chap. 11 has shown that SWTs support various aspects of production system's engineering being primarily employed to solve technical tasks such as *model integration*, *model consistency management*, and to a lesser extent, the more complex tasks of *flexible comparison*. Yet, from the perspective of the four typical *Industrie 4.0* scenarios identified in Chap. 2, it appears that those two that refer to run-time systems have been addressed only to a limited extent by work described in this book. At the same time, chapters in this book describe IEAs that do not fall under any of the four scenarios. We therefore conclude that future work could further diversify the scenarios addressed with SWTs, and also explore the application of SWTs to currently weakly addressed scenarios. For example, Chap. 10 envisions two new advances in the simulation model design area: (a) using Bond graphs for dealing with transformation of energy between various physical

[1]RDF2RDB: https://www.netestate.de/en/software-development/rdf2rdb/.

disciplines, such as between electrical and mechanical systems, and (b) the validation of designed simulation models with respect to the components from which the simulation has been assembled.

Developments in the research environment should be dovetailed by **ensuring successful SWT uptake by practitioners**. Much can be done in this respect by improving the accessibility and usability of SWTs. Various chapters in this book offer concrete ideas to achieve this overall goal. The lack of knowledge acquisition interfaces that are easy to use by engineers is seen as a major issue and in some chapters this was addressed by acquiring ontologies through transformations from modelling languages familiar to engineers (e.g., SysML). Chapter 5 identifies the need for supporting both industry adopters and Semantic Web experts in finding existing ontologies with (a) *ontology classification schemes*, which can be understood by both stakeholder groups, and (b) *surveys of engineering ontologies*, thus facilitating ontology reuse. For supporting ontology modelling, ontology design patterns (ODPs) should be brought closer to creators of engineering ontologies, for example, with a *catalogue of frequently emerging modelling needs and guidelines for solving these* with ODPs adapted to the engineering domain. In the area of semantic data creation from legacy data sources, Chap. 5 states that practitioners would highly benefit from the availability of *tool evaluation and selection frameworks* that support practitioners in finding the most suitable tools for their context. Chapter 14 finds that practitioners should be offered *support in resolving inconsistencies*. Besides the visualization of detected inconsistencies, support is needed for tracing and deciding on how an inconsistency should be resolved. Tools aimed at engineers should also explain results that were derived by virtue of the OWA. Last but not least, Oberle's (2014) points discussed in Sect. 15.2.4 remain valid: (1) support is needed for the *technical integration of SWTs with existing enterprise systems* and (2) *training engineers and developers* to better understand SWTs.

The chapters in this book also identified **new challenges and potential technical developments for SWTs**. IEA's that will support the *Industrie 4.0* vision should be based on high-performance tools that can deal with large, diverse, and rapidly changing datasets. It follows that Semantic Web tools should be mature enough to efficiently deal with engineering data. First, capabilities are needed to cater for frequent updates of the used ontologies to reflect system, technology, or market developments. To that end, Chap. 7 envisions a Semantic Web-based implementation of a generic infrastructure for *knowledge change management* in multidisciplinary engineering environments. As detailed in Sect. 15.2.1, research advances are also expected in the area of *managing dynamic engineering* data, i.e., streaming data.

Second, open topics in the area of data integration include: (1) the *automatic identification of semantic overlaps* between engineering models using probabilistic reasoning approaches or background domain knowledge (Chap. 14); (2) a comparison with languages and techniques currently applied by engineers to link different models across engineering disciplines or used within the Model-Driven Engineering field (Chap. 6); (3) *extending SHACL* (Chap. 13): (a) with rule support by applying the theory of Triple Graph Grammars; (b) to support one-to-many and many-to-many mappings and (c) to deal with structural heterogeneities between modeling languages and viewpoints.

Third, additional *evaluations of possible software architectures to achieve semantic integration* are needed especially tuned to the needs of industrial scenarios. In this direction, Chap. 8 investigated four ontology-based software architectures (ontology store, relational database, graph database store, and versioning management system) from the perspective of meeting requirements of typical multi-disciplinary engineering projects: querying of common concepts, transformation between local and common concepts, and versioning of engineering data. The results showed good usability and maintenance for ontology storage systems but also found that these (a) lag behind the software architecture using relational database in terms of scalability, specifically for insert performance, memory and disk usage; (b) are outperformed by the software architecture using graph databases in terms of query execution performance. Further evaluations of technical architectures are needed in engineering settings to identify weak points of SWTs that should be tackled by research but also to propose hybrid software architectures to overcome those.

As already mentioned before, **a closer collaboration between Semantic Web and Model-Driven Engineering technologies** seems a promising future research avenue. As discussed in Sect. 15.2.5, model-driven and Semantic Web technologies have compatible strengths that could be leveraged in hybrid solutions. Indeed, Chaps. 13 and 14 already demonstrate such hybrid solutions in this trend where SysML/UML are used as front end to engineers, models created in these languages are translated to ontologies, and SWTs are employed in the backend for data integration, constraint checking, and data analytics. Oberle (2014) proposes ways to bridge between MDE and Semantic Web models.

Exploring Linked Data in engineering. The IEAs described in this book show a weak uptake of Linked Data technologies and primarily focus on the use of classical SWTs. Although one of the strengths of SWTs is the combination of traditional knowledge representation and reasoning techniques with Web compliance features, there is a clear tendency, in the papers reviewed in Chap. 11 and IEAs reported in this book, to primarily explore the semantic features of these technologies as opposed to those related to Web compliance, in particular, C3 (see Sect. 15.2.3). We therefore see an opportunity in exploring and better understanding the benefits of Linked Data in the engineering domain, and more broadly in *Industrie 4.0*.

Acknowledgments We thank Peter Wetz and Manuel Wimmer for feedback on draft versions of this chapter. This work was supported by the Christian Doppler Forschungsgesellschaft, the Federal Ministry of Economy, Family and Youth, and the National Foundation for Research, Technology and Development in Austria.

References

Balduini, M., Della Valle, E., Dell'Aglio, D., Tsytsarau, M., Palpanas, T., Confalonieri, C.: Social listening of city scale events using the streaming linked data framework. In: Proceedings of the 12th International Semantic Web Conference—Part II (ISWC '13), Springer, New York, pp. 1–16 (2013)

Barbieri, D.F., Braga, D., Ceri, S., Della Valle, E., Grossniklaus, M.: C-SPARQL: a continuous query language for RDF data streams. Int. J. Semant. Comput. **4**(1), 3–25 (2010)

Biffl, S., Mordinyi, R., Moser, T.: Anforderungsanalyse für das integrierte Engineering - Mechanismen und Bedarfe aus der Praxis. ATP edition Automatisierungstechnische Praxis **54** (5), 28–35 (2012)

Biffl, S., Mordinyi, R., Steininger, H., Winkler, D.: Prozessunterstützung durch eine Integrationsplattform für anlagenmodellorientiertes Engineering – Bedarfe und Lösungsansätze. In: Vogel-Heuser, B., Bauernhansl, T., ten Hompel, M. (eds.) Handbuch Industrie 4.0, 2. Auflage (2015)

Brambilla, M., Cabot, J., Wimmer, M.: Model-Driven Software Engineering in Practice, p. 182. Morgan & Claypool (2012)

Compton, M., Barnaghi, P., Bermudez, L., García-Castro, R., Corcho, O., Cox, S., Graybeal, J., Hauswirth, M., Henson, C., Herzog, A., Huang, V., Janowicz, K., Kelsey, W.D., Le Phuoc, D., Lefort, L., Leggieri, M., Neuhaus, H., Nikolov, A., Page, K., Passant, A., Sheth, A., Taylor, K.: The SSN ontology of the W3C semantic sensor network incubator group. Web Semant. Sci. Serv. Agents World Wide Web **17**, 25–32 (2012)

Della Valle, E., Ceri, S., van Harmelen, F., Fensel, D.: It's a streaming world! Reasoning upon rapidly changing information. IEEE Intell. Syst. **24**, 83–89 (2009)

Fay, A., Biffl, S., Winkler, D., Drath, R., Barth, M.: A method to evaluate the openness of automation tools for increased interoperability. In: Proceedings of the 39th Annual Conference of the IEEE Industrial Electronics Society (IECON), Vienna, Austria (2013). doi:10.1109/IECON.2013.6700266

Gabrilovich, E., Markovitch, S.: Computing semantic relatedness using Wikipedia-based explicit semantic analysis. IJCAI **7**, 1606–1611 (2007)

Kernschmidt, K., Vogel-Heuser, B.: An interdisciplinary SysML based modeling approach for analyzing change influences in production plants to support the engineering, In: Proceedings of the IEEE International Conference on Automation Science and Engineering (CASE), pp. 1113–1118 (2013)

Kharlamov, E., Solomakhina, N., Özçep, Ö.L., Zheleznyakov, D., Hubauer, T., Lamparter, S., Roshchin, M., Soylu, A., Watson, S.: How semantic technologies can enhance data access at siemens energy. In: Proceedings of the 13th International Semantic Web Conference - Part I (ISWC'14), pp. 601–619. Springer, New York (2014)

Le-Phuoc, D., Dao-Tran, M., Xavier Parreira, J., Hauswirth, M.: A native and adaptive approach for unified processing of linked streamsand linked data. In: The Semantic Web—ISWC 2011. Lecture Notes in Computer Science, vol. 7031, pp. 370–388. Springer, Berlin (2011)

Margara, A., Urbani, J., van Harmelen, F., Bal, H.: Streaming the Web: Reasoning over dynamic data. J. Web Semant. Sci. Serv. Agents World Wide Web **25**, 24–44 (2014)

Oberle, D: Ontologies and Reasoning in Enterprise Service Ecosystems, In: Informatik Spektrum 37/4 (2014)

Paulheim, H., Oberle, D., Plendl, R., Probst, F.: An architecture for information exchange based on reference model, In: Sloane, A.M., Aβmann, U. (eds.) Revised Selected Papers of the 4th International Conference on Software Language Engineering (SLE). Lecture Notes in Computer Science, vol. 6940, pp. 160–179 (2011)

Pan, J.Z.: "Closing" some doors for the open Semantic Web. In: Proceedings of the 2nd International Conference on Web Intelligence, Mining and Semantics (WIMS'12). ACM, New York, Article 2, p. 2 (2012)

Rahmani, T., Oberle, D., Dahms, M.: An adjustable transformation from OWL to Ecore. In: Proceedings of the 13th International Conference, MODELS 2010, Part 2, Oslo, Norway. Lecture Notes in Computer Science, vol. 6395, pp. 243–257. Springer (2010)
Ren, Y., Pan, J.Z., Zhao, Y.: Closed world reasoning for OWL2 with NBox. J. Tsinghua Sci. Technol. **15**(10), 692–701 (2010)
Sabou, M., Ekaputra, F.J., Kovalenko, O., Biffl, S.: Supporting the engineering of cyber-physical production systems with the AutomationML analyzer. In: Proceedings of the CPPS Workshop, at the Cyber-Physical Systems Week, Vienna, (2016)
Seylan, I., Franconi, E., De Bruijn, J.: Effective query rewriting with ontologies over DBoxes. In: Proceedings of IJCAI'09, pp. 923–929 (2009)
Wache, H., Voegele, T., Visser, U., Stuckenschmidt, H., Schuster, G., Neumann, H., Hubner, S.: Ontology-based integration of information—a survey of existing approaches. In: Stuckenschmidt, H. (ed.) Proceedings of IJCAI Workshop: Ontologies and Information, pp. 108–117 (2001)
Weilkiens, T., Lamm, J.: Model-Based Systems Architecture. Wiley Series in Systems Engineering and Management, 384p. John Wiley & Sons Inc. (2015). ISBN 978-1118893647
Wenzel, K., Riegel, J., Schlegel, A., Putz, M.: Semantic web based dynamic energy analysis and forecasts in manufacturing engineering. In: Proceedings of the 18th CIRP International Conference on Life Cycle Engineering, pp. 507–512. Springer (2011)
Winkler, D., Biffl, S.: Improving quality assurance in automation systems development projects. In: Quality Assurance and Management, Intec Publishing (2012). ISBN 979-953-307-494-7, doi:10.5772/33487

Index

S. Biffl and M. Sabou (eds.), *Semantic Web Technologies for Intelligent Engineering Applications*, DOI 10.1007/978-3-319-41490-4

V

W

Printed in the United States
By Bookmasters